KINETIC THEORY AND TRANSPORT PHENOMENA

OXFORD MASTER SERIES IN PHYSICS

The Oxford Master Series is designed for final year undergraduate and beginning graduate students in physics and related disciplines. It has been driven by a perceived gap in the literature today. While basic undergraduate physics texts often show little or no connection with the huge explosion of research over the last two decades, more advanced and specialized texts tend to be rather daunting for students. In this series, all topics and their consequences are treated at a simple level, while pointers to recent developments are provided at various stages. The emphasis is on clear physical principles like symmetry, quantum mechanics, and electromagnetism which underlie the whole of physics. At the same time, the subjects are related to real measurements and to the experimental techniques and devices currently used by physicists in academe and industry. Books in this series are written as course books, and include ample tutorial material, examples, illustrations, revision points, and problem sets. They can likewise be used as preparation for students starting a doctorate in physics and related fields, or for recent graduates starting research in one of these fields in industry.

CONDENSED MATTER PHYSICS

1. M.T. Dove: *Structure and dynamics: an atomic view of materials*
2. J. Singleton: *Band theory and electronic properties of solids*
3. A.M. Fox: *Optical properties of solids, second edition*
4. S.J. Blundell: *Magnetism in condensed matter*
5. J.F. Annett: *Superconductivity, superfluids, and condensates*
6. R.A.L. Jones: *Soft condensed matter*
17. S. Tautz: *Surfaces of condensed matter*
18. H. Bruus: *Theoretical microfluidics*
19. C.L. Dennis, J.F. Gregg: *The art of spintronics: an introduction*
21. T.T. Heikkilä: *The physics of nanoelectronics: transport and fluctuation phenomena at low temperatures*
22. M. Geoghegan, G. Hadziioannou: *Polymer electronics*
25. R. Soto: *Kinetic theory and transport phenomena*

ATOMIC, OPTICAL, AND LASER PHYSICS

7. C.J. Foot: *Atomic physics*
8. G.A. Brooker: *Modern classical optics*
9. S.M. Hooker, C.E. Webb: *Laser physics*
15. A.M. Fox: *Quantum optics: an introduction*
16. S.M. Barnett: *Quantum information*
23. P. Blood: *Quantum confined laser devices*

PARTICLE PHYSICS, ASTROPHYSICS, AND COSMOLOGY

10. D.H. Perkins: *Particle astrophysics, second edition*
11. Ta-Pei Cheng: *Relativity, gravitation and cosmology, second edition*
24. G. Barr, R. Devenish, R. Walczak, T. Weidberg: *Particle physics in the LHC era*

STATISTICAL, COMPUTATIONAL, AND THEORETICAL PHYSICS

12. M. Maggiore: *A modern introduction to quantum field theory*
13. W. Krauth: *Statistical mechanics: algorithms and computations*
14. J.P. Sethna: *Statistical mechanics: entropy, order parameters, and complexity*
20. S.N. Dorogovtsev: *Lectures on complex networks*

Kinetic Theory and Transport Phenomena

Rodrigo Soto

Physics Department
Universidad de Chile

OXFORD
UNIVERSITY PRESS

Great Clarendon Street, Oxford, OX2 6DP,
United Kingdom

Oxford University Press is a department of the University of Oxford.
It furthers the University's objective of excellence in research, scholarship,
and education by publishing worldwide. Oxford is a registered trade mark of
Oxford University Press in the UK and in certain other countries

© Rodrigo Soto 2016

The moral rights of the author have been asserted

First Edition published in 2016

Impression: 1

Published in the United States of America by Oxford University Press
198 Madison Avenue, New York, NY 10016, United States of America

British Library Cataloguing in Publication Data
Data available

Library of Congress Control Number: 2015957821

ISBN 978–0–19–871605–1 (hbk.)
ISBN 978–0–19–871606–8 (pbk.)

Printed and bound by
CPI Group (UK) Ltd, Croydon, CR0 4YY

Preface

One of the questions about which humanity has wondered is the arrow of time. Why do we perceive that time goes forward? Why does temporal evolution seem irreversible? That is, we often see objects break into pieces, but we never see them reconstitute spontaneously. The observation of irreversibility was first put into scientific terms by the so-called second law of thermodynamics, which states that, in closed systems, the entropy never decreases. Note, however, that the second law does not explain the origin of irreversibility; it only quantifies it. With the emergence of classical mechanics in Hamiltonian form and, later, with quantum mechanics, it was evident that something was missing. Indeed, the classical and quantum equations of motion for atoms and molecules are time reversible: inverting all particles' velocities leads to exactly reversed trajectories. When we observe the motion of a few atoms, it is impossible to say whether the movie is running forward or backward. However, as we said, in thermodynamic systems, the evolution is clearly irreversible. The difference, as you will have noticed, lies in the number of degrees of freedom. We observe reversible dynamics with few particles, but irreversibility is observed when the number of degrees of freedom is large.[1]

Consider the diffusion process in a box. It can be described by a collection of random walkers, each moving independently. At each time step, each particle can move left or right with equal probability. If we initially place more particles on the left, simply because of the extra abundance on this side, there will be a net average flux to the right. This flux will end when an equal number of particles sit on both sides. Note that this is not a static state, as particles permanently cross from one side to the other. If the number of particles is small, there is a high chance that, due to fluctuations, particles will move to one side and create a noticeable mass imbalance. We would say that, in this case, the system dynamics is not irreversible, as asymmetric particle distributions are created and destroyed all the time. However, the probability of creating large mass imbalances decreases exponentially with the number of particles. Finally, with an infinite number of particles, it is impossible to spontaneously create mass imbalances and the system has a well-defined irreversible dynamics.

In the previous example, two ingredients were used. First, real irreversibility is only obtained in the thermodynamic limit. In this case, the system dynamics, which in our example was probabilistic, became deterministic. Second, in the thermodynamic limit, the system will evolve

[1] Normally, it is said that irreversibility is observed with $N \sim 10^{23}$ degrees of freedom. However, computer simulations show that in fact a much smaller number is needed and a few hundred are typically sufficient.

to a symmetric state. Hence, to observe irreversible behaviour, it is necessary to create initial conditions that are not symmetric. The origin of the observed irreversibility is the overwhelming majority of symmetric compared with asymmetric configurations.

Boltzmann noticed that the same two arguments could also be used in the context of classical mechanics for atoms and molecules. Here, the individual particles obey deterministic equations of motion and interact through collisions, making the analysis much more complex than in the random walk process. However, he first argued that, equally in this case, the number of configurations with atoms distributed evenly in the box is much larger than the corresponding number of configurations with atoms placed with any other spatial distribution. Detailed calculations using equilibrium statistical mechanics show that this is indeed the case. However, a giant step forward was later made by Boltzmann himself, in 1872, when he presented his now famous kinetic equation. He showed that gas dynamics is irreversible if one considers classical mechanics and applies a statistical analysis to the molecular interactions. Again, this statistical analysis becomes valid when the number of particles is large. Later, using the same equation, it was possible to obtain the dynamical laws that govern the irreversible evolution to equilibrium. It has been shown, for example, that temperature inhomogeneities are relaxed by heat transport and that the thermal conductivity can be computed in terms of the scattering properties of the molecular interactions. The Boltzmann equation provides a conceptual explanation for various phenomena and is also quantitatively predictive.

Another key question of nonequilibrium statistical mechanics is that of universality. It is common for quite different systems to present similar qualitative behaviour. For example: why is diffusion present in so many different systems? Why can all solids be classified as metals or insulators (semiconductors being simply poor insulators)? Or, why can all fluids—gases and liquids—be described by the same, Navier–Stokes hydrodynamic equations? Let us consider this last problem. In their motion, atoms collide with others with the exchange of momentum and energy, providing a mechanism for velocity relaxation at the collisional temporal scale. However, mass, momentum, and energy are conserved in collisions and can thus only relax by their transfer from one region to another, i.e. by transport processes, which are much slower than the collisional dynamics. This separation of temporal scales is key to understanding the question. Indeed, the hydrodynamic equations are precisely those that describe the evolution of the conserved fields at the slow scale that we measure, instead of the collisional scale of picoseconds. Kinetic theory provides the tools to deal with this scale separation, enabling us to derive, for example, the hydrodynamic equations from kinetic models. Moreover, it relates the evolution of the slow modes to the rapid collisional dynamics, closing the description.

The seminal approach to irreversibility by Boltzmann, called the kinetic theory of gases, rapidly led to application of similar approaches to other systems. Using the same idea of treating motion and interactions in a statistical way instead of computing the full motion of all the particles, it has been possible to understand the irreversible dynamics, derive transport laws, and compute the associated transport coefficients of systems as diverse as plasmas, colloidal particles suspended in liquids, electrons in metals or insulators, quantised vibrations in a solid, etc. In all these cases, which we present in this book, the kinetic theory is built based on the elementary dynamics of the system constituents, including the relevant interactions in play. Based on the different kinetic models, we will be able to explain macroscopic phenomena and make quantitatively accurate predictions. Kinetic theory is an approach to nonequilibrium physics that provides intuitive explanations of transport processes as well as tools for quantitative prediction. For example, when studying electron transport in solids, kinetic theory can explain the origin of the difference between conductors and insulators, but also gives the value of the electrical conductivity as a function of the quantum scattering amplitude of electrons on crystal defects.

It must be said that kinetic theory is an approximate approach to nonequilibrium physics. Many of the approximations and models are well justified, but some are applied only for convenience and must be checked against other theories or experiments. We still do not have a complete theory for nonequilibrium systems, and different complementary approaches should be used. Among them, we invite the reader to explore the methods provided by stochastic processes, nonlinear physics, dynamical systems, the renormalisation group, hydrodynamic or similar equations for relevant order parameters and fields, or the fluctuation theorems. To give the reader a broader view of the different approaches, appropriate books are referred to as additional reading at the end of the relevant chapters.

This book is based on lectures given at the School of Mathematical and Physical Sciences of the Universidad de Chile over several years. It aims to present a broad view of nonequilibrium statistical mechanics using the kinetic theory approach. The objective is to balance the mathematical formalism, qualitative analysis, and detailed description of different phenomena that are explained and quantified by kinetic theory. In any theory, calculations become rapidly intractable analytically as the models increase in detail. Kinetic theory is no exception, and bearing this in mind, we use simple models whenever we need to obtain explicit results. For example, hard sphere models will be used to compute the viscosity in gases or parabolic bands will be considered to obtain the electronic conductivity in metals. However, a general basis will be given for readers who need to go beyond the simple models. Also, a chapter on numerical methods is included at the end to present the main techniques that can be used to solve kinetic models computationally.

The first two chapters of the book present the general concepts and tools of kinetic theory. Chapter 1 provides a basic presentation of the one-particle distribution and the mean free path approach, while Chapters 2 presents a formal description of many-body distribution functions. The subsequent chapters each deal with a particular system or class of systems, described using specific kinetic models. Chapter 3 presents the Lorentz model for charge transport; this classical model is presented before the Boltzmann equation that inspired it, because of its comparative simplicity. The Boltzmann equation for classical gases is analysed in Chapter 4. Brownian motion and the corresponding Fokker–Planck equation are presented in Chapter 5, in which we describe diffusive motion. In Chapter 6, the Vlasov equation for plasmas is derived and analysed. It describes systems with long-range interactions and hence can also be applied to self-gravitating systems such as galaxies. In Chapter 7, the Boltzmann approach is extended to quantum gases, where the Pauli exclusion principle for fermions and stimulated emission for bosons are considered as essential ingredients of the kinetic equation. The case of phonons and electrons is considered in detail. Electrons moving in a crystalline solid are studied in Chapter 8. Here, besides the Pauli principle, the existence of bands qualitatively and quantitatively explains the difference between insulators and conductors. The special case of semiconductors is considered in the short Chapter 9. The final Chapter 10 presents numerical methods that can be used to solve all the kinetic models presented in the book. Finally, the book is complemented with some appendices that present tools used in different chapters. Specifically, appendices are devoted to some mathematical complements, perturbation methods for eigenvalue calculations, tensor analysis, the classical and quantum descriptions of scattering, and finally the electronic structure in solids.

At the end of most chapters, applications are given in special sections. The objective is to show how the concepts presented in that chapter can be applied to various situations as diverse as light diffusion, the quark–gluon plasma, bacterial suspensions, granular matter, the expanding universe, and graphene, among others. Finally, each chapter is complemented by recommended reading and a series of exercises, with solutions in a separate book. These exercises, tested over the years, solve specific problems, model new phenomena, deal with some theoretical issues, and finally aim to provide insight into various physical systems.

Whenever possible, standard notation and symbols are used, so that readers can switch easily to other books. Also, the international system of units (SI) is used everywhere, including electromagnetism. The only exception is in the description of the quark–gluon plasma where natural units are used ($\hbar = c = 1$). Finally, scalar quantities are presented with the standard italic font (e.g. $p = nk_{\mathrm{B}}T$), vectors are in boldface (e.g. $\mathbf{J} = \sigma\mathbf{E}$), and tensors and matrices in blackboard fonts (e.g. $\mathbb{P} = p\mathbb{I}$).

R.S., Santiago, Chile, 2016

Acknowledgements

I would like first to thank Jarosław Piasecki who read the early table of contents of this book, encouraging me to persevere with the project and giving me some interesting ideas. This book has been greatly improved by the corrections, suggestions, and comments of numerous friends and colleagues. Ricardo Brito, Marcos Flores, Víctor Fuenzalida, Andrés Meza, Gonzalo Palma, Alvaro Nuñez, and Mario Riquelme have read the manuscript drafts contributing with their experience and helping to identify numerous errors. I appreciate their suggestions on new topics, books, and articles in all those areas in which I am not expert. In this last aspect I am profoundly indebted to all those students worldwide who have performed a wonderful work in summarising novel results in their theses, which have allowed me to learn new topics and discover the relevant bibliographies. I would like to thank all of the students who contributed to the courses on which this book is based, with their questions and suggestions and also to Carmen Belmar who made the first typesetting of some of the chapters of the manuscript. A special acknowledgement is given to Patricio Cordero, Dino Risso, and Rosa Ramírez during my PhD and later to Michel Mareschal and Jarosław Piasecki in my discovery of the beauty and power of kinetic theory.

I acknowledge the Physics Department and the Faculty of Physical and Mathematical Sciences of the Universidad de Chile for their support and for providing a creative and stimulating space for research and teaching. Also, I thank my colleagues and students for their understanding and patience during the writing period of this book.

Lastly, I am profoundly grateful to Elena for her support and encouragement during this project.

Contents

Basic concepts

1.1   Velocity distribution function

In this book we will study the dynamics of different classes of systems, both classical and quantum. Each of them has its own peculiarities and, consequently, will be studied in detail by applying specific tools and concepts. However, the distribution function is a concept that is ubiquitous in the study of kinetic theory. It will be of quite general applicability when properly defined for specific situations. For concreteness, consider a gas composed of molecules at ambient conditions. In this context, the relevant object is the velocity distribution function. To make the description simpler, we will consider first the case of a gas with molecules of a single species, say N_2, for example. The gas occupies a volume $\mathcal{V}$ that is divided into subvolumes that are small enough such that in their interior the gas can be considered as homogeneous, but large enough to include a large number of molecules. Inside each subvolume, the molecules move in all directions with different velocities, and our objective is to characterise them. We define the velocity distribution function, $f(\mathbf{r}, \mathbf{c}, t)$, such that

$$f(\mathbf{r}, \mathbf{c}, t)\, \mathrm{d}^3 r\, \mathrm{d}^3 c \qquad (1.1)$$

is the average number of molecules in the vicinity $\mathrm{d}^3 r$ of $\mathbf{r}$ with velocities in the vicinity $\mathrm{d}^3 c$ of the velocity $\mathbf{c}$[1] at time t.[2]

This function is the central object of study in this book. We will learn how physical properties in equilibrium and, more importantly, out of equilibrium can be computed from it. Also, different evolution equations that model specific physical systems will be derived and analysed.[3] By solving these equations, we will be able to answer a large number of questions concerning transport phenomena and the relaxation of out-of-equilibrium systems. For quantum systems, instead of velocities, we will study the distribution of wavevectors.

If the gas were composed of different kinds of molecules, the velocity distribution function for each species $f_i(\mathbf{r}, \mathbf{c}, t)$ would be defined in an analogous way. This will be the case when we describe the dynamics of plasmas, for which we will consider the distribution functions of both electrons and ions in Chapter 6. Also, in Chapter 8, two distributions will be used when computing the electrical resistivity in metals due to the interaction of electrons with phonons. For the moment, we continue to consider a single species of particles.

[1] Note that we will use $\mathbf{c}$ to refer to the particles' velocities. The symbol $\mathbf{v}$ is reserved for the average velocity.

[2] J.C. Maxwell introduced the concept of the velocity distribution function for the first time in 1866 to describe the properties of gases.

[3] The first and most famous of these is the Boltzmann equation, which describes the dynamics of dilute classical gases. It was introduced in 1872 by L. Boltzmann.

Kinetic Theory and Transport Phenomena, First Edition, Rodrigo Soto.
© Rodrigo Soto 2016. Published in 2016 by Oxford University Press.

There are several aspects that require detailed clarification before continuing with the study of the distribution function. First, to define this object, we divided the phase space of positions and velocities into small parcels of size $d^3r\, d^3c$ and counted the number of particles in them. These parcels should be of mesoscopic size, being large enough to contain a significant number of molecules. Later, we will use $d^3r\, d^3c$ as a mathematical differential, so that integral calculus will be used to compute global quantities. The question that arises is whether it is possible to find such small-but-large parcels. Indeed it is. In a gas, the average number density of molecules is[4] $n = 2.69 \times 10^{26}\,\mathrm{m}^{-3}$. Therefore, it is possible to consider subvolumes as small as $1\,(\mu\mathrm{m})^3$ that contain 2.7×10^8 molecules each. This number of molecules is large enough to allow statistical analysis and define the velocity distribution function. In normal situations, the gas will be homogeneous on larger scales and the number of molecules in each subvolume can become even larger.

The second aspect that requires consideration is the use of the term 'average' in the definition of f in (1.1). As molecules move, they continuously enter and leave any given subvolume. Also, they perform collisions that can change their velocities on very short times. We are not interested in describing such a fluctuating and erratic function. Therefore, the distribution function is defined by blurring slightly in space and time the microscopic snapshots that we should be taking to compute the distribution function. This procedure, commonly applied in nonequilibrium statistical mechanics, is called coarse graining. The objective of this blurring procedure is to average, resulting in a smooth distribution function. This procedure should be done with caution to avoid the introduction of spurious effects. We will come to this when we consider the so-called H theorem that quantifies the irreversible dynamics of gases in Section 4.3.2.

1.2 The Maxwell–Boltzmann distribution function

Consider a system of classical particles in thermal equilibrium in the absence of external fields, therefore being homogeneous. Classical statistical mechanics states that, regardless of the mutual interaction between particles, their velocities are given by the Maxwell–Boltzmann distribution. In a homogeneous system, we consider the distribution function f such that $f(\mathbf{c})\, d^3c$ is the number of particles, per unit volume, with velocities in a neighbourhood d^3c about $\mathbf{c}$. In thermal equilibrium, then

$$f(\mathbf{c}) = n\widehat{f}_{\mathrm{MB}}(\mathbf{c}) = n\left(\frac{m}{2\pi k_{\mathrm{B}}T}\right)^{3/2} e^{-mc^2/2k_{\mathrm{B}}T}, \qquad (1.2)$$

where $c^2 = \mathbf{c}\cdot\mathbf{c}$, n is the particle number density, m is the particle mass, T is the temperature, and $k_{\mathrm{B}} = 1.38 \times 10^{-23}\,\mathrm{J\,K^{-1}}$ is the Boltzmann constant.[5]

[4]This number can be easily obtained by remembering that a mole having $N_A = 6.022 \times 10^{23}$ molecules occupies a volume of $\mathcal{V} = 22.4$ litres at standard temperature and pressure.

[5]The distribution $\widehat{f}_{\mathrm{MB}}$ is referred to as either the Maxwell–Boltzmann distribution or simply the Maxwellian distribution. It was first proposed by James Clerk Maxwell in 1860 as the only distribution invariant under collisions, and later, in 1872, Ludwig Boltzmann derived it using his kinetic equation of gases. It was also subsequently derived from the principles of statistical mechanics in 1877 by Boltzmann himself.

Here, we will use the microscopic notation for the Maxwell–Boltzmann distribution using m and k_B, instead of the equivalent macroscopic expression that uses the molar mass $\mu = N_A m$ and the gas constant $R = N_A k_B$, i.e. $\widehat{f}_{MB}(\mathbf{c}) = \left(\frac{\mu}{2\pi RT}\right)^{3/2} e^{-\mu c^2/2RT}$, where N_A is the Avogadro number. The advantage of the microscopic expression will become clear when computing the transport properties of gases, which are directly expressed in terms of the microscopic quantities.

As the velocity distribution is isotropic, we define the distribution of speeds (absolute value of velocity) by the differential relation $\phi(c)\,dc = f(\mathbf{c})\,d^3c$, which gives the number of particles per unit volume with speed between c and $c + dc$. To obtain it we use the Jacobian from d^3c to dc: $d^3c = 4\pi c^2\,dc$. In equilibrium,

$$\phi_{MB}(c) = 4\pi n \left(\frac{m}{2\pi k_B T}\right)^{3/2} c^2 e^{-mc^2/2k_B T}, \qquad (1.3)$$

which is depicted in Fig. 1.1. Its maximum is located at $c_{th} = \sqrt{2k_B T/m}$ and represents the typical thermal speed of the particles.[6] The average speed is $\langle|\mathbf{c}|\rangle = \sqrt{8k_B T/\pi m} \approx 1.13 c_{th}$ (see Exercise 1.1).

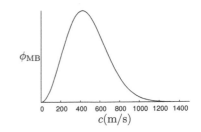

Fig. 1.1 Normalised Maxwellian distribution of speeds ϕ_{MB} for a gas of N_2 molecules at ambient temperature.

[6]In the atmosphere, essentially composed of N_2 molecules, the typical velocity under ambient conditions is $c_{th} = 422$ m/s.

1.3 Densities and fluxes

The distribution function is defined so as to give the number of particles per unit volume that have a particular velocity. Bearing this in mind, it is possible to compute the local particle density as

$$n(\mathbf{r}, t) = \int f(\mathbf{r}, \mathbf{c}, t)\,d^3c, \qquad (1.4)$$

which gives the number of particles per unit volume, irrespective of their velocities.

Consider a particular property φ that depends on the velocity of single particles; for example, this may be the kinetic energy $\varphi = mc^2/2$ or mass $\varphi = m$. We are interested in the local average of φ considering that the particles have different velocities. This is defined as

$$\langle\varphi\rangle(\mathbf{r}, t) = \frac{1}{n(\mathbf{r}, t)} \int \varphi(\mathbf{c})f(\mathbf{r}, \mathbf{c}, t)\,d^3c. \qquad (1.5)$$

In a similar way, the local density of this quantity is defined as

$$\rho_\varphi(\mathbf{r}, t) = \int \varphi(\mathbf{c})f(\mathbf{r}, \mathbf{c}, t)\,d^3c = n(\mathbf{r}, t)\langle\varphi\rangle(\mathbf{r}, t). \qquad (1.6)$$

The latter measures the content of φ in a unitary volume.

The local velocity field is then computed as the average of the molecular velocities,

$$\mathbf{v}(\mathbf{r}, t) = \frac{1}{n(\mathbf{r}, t)} \int \mathbf{c}f(\mathbf{r}, \mathbf{c}, t)\,d^3c. \qquad (1.7)$$

Next, according to the equipartition theorem, we might be tempted to compute the temperature from the average of the velocity squared.

However, we note that, due to Galilean invariance, the temperature of a sample does not depend on the mean velocity at which it moves, but rather on the fluctuations about this average velocity. The local temperature is then defined in three dimensions as

$$\frac{3}{2}k_{\mathrm{B}}T(\mathbf{r},t)=\left\langle\frac{m(\mathbf{c}-\mathbf{v})^{2}}{2}\right\rangle,\tag{1.8}$$

where $\mathbf{c}-\mathbf{v}$ are the particle velocities relative to the average value.[7]

When particles move, they transfer physical properties from one place to another. Associated to each quantity φ we can define a kinetic flux $\mathbf{J}_{\varphi}$ which measures the amount of this quantity that is transferred. Consider a mathematical surface ΔS with normal vector $\hat{\mathbf{n}}$, as shown in Fig. 1.2. In a time interval Δt a molecule with velocity $\mathbf{c}$ will cross the surface if it is located in the cylinder presented in the figure. The volume of this cylinder is $\Delta\mathcal{V}=\mathbf{c}\cdot\hat{\mathbf{n}}\Delta t\Delta S$. Therefore, the number of molecules with velocity in the vicinity $\mathrm{d}^{3}c$ of $\mathbf{c}$ that will cross the surface is $f(\mathbf{r},\mathbf{c},t)\mathbf{c}\cdot\hat{\mathbf{n}}\Delta t\Delta S\,\mathrm{d}^{3}c$. Considering that each molecule carries the quantity φ, the total amount transferred across ΔS is

$$\Delta\varphi=\Delta t\Delta S\int f(\mathbf{r},\mathbf{c},t)\varphi(\mathbf{c})\mathbf{c}\cdot\hat{\mathbf{n}}\,\mathrm{d}^{3}c.\tag{1.9}$$

We define the flux vector $\mathbf{J}_{\varphi}$ such that the total transfer per unit time is $\Delta\varphi/\Delta t=\mathbf{J}_{\varphi}\cdot\hat{\mathbf{n}}\Delta S$. Then,

$$\mathbf{J}_{\varphi}(\mathbf{r},t)=\int f(\mathbf{r},\mathbf{c},t)\varphi(\mathbf{c})\mathbf{c}\,\mathrm{d}^{3}c.\tag{1.10}$$

Note that, when computing the volume of the cylinder, we have retained the sign of $\mathbf{c}$. This is sensible, as it counts molecules moving parallel to $\hat{\mathbf{n}}$ as positive and molecules moving in the opposite direction as negative transfer. Hence, molecules moving in opposite directions are subtracted in the flux.

We proceed to compute some relevant fluxes. First, consider the mass flux. One can directly verify that $\mathbf{J}_{m}(\mathbf{r},t)=mn(\mathbf{r},t)\mathbf{v}(\mathbf{r},t)$, which corresponds also to the momentum density. Next, the kinetic energy flux is given by

$$\boldsymbol{J}_{e}(\mathbf{r},t)=\frac{m}{2}\int f(\mathbf{r},\mathbf{c},t)c^{2}\mathbf{c}\,\mathrm{d}^{3}c.\tag{1.11}$$

Molecules also transfer momentum as they move. In this case, momentum is a vectorial quantity. To clarify the interpretation of the associated flux, we will consider separately the Cartesian component $\varphi=mc_{i}$. The flux is also a vector, and we will compute its k-th component, $J_{k,c_{i}}=m\int fc_{i}c_{k}\,\mathrm{d}^{3}c$, which has two indices. It is a symmetric tensor[8] that we will denote by $\widetilde{P}_{ik}$,

$$\widetilde{P}_{ik}(\mathbf{r},t)=m\int f(\mathbf{r},\mathbf{c},t)c_{i}c_{k}\,\mathrm{d}^{3}c.\tag{1.12}$$

Its interpretation is such that $\widetilde{P}_{ik}n_{k}$[9] is the total momentum in the direction i that crosses a unitary surface oriented along the k direction per unit time. Note that, being a symmetric tensor, it is also the momentum in the direction k that crosses a surface oriented along i.

[7]Some books, to simplify notation, define the peculiar velocity $\mathbf{C}=\mathbf{c}-\mathbf{v}(\mathbf{r},t)$, in capitals, which depends explicitly on space and time through the mean velocity. In this book we will avoid its use. Although normally simplifying calculations, it can lead to confusion and notation problems.

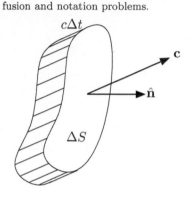

Fig. 1.2 Particles that have a velocity $\mathbf{c}$ will cross the surface ΔS in a time interval if they are inside the depicted volume. The normal vector to the surface is $\hat{\mathbf{n}}$, which is centred at $\mathbf{r}$.

[8]Relevant concepts of tensor analysis are introduced in Appendix B.

[9]Throughout the book, we will use the rule that repeated indices are summed, sometimes called Einstein notation. That is, for example, $a_{i}b_{i}=\sum_{i=1}^{3}a_{i}b_{i}$.

1.3.1   Stress tensor and energy flux

Both the kinetic energy flux and the momentum flux have a contribution
if the gas has a net velocity $\mathbf{v}$, being called the convective contributions.
It is useful to subtract these in (1.11) and (1.12) by defining

$$q_i(\mathbf{r},t) = \frac{m}{2}\int f(\mathbf{r},\mathbf{c},t)(\mathbf{c}-\mathbf{v})^2(c_i-v_i)\,\mathrm{d}^3c. \qquad (1.13)$$

$$P_{ik}(\mathbf{r},t) = m\int f(\mathbf{r},\mathbf{c},t)(c_i-v_i)(c_k-v_k)\,\mathrm{d}^3c, \qquad (1.14)$$

These quantities measure the flux of energy and momentum in a frame
comoving with the gas and are due to the molecular fluctuations around
the mean velocity. If there were no fluctuations, that is, if the temperature were zero, these fluxes would vanish. To interpret them, consider a
mathematical volume in the gas whose boundary S moves with the gas;
that is, the velocity of the points in the boundary is $\mathbf{v}(\mathbf{r})$. The total
mass that crosses the surface into the volume per unit time is

$$\frac{\mathrm{d}M}{\mathrm{d}t} = -\int\int_S f(\mathbf{r},\mathbf{c},t)m(\mathbf{c}-\mathbf{v})\cdot\mathrm{d}\mathbf{S}\,\mathrm{d}^3c. \qquad (1.15)$$

The minus sign appears because the normal vector is oriented toward
the exterior of the volume, and we have used that the rate at which
particles enter the volume is proportional to the relative velocity $\mathbf{c}-\mathbf{v}$.
One can directly verify, using the definition of $\mathbf{v}$ (eqn 1.7), that $\frac{\mathrm{d}M}{\mathrm{d}t}$
vanishes identically. Therefore, as expected for a volume that moves
with the fluid, it encloses a constant mass.

We now compute the momentum that enters the volume per unit time
as

$$\frac{\mathrm{d}p_i}{\mathrm{d}t} = -\int\int_S f(\mathbf{c})mc_i(\mathbf{c}-\mathbf{v})\cdot\mathrm{d}\mathbf{S}\,\mathrm{d}^3c. \qquad (1.16)$$

By the definition of $\mathbf{v}$, we can add a null term to obtain

$$\frac{\mathrm{d}p_i}{\mathrm{d}t} = -\int\int_S f(\mathbf{c})m(c_i-v_i)(\mathbf{c}-\mathbf{v})\cdot\mathrm{d}\mathbf{S}\,\mathrm{d}^3c = -\int_S P_{ik}\,\mathrm{d}S_k; \quad (1.17)$$

That is, the enclosed momentum changes by the action of the tensor $\mathbb{P}$
(with components P_{ik}) at the surface. However, as the total mass inside
the volume is constant, we can apply Newton's law to interpret $-P_{ik}\,\mathrm{d}S_k$
as the force acting on the volume due to the gas in the exterior. By the
action–reaction principle, $P_{ik}\,\mathrm{d}S_k$ is the force exerted by the gas on the
exterior through the surface element $\mathrm{d}\mathbf{S}$. Based on this interpretation,
P_{ik} is called the stress tensor.

Using a similar analysis, we can study the energy flux through the
comoving surface:

$$\frac{\mathrm{d}E}{\mathrm{d}t} = -\int\int_S f(\mathbf{c})\frac{mc^2}{2}(\mathbf{c}-\mathbf{v})\cdot\mathrm{d}\mathbf{S}\,\mathrm{d}^3c. \qquad (1.18)$$

Writing $c^2 = (\mathbf{c}-\mathbf{v})^2 + 2(\mathbf{c}-\mathbf{v})\cdot\mathbf{v} + v^2$, we obtain after some simple
manipulations that

$$\frac{\mathrm{d}E}{\mathrm{d}t} = -\int_S \mathbf{q}\cdot\mathrm{d}\mathbf{S} - \int_S v_i P_{ik}\,\mathrm{d}S_k. \qquad (1.19)$$

The first term accounts for the energy change in the volume due to the heat flux that enters through the surface. The second term is the mechanical work done per unit time by the external fluid on the volume (mechanical power). Indeed, $dF_i = -P_{ik}\,dS_k$ is the force on a surface element, and the mentioned term is the integral of $-\mathbf{v}\cdot d\mathbf{F}$. Therefore, this expression corresponds to the first law of thermodynamics, and $\mathbf{q}$ should be interpreted as the thermodynamic heat flux.

Finally, it is easy to show[10] that

$$\widetilde{P}_{ik} = mnv_iv_k + P_{ik}, \tag{1.20}$$

$$J_{ei} = n\left(\frac{mv^2}{2} + \frac{3k_\mathrm{B}T}{2}\right)v_i + P_{ik}v_k + q_i, \tag{1.21}$$

which relate the fluxes in the laboratory frame with those in the comoving frame. Equation (1.20) indicates that the momentum flux through a fixed surface has two contributions: the first term is the momentum that is merely transported by the flow (convective transport), while the second contribution is due to the fluctuations around the mean flow velocity, which is given by the stress tensor. Similarly, we can separate and identify the three contributions for the energy flux in eqn (1.21). The first term is the convective transport, given by the total kinetic energy transported by the average fluid motion. The second term is the mechanical power, and the last term is the heat flux associated with the velocity fluctuations.

1.3.2   Stress tensor and heat flux in equilibrium

In equilibrium, the velocity distribution is the Maxwellian (1.2). Using this distribution, one can directly verify that the heat flux (1.13) vanishes and the stress tensor (1.14) is an isotropic tensor,

$$P_{ik} = nk_\mathrm{B}T\delta_{ik}. \tag{1.22}$$

That is, the force that a gas in equilibrium exerts on a surface is normal to it, with intensity $p = nk_\mathrm{B}T$, independent of the direction, and therefore it can be interpreted as a pressure. Note that we have obtained the ideal gas equation of state, which we know from equilibrium statistical mechanics is only valid at low densities.[11] However, where did we make the assumption of low density? The answer is that, in the analysis of the momentum balance, we neglected the molecular interactions, which can transfer momentum through the surface (see Fig. 1.3). Indeed, particles in the exterior exert forces on particles in the interior. These forces are short ranged and therefore can be interpreted again as surface terms, making additional contributions to the stress tensor (the collisional transport). We will take these terms into account in the analysis of dense gases in Sections 2.6.3 and 4.8.1 and show that they are proportional to the collision frequency that is proportional to density.

Under nonequilibrium conditions, the distribution function will differ from a Maxwellian and consequently the stress tensor will not generally

[10]See Exercise 1.3.

[11]In fact, in equilibrium, the pressure can be expanded in a power series of density $p = nk_\mathrm{B}T(1 + B_2n + B_3n^2 + \ldots)$, the virial expansion, that is convergent in the gas state. The ideal gas law $p = nk_\mathrm{B}T$ is only valid at low densities.

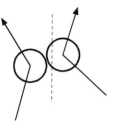

Fig. 1.3 Collisional transfer of momentum: particles interact through the surface, exchanging momentum between them and, therefore, transferring it from one side to the other.

be isotropic or given by eqn (1.22). It is customary to define the (scalar) pressure as the isotropic part of the tensor, which is computed in terms of its trace,

$$p = \frac{1}{3}P_{ii}. \tag{1.23}$$

Note that, according to the definition (1.14), we obtain the ideal gas equation of state $p = nk_BT$ under any nonequilibrium condition, because p is defined via eqn (1.8).

1.3.3 Flux distribution

Consider the mathematical surface depicted in Fig. 1.2. We want to determine the velocity distribution of the particles that cross the surface in the direction $+\hat{n}$. In a small time interval Δt, the particles with velocity $\mathbf{c}$ that cross the surface are those that are inside the cylinder shown in the same figure. Its volume is $\Delta \mathcal{V} = \mathbf{c} \cdot \hat{n}\Delta t\Delta S$, which is proportional to $\mathbf{c}\cdot\hat{n}$. As expected, the fastest particles will cross the surface more frequently that those with smaller velocities. The distribution of particles crossing the surface is then given by the so-called flux distribution,

$$f_{\text{flux}}(\mathbf{c}) = f(\mathbf{c})\mathbf{c} \cdot \hat{n}, \tag{1.24}$$

which is anisotropic. This distribution will be used for imposing boundary conditions or in the effusion problem.[12]

[12]See Exercise 1.5.

1.4 Collision frequency

In a gas, the molecules experience collisions. We will compute the average rate at which each molecule is collided with by others. Consider a target molecule with velocity $\mathbf{c}_1$ that can be collided with by another with velocity $\mathbf{c}_2$. In the reference frame of the target molecule, the second molecule moves with relative velocity $\mathbf{g} = \mathbf{c}_2 - \mathbf{c}_1$. If the total cross section is σ,[13] then for the second molecule to collide with the target in a time interval Δt, it must be in the collision cylinder shown in Fig. 1.4.

The volume of the cylinder is $\Delta \mathcal{V}_2 = \sigma|\mathbf{g}|\Delta t$, hence the number of projectiles with velocities in an interval d^3c_2 is $\Delta N_2 = f(\mathbf{c}_2)\Delta \mathcal{V}_2\, d^3c_2$. The number of targets within an interval of positions and velocities is $\Delta N_1 = f(\mathbf{c}_1)\, d^3r\, d^3c_1$. Finally, the number of collisions for the specified parameters is simply the number of targets multiplied by the number of projectiles $\Delta N_{\text{coll}} = \Delta N_1 \Delta N_2$.[14] The total number of collisions is obtained by integrating over all possible velocities and positions. Then, the total collision rate is

$$\frac{\Delta N_{\text{coll}}}{\Delta t} = \mathcal{V}\sigma \int f(\mathbf{c}_1)f(\mathbf{c}_2)|\mathbf{c}_2 - \mathbf{c}_1|\, d^3c_1\, d^3c_2. \tag{1.25}$$

The collision frequency is defined as the average number of collisions per particle, $\nu = \frac{1}{N}\frac{\Delta N_{\text{coll}}}{\Delta t}$, resulting in the following expression:

$$\nu = \frac{\sigma}{n}\int f(\mathbf{c}_1)f(\mathbf{c}_2)|\mathbf{c}_2 - \mathbf{c}_1|\, d^3c_1\, d^3c_2. \tag{1.26}$$

[13]In the case of spheres of equal radii R, the total cross section is $\sigma = \pi(2R)^2$, where the factor 2 appears because, to have a collision, the distance between the centres of the spheres must be $2R$.

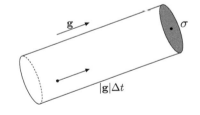

Fig. 1.4 Collision cylinder in which the molecules must be to collide with the target molecule in a time interval Δt. The base of the cylinder is the collision cross section σ, centred at the target molecule, and its length is $|\mathbf{g}|\Delta t$, where $\mathbf{g} = \mathbf{c}_2 - \mathbf{c}_1$ is the relative velocity.

[14]Here, we have assumed that the molecules are uncorrelated. As we will see in Chapter 4, this hypothesis is valid if the gas is rarefied.

As the distribution function is proportional to the particle density, the collision frequency turns out to be proportional to the total cross section and the particle density.

At equilibrium, the collision frequency can be computed using the Maxwellian distribution. This calculation is easily performed by changing integration variables to the centre of mass and relative velocities, $\mathbf{C} = (\mathbf{c}_1 + \mathbf{c}_2)/2$ and $\mathbf{g} = \mathbf{c}_2 - \mathbf{c}_1$ (with Jacobian 1), thus

$$
\nu = n\sigma \left(\frac{m}{2\pi k_\mathrm{B} T} \right)^3 \int e^{-mC^2/k_\mathrm{B}T} e^{-mg^2/4k_\mathrm{B}T} g \, \mathrm{d}^3 C \, \mathrm{d}^3 g,
$$

$$
\nu = 4n\sigma \sqrt{\frac{k_\mathrm{B}T}{\pi m}}.
$$

(1.27)

Associated to the collision frequency, we define the mean free time $\tau = 1/\nu$, which measures the average time between successive collisions.[15]

[15]In the atmosphere, under normal conditions, for molecules of radii $r_0 \approx 10^{-10}$ m, the collision frequency is $\nu \approx 5.7 \times 10^9 \, \mathrm{s}^{-1}$ and the mean free time $\tau \approx 1.8 \times 10^{-10}$ s.

1.5 Mean free path

The mean free path ℓ is defined as the average distance covered by a molecule between collisions (see Fig. 1.5). It is given by the average speed times the mean free time, $\ell = \langle |\mathbf{c}| \rangle \tau$. In equilibrium, it is given by

$$
\ell = \frac{1}{\sqrt{2}n\sigma}.
$$

(1.28)

The factor $\sqrt{2}$ appears due to the fact that the collision rate depends on the relative velocity rather than on the velocity of a single molecule. Indeed, if we consider the Lorentz model, described in detail in Chapter 3, in which point particles collide with a series of fixed scatters with cross section σ and number density n, the mean free path for the point particles will be $\ell = 1/(n\sigma)$, without the factor $\sqrt{2}$.[16]

At standard pressure, the mean free path of the gas molecules in the atmosphere is of the order of 10^{-7} m.

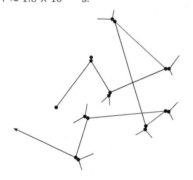

Fig. 1.5 A molecule in a gas performs a random walk as a consequence of collisions with other molecules. Between collisions, they move in straight lines at constant speed. The lengths of those segments follow a Poisson distribution whose average value is the mean free path ℓ. For this picture we had hard spheres in mind for which collisions are abrupt events.

[16]See Exercise 1.7.

1.6 Transport properties in the mean free path approximation

The mean free path is the average distance travelled by a molecule between successive collisions. If we consider any point in the gas, the molecules that arrive there come from regions that are, roughly, one mean free path away (see Fig. 1.6). At any point, then, the properties of the gas are a mixture of the gas properties in a neighbourhood of radius ℓ. This mixing process does not produce any effect in equilibrium, but if the system is out of equilibrium, it will lead to the transport processes that are dominant in gases. In this section, we will present a rough derivation of some of those transport processes and estimate the associated transport coefficients using the idea mentioned above. Although

Fig. 1.6 At any point in the gas, the molecules come from regions that are, on average, one mean free path away.

this is not accurate and does not constitute a full theory of transport, this model will provide us with intuition and order-of-magnitude estimates. In Chapter 4, we will develop the full theory of the transport processes in gases by means of the Boltzmann equation.

1.6.1   Thermal conductivity

Consider a gas that is in contact with two parallel and infinite walls, separated by a distance L and having fixed temperatures T_- and T_+ (Fig. 1.7). The system has been allowed to relax, and it is stationary and static. Under this condition, the gas develops a temperature profile that depends only on the vertical coordinate z.

Imagine a mathematical surface parallel to the walls as indicated in the figure, located at a fixed height z. To compute the heat flux at z we notice that, as the mean free path is finite, particles that cross the surface come from a neighbourhood of radius ℓ. To simplify the analysis, we consider that there are only six possible directions of motion: $\pm\hat{\mathbf{x}}$, $\pm\hat{\mathbf{y}}$, and $\pm\hat{\mathbf{z}}$. Only the latter is relevant for energy transport through the surface. Then, $1/6$ of the particles at z come from $z+\ell$ and another $1/6$ from $z-\ell$, each of them crossing the surface carrying kinetic energy $3k_BT(z\pm\ell)/2$. Finally, the flux of particles crossing the surface equals $n\langle|\mathbf{c}|\rangle$. Therefore, the heat flux at z is estimated by

Fig. 1.7 Heat transfer between two parallel walls kept fixed at different temperatures. The gas inside the system develops the temperature profile $T(z)$.

$$q_z = n\frac{\langle|\mathbf{c}|\rangle}{6}\left[\frac{3}{2}k_BT(z-\ell) - \frac{3}{2}k_BT(z+\ell)\right], \qquad (1.29)$$

where the signs indicate that particles from $z-\ell$ move upward, while the opposite happens for particles moving from $z+\ell$. Applying a Taylor expansion for small ℓ, we obtain

$$q_z = -\left(\frac{n\langle|\mathbf{c}|\rangle k_B\ell}{2}\right)\frac{dT}{dz}. \qquad (1.30)$$

Fourier's law of thermal conductivity, $\mathbf{q} = -\kappa\nabla T$ is recovered, and the thermal conductivity is given by

$$\kappa = \sqrt{\frac{k_BT}{\pi m}}\frac{k_B}{\sigma}, \qquad (1.31)$$

where the expressions for the mean free path and the average speed, $\langle|\mathbf{c}|\rangle = \sqrt{8k_BT/\pi m}$, have been used. This result is somehow surprising: the thermal conductivity does not depend on density.[17] Indeed, this result is true for gases: the heat transport is proportional to the density (more particles transport more energy) and to the mean free path (the longer the mean free path, the larger the region the molecules explore, coming therefore from regions with greater differences in temperature). The cancellation is due to the inversely proportional dependence of the mean free path on density. We also note that the thermal conductivity is proportional to the mean speed, a condition that can be verified experimentally by changing the temperature or the molecular mass, for example, by using isotopic mixtures.

[17]In 1866, Maxwell obtained, by a similar argument, that the viscosity for low density gases is also independent of density. To verify this surprising result he conducted experiments that confirmed this.

In Chapter 4 we will develop a full consistent theory for gases, and the transport coefficients that stem from it will be obtained. Rather surprisingly, we will see that the mean free path model is also quantitatively accurate. However, it must be remembered that, in general, the mean free path model only allows us to make estimations and is not a quantitative theory.

1.6.2 Viscosity

Consider a gas between two parallel walls that move relative to each other. After a while, if the relative velocity is not large, the gas will reach a stationary state in which the average velocity varies continuously and monotonically in space to match the wall velocities. Under ideal conditions (isothermal gas, pure gas, and no gravity) the average velocity varies linearly. This is what we call Couette flow. Specifically, if we consider the geometry presented in Fig. 1.8, the average gas velocity is

$$\mathbf{v} = V'y\hat{\mathbf{x}}, \tag{1.32}$$

where $V' = V_0/L$ is the imposed velocity gradient.

Consider now the mathematical surface, parallel to the flow, presented in the figure, which is located at a distance y from the bottom wall and moves with the fluid. Particles arrive there from regions that are one mean free path away. In the x and z directions the system is homogeneous, therefore one only has to consider the y dependence of the distribution function.

Using the same procedure as in Section 1.6.1, we now estimate the total momentum along the x direction that crosses the surface per unit time. Particles moving up and down (coming from regions at $y - \ell$ and $y + \ell$, respectively) carry horizontal momentum and make a contribution to the stress tensor,

$$P_{xy} = \frac{n\langle|\mathbf{c}|\rangle}{6}\left[mv_x(y - \ell) - mv_x(y + \ell)\right], \tag{1.33}$$

where we have used that, in a Couette flow, both n and T are uniform. Applying a Taylor expansion, we find

$$P_{xy} = -\left(\frac{mn\langle|\mathbf{c}|\rangle\ell}{3}\right)\frac{dv_x}{dy}. \tag{1.34}$$

Newton's law of viscous transport, $P_{xy} = -\eta\frac{dv_x}{dy}$, is recovered with a viscosity given by

$$\eta = \frac{m\langle|\mathbf{c}|\rangle}{3\sqrt{2}\sigma} = \sqrt{\frac{k_{\mathrm{B}}T}{\pi m}}\frac{2m}{3\sigma}, \tag{1.35}$$

which is again independent of density.

Having obtained the viscosity and the thermal conductivity, we can compute the Prandtl number Pr, which is of importance in fluid mechanics. It is defined as $\mathrm{Pr} = c_p\eta/\kappa$, where c_p is the heat capacity per unit

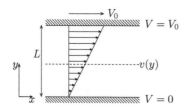

Fig. 1.8 Couette geometry used to study viscous flows. The gas is bounded between two parallel walls: the bottom one is kept fixed, while the top wall moves at a constant speed V_0. The gas develops a velocity profile $\mathbf{v} = \frac{V_0}{L}y\hat{\mathbf{x}}$.

mass at constant pressure. Using the expressions for the transport coefficients (1.31) and (1.35) in the mean free path approximation, $Pr = \gamma$, where $\gamma = c_p/c_V$ is the ratio of the specific heats, which for monoatomic gases is $\gamma = 5/3$. In real gases, its value is rather close to $Pr = 2/3$ and $Pr = 5/7$ for monoatomic and diatomic gases, respectively, as can be seen from Table 1.1. This discrepancy is due to the crude approximations in the mean free path description, and we will obtain the correct value when working with the Boltzmann equation in Chapter 4.

Table 1.1 Prandtl number (Pr) for various gases at 20 °C.

		Pr
Monoatomic gases	He	0.69
	Ne	0.66
	Ar	0.67
Diatomic gases	N_2	0.72
	O_2	0.72
	NO	0.75
	CO	0.75

1.6.3   Wall slip

In the previous subsection we said that, in a Couette flow, the average gas velocity matches the wall speed. Here, we explain the origin of this phenomenon and show that, in rarefied gases, this is not exactly true. Indeed, consider the gas near a solid static wall, which to be concrete we consider to be horizontal. After the particles collide with the wall, they are thermalised at the temperature of the wall and emitted back into the gas with a flux Maxwellian distribution (1.24). The average horizontal velocity of the emitted particles is thus zero. However, those are only half of the particles at the wall, the other half being those that approach it. They come from one mean free path away. Therefore, the average horizontal velocity at the wall is $v(y = 0) = [0 + v(y = \ell)]/2$. Applying a Taylor expansion, we can obtain the following relation at the wall:

$$v(0) = v'(0)\ell, \tag{1.36}$$

where $v' = \mathrm{d}v/\mathrm{d}y$. This relation should be used as the boundary condition instead of the usual nonslip boundary condition $[v(0) = 0]$.

Locally, keeping only linear terms in the Taylor expansion, the velocity profile results as $v(y) = (y+\ell)v'(0)$. More precise calculations using the Boltzmann equation (see Chapter 4) give

$$v(y) = (y + \beta)v'(0), \tag{1.37}$$

where β, called the slip length, is proportional to the mean free path for small velocity gradients and can be interpreted as shown in Fig. 1.9 as the fictitious position inside the wall where the velocity should vanish. In the mean free path model we have obtained $\beta = \ell$. Note that, if the mean free path is much shorter than the distance over which the velocity field varies, the nonslip boundary condition is effectively recovered. On the contrary, in the molecular or Knudsen regime, where the mean free path is large, slip effects appear.

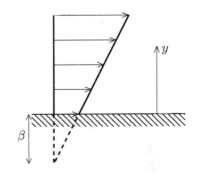

Fig. 1.9 Interpretation of the slip length β. In a shear flow, the velocity profile is linear and the velocity does not vanish at the wall. Rather, it is as if it vanishes at a distance β inside the wall. In a dilute gas, the slip length is proportional to the mean free path.

1.6.4   Self-diffusion

When we take a gas in equilibrium and look at the motion of a given molecule (tagged but otherwise identical to the others), we obtain that it follows straight trajectories interrupted by collisions that change its direction of motion. After a few collisions, the molecule has completely

[18]This should be distinguished from the diffusion process, in which particles or molecules of a different kind are placed in the gas. This process will be studied in detail in Chapter 5.

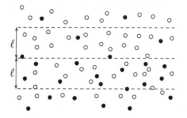

Fig. 1.10 Self-diffusion process. In a homogeneous gas, some molecules—identical to the others—are tagged. The flux of tagged particles follows Fick's law, $\mathbf{J}_T = -D\nabla n_T$, where n_T is the density of tagged molecules.

forgotten its original velocity, and we say that it has decorrelated from the initial state. The tagged molecule is said to follow a random walk. This is the self-diffusion process.[18]

If many molecules in a region are tagged, due to their random walks, they will diffuse into the system and ultimately mix homogeneously with the others. The rate at which this process is accomplished is given by the self-diffusion coefficient D. To compute it, consider a situation where the tagged molecules are almost evenly distributed with some small gradient in the z direction, as indicated in Fig. 1.10. At the mathematical surface shown in the figure, there is a flux of particles coming from $z + \ell$ and from $z - \ell$, with values given by the local density of tagged particles n_T,

$$J_T = [\langle|\mathbf{c}|\rangle n_T(z - \ell) - \langle|\mathbf{c}|\rangle n_T(z + \ell)]. \tag{1.38}$$

Applying, as usual, a Taylor expansion and writing in vectorial form, we obtain

$$\mathbf{J}_T = -2\ell\langle|\mathbf{c}|\rangle \nabla n_T. \tag{1.39}$$

We recover Fick's law of diffusion, $\mathbf{J}_T = -D\nabla n_T$, where the self-diffusion coefficient in the mean free path approximation, is $D = 2\ell\langle|\mathbf{c}|\rangle$.

1.7   Drude model for electric transport

As a first model of electric transport, charges are modelled as classical particles that collide with ions in the crystal lattice. This model, which we will work out in detail in Chapter 3, is helpful in providing some intuition on charge transport but is unable to explain many of the features of this process, mainly the qualitative difference between conductors and insulators. The origin of this failure is the use of a classical description for the electrons, instead of a quantum model that will be presented in Chapter 8 (Ashcroft and Mermin, 1976).

In the classical model, we can apply a mean free path description of the transport process of charges, known as the Drude model. Here, electrons in the presence of an electric field $\mathbf{E}$ are accelerated and, when they collide with an ion, instantly thermalised and emerge with a Maxwellian distribution at the temperature of the ion. After that, they start to be accelerated again. The mean free time between collisions with ions is $\tau = \ell/\langle|\mathbf{c}|\rangle$. Therefore, the mean electron velocity is

$$\langle\mathbf{c}\rangle = \frac{\tau q\mathbf{E}}{2m}, \tag{1.40}$$

where q and m are the electron charge and mass, and the factor $1/2$ comes from the average between the null mean velocity after the collision and the maximum speed just before the collision. The electric current $\mathbf{J} = nq\langle\mathbf{c}\rangle$ is therefore

$$\mathbf{J} = \frac{nq^2\tau}{2m}\mathbf{E}, \tag{1.41}$$

and we recover Ohm's law, $\mathbf{J} = \sigma\mathbf{E}$, with an electrical conductivity given by $\sigma = nq^2\tau/2m$.

Exercises

(1.1) **Average speed.** Compute the average speed $\langle|\mathbf{c}|\rangle$ for a gas in equilibrium at temperature T.

(1.2) **Collision frequency.** Derive eqn (1.27) for the collision frequency in a gas of density n and temperature T, where the collision cross section is σ.

(1.3) **Stress tensor and energy flux.** Prove the relations for the fluxes (1.20) and (1.21).

(1.4) **Kinetic pressure on a wall.** An ideal gas at equilibrium with density n and temperature T is placed in a container. We want to compute the force $\mathbf{F}$ that the molecules of the gas exert on the container walls. Consider a surface of area A oriented along the normal vector $\hat{\mathbf{n}}$ and use a similar procedure as in Section 1.3.

    (a) Consider a small time interval Δt. Determine the number of molecules with velocity $\mathbf{c}$ that will collide with the surface in that period.

    (b) In a collision, the molecules rebound elastically. Compute the momentum transferred to the wall by a molecule with velocity $\mathbf{c}$ that collides with it.

    (c) Integrate now over all possible velocities to compute the total transferred momentum per unit time in order to determine the force on the wall. Note that only molecules coming from inside can collide with the wall. Here, it could be useful to use an explicit coordinate system, for example with $\hat{\mathbf{n}}$ pointing along the z axis.

    (d) Show that the force points along $\hat{\mathbf{n}}$ and obtain the pressure p with the usual definition, $\mathbf{F} = pA\hat{\mathbf{n}}$. Show that p is given by the ideal gas equation of state.

(1.5) **Effusion.** An ideal gas at equilibrium with density n and temperature T is placed in a container. At one side of the container there is a small orifice, through which particles can escape. The area A of the orifice is small enough that the gas inside always remains at equilibrium.

    (a) Compute the particle flux through the orifice, that is, compute the number of particles that escape per unit time and area. Use the same approach as in Exercise 1.4.

    (b) Compute now the energy flux through the orifice.

    (c) Knowing that the energy in an ideal gas is $E = \frac{3}{2}Nk_BT$, use the previous results to determine the time derivative of the temperature. Does the gas cool down, heat up, or remain at constant temperature? Explain this result.

(1.6) **Black body radiation.** Consider a box that contains electromagnetic radiation in thermal equilibrium. The radiation is quantised in the form of photons with energies $\varepsilon = \hbar\omega$, where ω is the photon frequency. In a cubic box of length L, the possible electromagnetic modes are standing waves with wavevector $\mathbf{k} = \pi\mathbf{n}/L$, where the vector $\mathbf{n} = (n_x, n_y, n_z)$ has non-negative integer components. Finally, $\omega = c|\mathbf{k}|$, with c being the speed of light. According to the Planck distribution, the number of photons in each mode is

$$n_{\mathbf{k}} = \frac{1}{e^{\hbar\omega(\mathbf{k})/k_B T} - 1},$$

which we have to multiply by 2 to take into account the two polarisations of light. Consider an aperture of size A in one of the borders of the box. Compute the total energy flux per unit area through the aperture J_{total} and recover the Stefan–Boltzmann law,

$$J_{\text{total}} = \sigma T^4,$$

with

$$\sigma = \frac{2\pi^5 k_B^4}{15c^2 h^3} A.$$

(1.7) **Lorentz model.** Compute the collision frequency and mean free path of point particles with a Maxwellian distribution that collide with fixed obstacles of number density n when the collision is described by a total cross section σ.

(1.8) **Reaction rates.** Consider a gas composed of a mixture of two molecular species A and B that can perform the chemical reaction $A + B \rightarrow C + D$. When two molecules meet, not all collisions produce the reaction but only those that overcome a reaction barrier. Specifically, only those reactions in which the relative energy, $E_{\text{rel}} = \mu g^2/2$, is larger than a threshold ε are effective, where $\mu = m_A m_B/(m_A + m_B)$ is the reduced mass. Both species are at equilibrium at the same temperature and the mutual total cross section is σ.

Employing the same argument as used to obtain the collision frequency, compute the reaction rate, that is, the number of reactions per unit time and volume.

(1.9) **Hall effect.** The electron transport in a metal, described in Section 1.7, is modified when a magnetic field **B** is added on top of the imposed electric field $\mathbf{E}_x = E_x\hat{\mathbf{x}}$. In the Hall geometry, shown in Fig. 1.11, these two fields are perpendicular. The magnetic field generates a Lorentz force on the electrons that will deviate them along the y direction. If the conductor is finite in this direction, electrons will accumulate at one border, generating in turn an electric field $\mathbf{E}_y = E_y\hat{\mathbf{y}}$. This accumulation process will end when the y component of the total force on the electrons $\mathbf{F} = q(\mathbf{E} + \mathbf{c} \times \mathbf{B})$ vanishes. Compute the value of E_y. The Hall resistance R_H is defined by the relation $E_y = -R_\mathrm{H}BJ_x$, where J_x is the electric current. Using the Drude model compute R_H and show that it depends on the sign of the carrier charge.

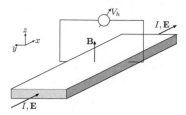

Fig. 1.11 Geometry used to measure the Hall effect.

(1.10) **Temperature drop.** When a gas is placed in contact with a wall at a given temperature it will evolve to equilibrate with it and have the same temperature as the wall. However, if the gas is simultaneously in contact with another wall at a different temperature, it will not reach a global equilibrium, but instead, a temperature profile will develop. Usually, the temperature profile is obtained by solving the heat equation with the boundary conditions that the gas has the same temperature as the walls. By a similar argument as the one presented in Section 1.6.3, a nonuniform gas in contact with a thermal wall will exhibit a temperature drop and its temperature in contact with the wall will be different. Using the mean free path approximations, compute this temperature drop in terms of the temperature gradient. Finally, considering the

case of a gas of density n between two walls at temperatures T_1 and T_2 and separated by a distance L, obtain the temperature profile.

(1.11) **Molecular heat transport.** A particular regime of heat transport occurs when the gas is so dilute that the mean free path is much larger than the system dimensions. The molecules then move freely from one side to another of the container, carrying without perturbation the information of the wall with which they had their last collision. In this regime, consider a gas of density n that is placed between two walls, separated by a distance L, that are kept at temperatures T_1 and T_2 ($T_2 > T_1$). The distribution function is such that, at each wall, particles emerge with a flux Maxwellian distribution at the respective temperature and densities n_1 and n_2. In the bulk, the distribution will be discontinuous on the velocities: for particles going to the right it is a Maxwellian with T_1, while for particles going to the left the temperature is T_2.

(a) Write down the full velocity distribution in terms of $T_{1/2}$ and $n_{1/2}$.

(b) Impose that the system is in steady state, meaning that there is no net mass flux, to determine $n_{1/2}$ in terms of $T_{1/2}$ and the global density n.

(c) Compute the energy flux q and derive the thermal conductivity $\kappa = -Lq/(T_2 - T_1)$.

(1.12) **Grad distribution.** As shown in the text, if the distribution function is Maxwellian, the pressure tensor is diagonal and the heat flux vanishes. This result indicates that the Maxwellian distribution cannot describe all physical situations, in particular those nonequilibrium conditions in which there are stresses or heat flux. Considering this, Grad argued in 1949 (Grad, 1949; Grad, 1958) that, in conditions close to equilibrium, one should use instead,

$$f(\mathbf{c}) = f_0 \left[1 + (\alpha C^2 - \gamma)\mathbf{C} \cdot \mathbf{q} + \delta p_{ij}\mathbf{C}_i\mathbf{C}_j\right],$$

where

$$f_0 = n(\mathbf{r}, t)\left(\frac{m}{2\pi k_B T(\mathbf{r}, t)}\right)^{3/2} e^{-mC^2/2k_B T(\mathbf{r}, t)}$$

is the local Maxwellian distribution, $\mathbf{C} = \mathbf{c} - \mathbf{v}(\mathbf{r}, t)$ is the peculiar velocity, and n, T, $\mathbf{v}$, α, γ, and δ can depend on position and time.

Find the constants α, γ, and δ such that f is properly normalised and that, when the heat flux and the stress tensor are computed, result in $\mathbf{q}$ and $P_{ij} = nk_\mathrm{B}T\delta_{ij} + p_{ij}$, respectively.

Distribution functions

2

2.1 Introduction

In the previous chapter we made use of the velocity distribution function to characterise the properties of the molecules in a gas or the charges in a classical model for electrons in a metal. Using this concept it was possible to derive transport laws and estimate the values of the associated transport coefficients. This was done using the Maxwellian distribution function and the simple mean free path approximation. However, we did not give any evidence for how different systems reach the Maxwellian distribution and whether the methodology proposed in that chapter is still valid out of equilibrium, as in the cases of a shear or electric flow, where we expect that the distributions will not be Maxwellian.[1] Besides, the mean free path approximation is not completely consistent. To compute the transport coefficients in this approximation, we use that at each point particles come from a neighbourhood with a radius equal to the mean free path, but, at the same time, we use that the velocity distribution is Maxwellian. However, the mixture of particles coming from regions with different properties does not produce a Maxwellian distribution. Finally, the calculations, although intuitive, very useful, and quite accurate, do not provide a route to extensions or generalisations. Rather, they appear to be ad hoc for each studied problem.

In this chapter we provide the tools necessary to build a systematic theory of transport processes. At the end of the chapter, we present a hierarchy of equations for the distribution functions where the need for approximations in order to have a closed kinetic description will become evident. This will be the task of subsequent chapters, where approximations valid for different systems will be justified and constructed.

In this chapter we consider the case of classical systems, whereas in Chapters 7, 8, and 9 we will extend the description to quantum systems in an appropriate manner. The description will be done for classical particles following conservative Hamiltonian dynamics, which is the relevant case for modelling most systems. The extension to nonconservative systems will be worked out at the end of this chapter, in Section 2.11.

To achieve a more systematic derivation of the kinetic theory formalism, in this chapter we will use the Hamiltonian description of classical mechanics.[2] The Hamiltonian formalism will be used only rarely in subsequent chapters, and readers not familiar with it can concentrate on the results and discussions, leaving the more technical details for later study.

[1]See Exercise 1.12.

[2]This was established by W.R. Hamilton in 1833, who started from the Lagrangian mechanics, introduced by J.L. Lagrange in 1788.

Kinetic Theory and Transport Phenomena, First Edition, Rodrigo Soto.
© Rodrigo Soto 2016. Published in 2016 by Oxford University Press.

2.2   Hamiltonian dynamics

[3]In this chapter we will adopt the Hamiltonian notation instead of using position and velocities because the former description will allow us to demonstrate important statistical properties.

In classical mechanics, the state of a system is completely specified by giving all the generalised coordinates and momenta q_a and p_a, where the index a runs from one to the number of degrees of freedom.[3] The dynamics of the system is specified via the Hamiltonian H that depends on the coordinates and momenta such that the equations of motion are

$$\dot{q}_a = \frac{\partial H}{\partial p_a}, \qquad\qquad \dot{p}_a = -\frac{\partial H}{\partial q_a}, \qquad (2.1)$$

termed the Hamilton equations. Note that these are differential equations of first order, which are equivalent to the second-order equations in the Newton or Lagrange formulations of classical mechanics.[4] To simplify the notation and later find solutions for the dynamics of systems with many degrees of freedom, it is convenient to define the Poisson bracket between two functions of the coordinates and momenta, f and g, as[5]

[4]For example, consider a single particle moving in one dimension in an external potential $U(x)$. The Hamiltonian is $H = p^2/2m + U(x)$, and the Hamilton equations read in this case $\dot{x} = p/m$ and $\dot{p} = -U'(x)$. Combining these two equations we recover the usual Newton equation, $m\ddot{x} = -U'(x)$.

[5]In this book we use the sign convection chosen in the series of books of Landau and Lifshitz. The book by Goldstein *et al.* (2013) uses the opposite sign convention.

$$\{f, g\} = \sum_a \left(\frac{\partial f}{\partial p_a} \frac{\partial g}{\partial q_a} - \frac{\partial g}{\partial p_a} \frac{\partial f}{\partial q_a} \right). \qquad (2.2)$$

With this definition, the Hamilton equations can be written as $\dot{q}_a = \{H, q_a\}$ and $\dot{p}_a = \{H, p_a\}$. Finally, the state vector of the system is represented as $\Gamma = (q_1, p_1, q_2, p_2, \dots)$, which sits in the phase space. For a three-dimensional system composed of N particles, the phase space has $6N$ dimensions. The whole system is represented by a single point in phase space, and the system evolution is represented by a trajectory in the phase space. Using this notation, the Hamilton equations reduce to

$$\dot{\Gamma} = \{H, \Gamma\}, \qquad (2.3)$$

and we define the flux vector $\Phi = \{H, \Gamma\}$, which gives the generalised velocity of the system in phase space. Note that the equations of motion $\dot{\Gamma} = \Phi(\Gamma)$ are of first order, and therefore the trajectories do not cross in phase space (see Fig. 2.1).

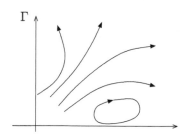

Fig. 2.1 Representation of the phase space. The state of the system at any instant is given by a point. Different initial conditions lead to different trajectories, which are shown in the figure.

2.3   Statistical description of the phase space

When a system is composed of many particles and the number of degrees of freedom becomes large, it is in practice impossible to give in detail its state. The coordinates and momenta present uncertainties, and for example, the initial condition is not given as an infinitesimal point in phase space but rather as a region of size $\Delta\Gamma$, as represented in Fig. 2.2. Note that this region need not be connected. For example, if the particles are indistinguishable, the uncertainty in position should be included in all possible permutations of the particles. Also, if initially other global quantities are measured with some error, for example the energy, instead of the positions and momenta, the initial conditions in phase space are all those that are compatible with the measurements, for example in a range of energies.

An important question that emerges is how this uncertainty will evolve in time. In going from time t to time $t+\Delta t$, the system moves from $\Gamma(t)$ to $\Gamma(t+\Delta t)$ according to the Hamilton equations. Then, the size of the uncertainty region is modified by the change of variable $\Delta\Gamma(t+\Delta t) = J\Delta\Gamma(t)$, where J is the Jacobian of the transformation,

$$J = \begin{vmatrix} \frac{\partial q(t+\Delta t)}{\partial q(t)} & \frac{\partial q(t+\Delta t)}{\partial p(t)} \\ \frac{\partial p(t+\Delta t)}{\partial q(t)} & \frac{\partial p(t+\Delta t)}{\partial p(t)} \end{vmatrix}. \qquad (2.4)$$

In the small time step Δt, the Hamilton equations can be integrated to give $q(t+\Delta t) = q(t) + \frac{\partial H}{\partial p}\Delta t + \mathcal{O}(\Delta t^2)$ and $p(t+\Delta t) = p(t) - \frac{\partial H}{\partial q}\Delta t + \mathcal{O}(\Delta t^2)$. On substituting these expressions into eqn (2.4), we find that the Jacobian equals one plus small corrections of order Δt^2. Hence, the uncertainties in phase space remain constant under Hamiltonian dynamics. This result, known as the Liouville theorem, should be reconciled with the fact that generically any system with more than three degrees of freedom is chaotic; that is, the evolution of two nearby initial conditions will lead to trajectories that separate exponentially in time. We have, on the one hand, that the volume occupied by different trajectories that started in the region of uncertainty $\Delta\Gamma$ remains constant, but, on the other, that those trajectories separate exponentially fast. The solution to this conundrum is to admit that the initial region of phase space will deform, with some directions that expand while others contract, such as to keep the volume constant. In complex systems, these directions change over the course of time, and consequently, the region of initial conditions becomes extremely deformed. If we follow the evolution of the system for a long time, the set of phase points that were in the region of the initial conditions will spread throughout the phase space, reaching regions compatible with the global conservations (normally the energy is the only conserved quantity, and therefore the available phase space is the surface of equal energy).

Under these circumstances, it becomes natural to question the relevance of giving the system state by a single point Γ. Furthermore, we will normally be interested in measurable quantities. However, any measurement process takes a finite (albeit short) time to achieve and, consequently, averages over different states. If we make predictions based on some measurement, the errors introduced by this blurring effect will persist, as the Liouville theorem shows. Besides, if we were interested in microscopic details, such as the position of a given particle, the error associated with this value would grow exponentially.

It is appropriate then to appeal to a statistical approach. Instead of giving the precise state of the system, we give the probabilities that the system is in given states. We define the distribution function $F(\Gamma, t)$ such that $F(\Gamma, t)\,\mathrm{d}\Gamma$ is the probability that the system is in the vicinity $\mathrm{d}\Gamma$ of Γ at the instant t.[6] By the definition of probability, it is normalised

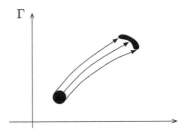

Fig. 2.2 When the initial condition has uncertainties, this is represented by a region in phase space, such that all points inside the region are compatible with the initial condition. The different possible initial conditions evolve, deforming this region.

[6]The distribution function can also be defined in the context of ensemble theory. Consider an ensemble of identical systems (with the same Hamiltonian but different initial conditions). Then, F gives the fraction of those systems that are in the specified state at a given instant.

as

$$\int F(\Gamma, t)\, d\Gamma = 1. \tag{2.5}$$

We will show in subsequent sections that the time evolution of any observable can be computed in terms of F and, therefore, that F contains all the necessary information regarding the system. We then aim to find an equation that describes the evolution of the distribution function in such a way that we can determine the probability of finding the system in some configuration given the initial conditions. To obtain the dynamical equation for F, we follow a similar procedure as in continuum mechanics to find the conservation equation of fluids or the charge conservation equation in electromagnetism. Here the motion takes place in phase space instead of real space. We define a volume Ω in the phase space. The probability of finding the system in this volume at a given instant is

$$P_\Omega(t) = \int_\Omega F(\Gamma, t)\, d\Gamma. \tag{2.6}$$

Its time derivative is simply

$$\dot{P}_\Omega(t) = \int_\Omega \frac{\partial F(\Gamma, t)}{\partial t}\, d\Gamma. \tag{2.7}$$

On the other hand, because probability is conserved, this time derivative equals the probability flux across the boundary of Ω. At each point of the boundary, the probability flux equals the probability that the system is in this configuration times the velocity in phase space Φ. Using the standard convention of normal vectors pointing outward, we obtain

$$\dot{P}_\Omega(t) = -\int_S F(\Gamma, t)\Phi(\Gamma) \cdot dS. \tag{2.8}$$

Using the divergence theorem and noting that eqns (2.7) and (2.8) should be equal for any volume Ω, we obtain

$$\frac{\partial F}{\partial t} = -\nabla_\Gamma \cdot (F\Phi), \tag{2.9}$$

which is the usual conservation or continuity equation. This equation can be further simplified by expanding the divergence in the right-hand side and using the properties of the Hamilton equations:

$$\begin{aligned}
\nabla_\Gamma \cdot (F\Phi) &= (\nabla_\Gamma \cdot \Phi)\, F + (\nabla_\Gamma F) \cdot \Phi \\
&= \sum_a \left[\frac{\partial}{\partial q_a}\left(\frac{\partial H}{\partial p_a}\right) + \frac{\partial}{\partial p_a}\left(-\frac{\partial H}{\partial q_a}\right) \right] F \\
&\quad + \sum_a \left[\frac{\partial F}{\partial q_a}\frac{\partial H}{\partial p_a} + \frac{\partial F}{\partial p_a}\left(-\frac{\partial H}{\partial q_a}\right) \right],
\end{aligned} \tag{2.10}$$

where we have used that the flux vector has components $\Phi_a = (\frac{\partial H}{\partial p_a}, -\frac{\partial H}{\partial q_a})$. The first term vanishes identically, while the second equals the Poisson bracket $\{H, F\}$. We have thus obtained the Liouville equation for the distribution function, valid for Hamiltonian dynamics,

$$\frac{\partial F}{\partial t} = -\{H, F\}. \tag{2.11}$$

The previous discussion of the deformation of the set of possible initial conditions by the combination of the Liouville theorem and the chaotic dynamics can be reinterpreted using the language of the distribution function. The condition represented in Fig. 2.2 means that $F(\Gamma, 0)$ is constant inside a small region and zero outside. This initial discontinuity of F is preserved subsequently, leading to the development of a highly elongated and tortuous distribution function. However, if instead of a discontinuous initial distribution function, we adopt a smooth function (e.g. a Gaussian distribution around a central point, compatible with the measured values), then the distribution function will evolve to a smooth function that will cover the phase space (Fig. 2.3).

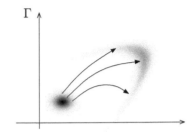

Fig. 2.3 Evolution of a distribution function with a smooth initial condition. The chaotic stretching and compression of the trajectories will lead to a smooth coverage of the phase space.

The Liouville equation (2.11) was derived assuming that the trajectories in phase space were continuous. This is the case in most situations, but in some models, discontinuous trajectories appear. The simplest case consists of a gas confined by a rigid wall. When the particles hit the wall, their momentum is instantaneously reversed and, hence, the trajectory jumps in phase space. This case can be easily incorporated into the Liouville equation through boundary conditions. More complex is the hard sphere model for gases and liquids, in which particles change their momenta instantaneously at collisions. In this case, the evolution of the distribution function can be derived by noting that the probability of finding the system in a state Γ equals the probability that the system was previously in a state Γ', such that the temporal evolution of Γ' after an interval Δt is Γ. Taking the limit of small time intervals, a pseudo-Liouville equation is obtained.[7] In this book, however, we will consider only continuous dynamics described by the Liouville equation (2.11).

[7]It is called pseudo-Liouville because the differential equation presents fewer analytical properties as compared with the Liouville equation.

2.4 Equilibrium distribution

We know from equilibrium statistical mechanics that, in thermodynamic equilibrium, the distribution function does not depend on time, and for example, if the system is in contact with a thermal bath, it is given by the equilibrium canonical distribution, $F_{\text{eq}} = Z^{-1} e^{-H/k_B T}$, where Z is the partition function. But, is this description compatible with the Liouville equation (2.11)? Can we derive this equilibrium distribution from the Liouville equation?

If we impose that the distribution function is stationary, the Liouville equation reduces to the condition

$$\{H, F_0\} = 0. \tag{2.12}$$

We know from classical mechanics that this equation admits as solution any function of the Hamiltonian $F_0 = f(H)$. Indeed, it can be easily

[8]See Exercise 2.2.

[9]Indeed, a caveat should be made here. If the system, besides energy, presents other conserved quantities $X_1, X_2, \ldots$ (for example, the total linear and angular momenta in a galaxy, or the spin in a nonmagnetic medium), the stationary distributions are of the form, $F_0 = f(H, X_1, X_2, \ldots)$. The initial condition fixes the probabilities of the conserved quantities, and the stationary distribution is uniform on the iso-conserved-quantities hypersurfaces.

verified by applying the chain rule that $\{H, f(H)\} = 0$.[8] That is, the condition of stationarity admits the canonical distribution as one possible solution, but many other solutions are possible. How then is the canonical distribution selected among this myriad of possibilities? As a first answer, let us say that, in the Liouville equation, we did not incorporate a bath or how the system interacts with it, so it is not surprising that we do not get the canonical distribution as a unique solution. However, before proceeding to the main point, let us note that the stationary solutions $F_0 = f(H)$ give equal probabilities to any state with the same energy; that is, on isoenergy hypersurfaces, the distributions are uniform. This completes our discussion of how an initial condition with some finite uncertainty evolves. The initial region will extend into the phase space until it uniformly covers each isoenergy hypersurface. The probability of being on any of these isohypersurfaces depends on how probable it is to have different energies in the initial condition, as reflected in the function f. This is an effect of the dynamics being Hamiltonian, where energy is conserved and, hence, trajectories (and the associated probabilities) cannot cross isoenergy hypersurfaces.

So, we have found that, in a purely Hamiltonian dynamics, the final stationary distribution depends on the probability of having different energy values in the initial condition. All other information is lost, and the probability distribution is otherwise uniform.[9] This is the case for the microcanonical ensemble, in which the system is isolated and evolves conserving energy. If the initial condition is such that the energy E is fixed with high precision, the stationary distribution is then given by a Dirac delta function,

$$F_0(\Gamma) = \Omega^{-1}\delta(H(\Gamma) - E), \qquad (2.13)$$

where the normalisation constant Ω is the phase space volume of the isoenergy hypersurface. This is the well-known microcanonical distribution, being $S = k_B \ln \Omega$, the Boltzmann entropy.

To reach the canonical distribution, the system must be kept in contact with a thermal bath; that is, it must exchange energy with a larger system, in a way that we need not specify except that the interaction is sufficiently weak that the internal dynamics of the system is almost unperturbed. If we consider the bath and the studied system as a whole, the total system should be described by a big Hamiltonian H_T to which we can apply the same methods as before to conclude that it will reach the microcanonical distribution. Then, we can use the same arguments as in equilibrium statistical mechanics to derive that, for the smaller system under study, the stationary distribution is the canonical

$$F_0(\Gamma) = Z^{-1}e^{-H(\Gamma)/k_B T}, \qquad (2.14)$$

with a temperature given by $T = \left(\frac{\partial S_B}{\partial E_B}\right)^{-1}$, which is computed in terms of the bath entropy S_B and energy E_B. The normalisation factor is the partition function, $Z = \int e^{-H(\Gamma)/k_B T}\, d\Gamma$.

In both cases, we have shown that the Liouville equation admits as stationary solutions the microcanonical or canonical equilibrium distributions, depending on whether the system is isolated or in contact with a thermal bath. However, no information is given on the mechanism that allows it to reach equilibrium, the time scales associated with the process of equilibration, or the role played by different energy exchange mechanisms with the bath. The answer to these questions, which is the main purpose of this book, depends strongly on the type of system one considers. Different approaches will be needed depending on the nature of the particle dynamics and their interactions. Nevertheless, in the next section, we accomplish the derivation of a quite generic kinetic equation that will be the starting point for the different approximations made in the following chapters.

2.5   Reduced distributions

One should note that the distribution function F is a function of all the coordinates and momenta in the system. That is, for a macroscopic system, is has roughly 10^{24} variables and the Liouville equation is a first-order differential equation for 10^{24} variables. With all this detail, the Liouville equation is indeed equivalent to the Hamilton equations.[10] Hence, it contains the detailed information concerning all the particles, their precise motion, and interactions. This equivalence makes it an exact equation and, therefore, an excellent starting point for any theoretical development of the statistical mechanics of many-particle systems. However, being exact, it is rather useless, as there should be no hope of finding solutions to it in its present form, except for the equilibrium solutions already described. We need to apply some approximations; for example, in a gas, we will be interested in eliminating the interactions between distant particles. To facilitate the implementation of different approximation schemes, we will define reduced distribution functions and then find the equations that dictate their temporal evolution.

To be concrete, let us first consider a system of N identical particles. In Section 2.8, we will generalise this analysis to a mixture of different species. Also, we will use Cartesian coordinates such that the generalised coordinates and momenta are written $\mathbf{r}_a$ and $\mathbf{p}_a$, with $a = 1, \ldots, N$. It is natural to symmetrise the distribution function such that it does not distinguish between permutations of the identical particles. We define the symmetrised distribution function as

$$\widehat{F}(\Gamma, t) = \frac{1}{N!} \sum_P F(P\Gamma, t), \qquad (2.15)$$

where $P\Gamma$ permutes the positions and momenta of the particles and the sum runs over possible permutations P of N indices.

The reduced distribution functions of $1 \leq n \leq N$ particles are defined to be an extension of the one-particle distribution studied in detail in Chapter 1. Here, we retain the information of a subset of n particles,

[10]If we know the initial condition of the system Γ_0 exactly, the initial distribution function is $F(\Gamma, 0) = \delta(\Gamma - \Gamma_0)$. As the dynamics is deterministic, this distribution will evolve to $F(\Gamma, t) = \delta(\Gamma - \Gamma_t)$, where Γ_t is the evolved phase point. Therefore, the Liouville equation contains the solutions of the Hamilton equations, showing their equivalence.

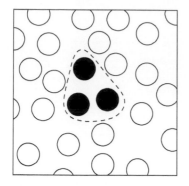

Fig. 2.4 The reduced distributions, in this case $F^{(3)}$, give information regarding a reduced subset of particles inside a large system.

while the other positions and momenta are disregarded (integrated out). These are called marginal distributions in probability theory. We then define

$$F^{(n)}(\mathbf{r}_1, \mathbf{p}_1, \ldots, \mathbf{r}_n, \mathbf{p}_n, t)$$
$$= \frac{N!}{(N-n)!} \int \widehat{F}(\Gamma, t) \, \mathrm{d}^3 r_{n+1} \, \mathrm{d}^3 p_{n+1} \ldots \mathrm{d}^3 r_N \, \mathrm{d}^3 p_N, \quad (2.16)$$

where the advantage of the factorial prefactor will become evident below.

We will use the compact notation, $a \equiv (\mathbf{r}_a, \mathbf{p}_a)$, such that the reduced distributions (Fig. 2.4) are

$$F^{(n)}(1, \ldots, n; t) = \frac{N!}{(N-n)!} \int \widehat{F}(1, \ldots, N; t) \, \mathrm{d}(n+1) \ldots \mathrm{d}N, \quad (2.17)$$

being normalised to

$$\int F^{(n)}(1, \ldots n; t) \, \mathrm{d}1 \ldots \mathrm{d}n = \frac{N!}{(N-n)!}. \quad (2.18)$$

Note that, by normalisation, these distributions are not probability densities; instead they give the densities of n-tuples of particles in the reduced phase space, regardless of the configuration of the others. Indeed, consider first the case $n = 1$. The distribution $F^{(1)}(\mathbf{r}, \mathbf{p})$ is normalised to N. It gives the number of particles in a neighbourhood of the specified position and momentum and, therefore, is equivalent to the one-particle distribution studied in Chapter 1, except that it depends on momentum. Hence, they are related by $f^{(1)}(\mathbf{r}, \mathbf{c}) = m^3 F^{(1)}(\mathbf{r}, \mathbf{p})$. The two-particle distribution, normalised to $N(N-1)$, gives the number of pairs with specified positions and momenta, and so on.

After neglecting any spatial dependence, the integral over momenta of the reduced distributions scale, for the first two, as $\int F^{(1)}(\mathbf{q}, \mathbf{p}) \, \mathrm{d}^3 p = N/\mathcal{V}$ and $\int F^{(2)}(\mathbf{r}_1, \mathbf{p}_1, \mathbf{r}_2, \mathbf{p}_2) \, \mathrm{d}^3 p_1 \, \mathrm{d}^3 p_2 = N(N-1)/\mathcal{V}^2$. In the thermodynamic limit ($N \to \infty$, $\mathcal{V} \to \infty$, while $N/\mathcal{V} \to n$, finite), we obtain the normalisations $\int F^{(1)}(\mathbf{r}, \mathbf{p}) \, \mathrm{d}^3 p = n$ and $\int F^{(2)}(\mathbf{r}_1, \mathbf{p}_1, \mathbf{r}_2, \mathbf{p}_2) \, \mathrm{d}^3 p_1 \, \mathrm{d}^3 p_2 = n^2$. Hence, the reduced distribution functions for finite n have well-defined thermodynamic limits. Note that this would be not the case if the factorial prefactor were not introduced in their definition in eqn (2.17). In that case, they would have vanished in the thermodynamic limit. It is worth mentioning that neither F is well defined in the thermodynamic limit; firstly, it would have an infinite number of variables, and secondly, it would be impossible to normalise to unity.

We have seen that the reduced distributions thus have two important advantages over the full distribution function F. They present the information of only a reduced subset of particles, making it easier to analyse and perform approximations on them, and they are well behaved in the thermodynamic limit.[11] In the next section, we will show that most of the average properties in a system can be computed in terms of a few reduced distribution functions. Finally, in Section 2.7, we will derive the equations that describe their time evolution.

[11] Under nonequilibrium conditions, to observe irreversible behaviour, we should first take the thermodynamic limit and later the observation times must be large. Otherwise, for finite systems, quasiperiodic motion would be observed at large times, which is not irreversible. This order of the limiting processes is difficult to perform with the full distribution, but we will see that it is simpler in the reduced distribution language. See the discussion in Section 4.3.3.

2.6   Microscopic and average observables

For any system under study, we will be interested in the temporal evolution, spatial distribution, and/or average values of different observables. Those observables may be global quantities such as the total energy, space-dependent fields such as the mass density, or fine-grained quantities such as the position of a tagged particle. In all these cases, we can define microscopic observables $A(\Gamma)$ as functions of the phase space state, $\Gamma = (\mathbf{r}_1, \mathbf{p}_1, \mathbf{r}_2, \mathbf{p}_2, \ldots, \mathbf{r}_N, \mathbf{p}_N)$, that give the value of the desired quantity; for example, the total momentum is simply $\mathbf{P}(\Gamma) = \sum_a \mathbf{p}_a$. For the statistical state of the system described by the distribution function, the (phase space) average of $A(\Gamma)$ is

$$\langle A \rangle(t) = \int F(\Gamma, t) A(\Gamma) \, d\Gamma, \tag{2.19}$$

which will generally depend on time.

In the following subsections, some relevant observables will be presented and it will be shown how their averages can be computed in terms of the reduced distribution functions. The different observables are classified as global, densities, and fluxes, as explained below.

2.6.1   Global observables

This type of observable gives a single value that characterises some property of the system. For example, in a system described by a Hamiltonian of N identical particles with two-body interactions,

$$H = \sum_a \frac{p_a^2}{2m} + \sum_{a<b} \phi(\mathbf{r}_a - \mathbf{r}_b), \tag{2.20}$$

the microscopic observables that give the kinetic and potential energy are

$$K = \sum_a \frac{p_a^2}{2m}, \tag{2.21}$$

$$U = \sum_{a<b} \phi(\mathbf{r}_a - \mathbf{r}_b). \tag{2.22}$$

Their averages are

$$\begin{aligned}
\langle K \rangle &= \int F(\Gamma, t) \sum_a \frac{p_a^2}{2m} \, d\Gamma \\
&= N \int \widehat{F}(\Gamma, t) \frac{p_1^2}{2m} \, d\Gamma \\
&= \int F^{(1)}(\mathbf{r}_1, \mathbf{p}_1, t) \frac{p_1^2}{2m} \, d^3 r_1 \, d^3 p_1,
\end{aligned} \tag{2.23}$$

where we have used that the particles are identical and the definition of $F^{(1)}$. Analogously for the potential energy,

$$
\begin{aligned}
\langle U \rangle &= \int F(\Gamma, t) \sum_{a<b} \phi(\mathbf{r}_a - \mathbf{r}_b)\, d\Gamma \\
&= N(N-1) \int \widehat{F}(\Gamma, t)\phi(\mathbf{r}_1 - \mathbf{r}_2)\, d\Gamma \\
&= \int F^{(2)}(\mathbf{r}_1, \mathbf{p}_1, \mathbf{r}_2, \mathbf{p}_2, t)\phi(\mathbf{r}_1 - \mathbf{r}_2)\, d^3r_1\, d^3p_1\, d^3r_2\, d^3p_2,
\end{aligned}
\tag{2.24}
$$

which now depends on the two-particle distribution function. The resulting expressions for $\langle K \rangle$ and $\langle U \rangle$ are readily interpreted by recalling that $F^{(1)}$ and $F^{(2)}$ are the one- and two-particle distributions. Again, the advantage of the factorial prefactor in their definition is evident.

We have shown that the averages of simple global observables can be obtained using only the first two reduced distribution functions and there is no need to use the full distribution function. Other examples are presented in the exercise section (Exercises 2.4 to 2.7).

2.6.2 Densities

Besides global quantities, we will also be interested in density fields. In the case of fields where the property φ is carried by the particles themselves, as in the case, for example, of mass, charge, momentum, or energy densities, we say that the field is located at the position of the particles. We make use of the fact that the Dirac delta functions have units of $1/\text{length}^3$ to define the density field associated with φ as

$$
\rho_\varphi(\mathbf{r}) = \sum_a \varphi(\mathbf{r}_a, \mathbf{p}_a)\delta(\mathbf{r} - \mathbf{r}_a),
\tag{2.25}
$$

which, we should note, depends on the observation point $\mathbf{r}$ and also on the phase space through $\varphi_a = \varphi(\mathbf{r}_a, \mathbf{p}_a)$ and $\mathbf{r}_a$. It is common to consider also its Fourier transform,[12]

$$
\widetilde{\rho}_\varphi(\mathbf{k}) = \int e^{i\mathbf{k}\cdot\mathbf{r}} \rho_\varphi(\mathbf{r})\, d^3r = \sum_a \varphi_a e^{i\mathbf{k}\cdot\mathbf{r}_a}.
\tag{2.26}
$$

The phase space average is done analogously to the kinetic energy case in the previous section, i.e.

$$
\langle \rho_\varphi \rangle(\mathbf{r}, t) = \int F^{(1)}(\mathbf{r}, \mathbf{p}_1, t)\varphi(\mathbf{r}, \mathbf{p}_1)\, d^3p_1,
\tag{2.27}
$$

$$
\langle \widetilde{\rho}_\varphi \rangle(\mathbf{k}, t) = \int F^{(1)}(\mathbf{r}_1, \mathbf{p}_1, t)\varphi(\mathbf{r}_1, \mathbf{p}_1)e^{i\mathbf{k}\cdot\mathbf{r}_1}\, d^3r_1\, d^3p_1.
\tag{2.28}
$$

That is, they can be computed in terms of $F^{(1)}$ only.

2.6.3 Fluxes

If a density field is associated with a conserved quantity, it is expected that a flux field will exist, being related to the density via a conservation

[12]Note that the microscopic density is a highly irregular function that, for the purpose of computer calculations, should be coarse grained, for example using measurement cells or kernel smoothing functions. The Fourier transform, on the other hand, is a smooth and regular function, allowing the use of standard analytic or computer tools.

equation, $\frac{\partial}{\partial t}\rho_\varphi + \nabla \cdot \mathbf{J}_\varphi = 0$. This relation allows us to determine the associated flux field. We first consider the case of the mass density, which is quite direct, and then the momentum density will be analysed separately, as it deserves greater attention.

We compute the time derivative of the microscopic mass density, which depends implicitly on time though the particle coordinates,[13]

$$\frac{\partial}{\partial t}\rho(\mathbf{r}) = \frac{\partial}{\partial t}\sum_a m\delta(\mathbf{r} - \mathbf{r}_a) = -\sum_a m\delta'(\mathbf{r} - \mathbf{r}_a)\mathbf{c}_a$$
$$= -\sum_a \mathbf{p}_a \cdot \nabla\delta(\mathbf{r} - \mathbf{r}_a) = -\nabla \cdot \sum_a \mathbf{p}_a\,\delta(\mathbf{r} - \mathbf{r}_a), \qquad (2.29)$$

where we have used that $\dot{\mathbf{r}}_a = \mathbf{c}_a$ and that $\delta'(\mathbf{r} - \mathbf{r}_a) = \nabla\delta(\mathbf{r} - \mathbf{r}_a)$. Therefore, the mass flux can be identified as

$$\mathbf{J} = \sum_a \mathbf{p}_a\,\delta(\mathbf{r} - \mathbf{r}_a), \qquad (2.30)$$

which we note corresponds also to the momentum density. Its ensemble average is computed using the general expression for density fields as

$$\langle \mathbf{J}\rangle(\mathbf{r}, t) = \int F^{(1)}(\mathbf{r}, \mathbf{p}_1, t)\mathbf{p}_1\,\mathrm{d}^3 p_1. \qquad (2.31)$$

In Chapter 1 we showed that the flux associated with the momentum density is a tensor of rank two, called the stress tensor, and we derived an expression for it in the context of dilute gases [eqn (1.14)]. The momentum is also conserved in interacting systems, hence we should be able to derive an expression for the stress tensor for a system described by the Hamiltonian (2.20). To obtain it, we differentiate the momentum density (2.30) with respect to time, noting that the particle positions and momenta depend on time with $\dot{\mathbf{r}}_a = \mathbf{c}_a$ and $\dot{\mathbf{p}}_a = \mathbf{f}_a$, where $\mathbf{f}_a$ is the total force acting on particle a,

$$\frac{\partial \mathbf{J}}{\partial t} = -\sum_a m\mathbf{c}_a\,\mathbf{c}_b \cdot \nabla\delta(\mathbf{r} - \mathbf{r}_a) + \sum_a \mathbf{f}_a\delta(\mathbf{r} - \mathbf{r}_a). \qquad (2.32)$$

The first term is already in the form of a divergence; it remains to do so for the second term. We use the fact that the total force is obtained as a sum of pair forces $\mathbf{f}_a = \sum_{b\neq a}\mathbf{f}_{ab}$ to symmetrise the sum and obtain $\sum_a \mathbf{f}_a\delta(\mathbf{r} - \mathbf{r}_a) = \frac{1}{2}\sum_{a,b}[\mathbf{f}_{ab}\delta(\mathbf{r} - \mathbf{r}_a) + \mathbf{f}_{ba}\delta(\mathbf{r} - \mathbf{r}_b)] = \frac{1}{2}\sum_{ab}\mathbf{f}_{ab}[\delta(\mathbf{r} - \mathbf{r}_a) - \delta(\mathbf{r} - \mathbf{r}_b)]$, where the action–reaction principle $\mathbf{f}_{ba} = -\mathbf{f}_{ab}$ is used. Next, we write the difference of the Dirac delta functions in integral form as

$$\delta(\mathbf{r} - \mathbf{r}_a) - \delta(\mathbf{r} - \mathbf{r}_b) = \int_{\mathbf{r}_b}^{\mathbf{r}_a} \frac{d}{d\mathbf{s}}\delta(\mathbf{r} - \mathbf{s}) \cdot d\mathbf{s}$$
$$= -\nabla \cdot \int_{\mathbf{r}_b}^{\mathbf{r}_a} \delta(\mathbf{r} - \mathbf{s})\,d\mathbf{s}. \qquad (2.33)$$

Collecting all terms, we obtain the momentum conservation equation,

$$\frac{\partial \mathbf{J}}{\partial t} = -\nabla \cdot \mathbb{P}, \qquad (2.34)$$

[13]We recall that $\mathbf{r}$ is a parameter that indicates the observation point and, therefore, does not depend on time.

where the stress tensor $\mathbb{P}$ is defined by components as

$$P_{ik} = \sum_{a=1}^{N} m\, c_{a,i}\, c_{a,k}\, \delta(\mathbf{r} - \mathbf{r}_a) + \frac{1}{2} \sum_{a,b=1}^{N} f_{ab,k} \int_{\mathbf{r}_a}^{\mathbf{r}_b} \delta(\mathbf{r} - \mathbf{s})ds_i, \quad (2.35)$$

where $c_{a,i}$ represents the i-th Cartesian component of the velocity of particle a and analogously for $f_{ab,k}$.

The first contribution to the stress tensor is equivalent to the one we derived for dilute gases [eqn (1.14)] and corresponds to the momentum flux due to the particles that are moving, carrying their momentum along, i.e. the kinetic transfer. The second contribution is due to the interparticle interactions and indicates the momentum change (i.e. force according to the Newton equation) that is transmitted along the path that joins each pair of particles; that is, the interparticle force is delocalised in the region between the two particles (Fig. 2.5). In the original derivation of the stress tensor, made by Irving and Kirkwood in 1950, they noted that the integral representation in (2.33) is not unique. Indeed, any path going from $\mathbf{r}_a$ to $\mathbf{r}_b$ gives the same result. They argued that, for simplicity, one is tempted to choose the straight line joining the centres, but any other election is equally valid from a mathematical point of view. Physically, this implies that, having many equally valid ways of defining the stress tensor, it should not be a measurable quantity. Indeed, we never measure the stress tensor directly but rather its effects, which manifest in two ways: either as the integral over the surface of a volume to give the total force on the volume, or as a divergence to compute the rate of change of the momentum density. In both cases, the ambiguity in the definition disappears and any path gives the same physically measurable value. If the pair forces are central (i.e. $\mathbf{f}_{ab} = f_{ab}\hat{\mathbf{r}}_{ab}$), the choice of the straight-line path guarantees that the stress tensor is symmetric. This requirement is usually applied so that, in macroscopic equations (e.g. fluid equations), angular momentum is conserved at the level of the momentum equation and there is no need for a separate equation for angular momentum.

For a homogeneous system, the stress tensor can be averaged in space, simplifying to

$$\mathbb{P} = \frac{1}{\mathcal{V}} \int \mathbb{P}(\mathbf{r})\, d^3r = \frac{1}{\mathcal{V}} \left[\sum_{a=1}^{N} m\, \mathbf{c}_a\, \mathbf{c}_a + \frac{1}{2} \sum_{a,b=1}^{N} \mathbf{f}_{ab}\, \mathbf{r}_{ab} \right], \quad (2.36)$$

where $\mathbf{r}_{ab} = \mathbf{r}_a - \mathbf{r}_b$ and we recall that $\mathbf{f}_{ab}$ is the force that particle b exerts on particle a.

Finally, the pressure, defined as the isotropic part of the stress tensor (see discussion is Section 1.3.2), for a homogeneous system is

$$p = \frac{1}{3}\text{Tr}\mathbb{P} = \frac{1}{3\mathcal{V}} \left[\sum_{a=1}^{N} mc_a^2 + \frac{1}{2} \sum_{a,b=1}^{N} \mathbf{f}_{ab} \cdot \mathbf{r}_{ab} \right]. \quad (2.37)$$

This is the known virial expression for the pressure. The first term is the usual kinetic contribution that, for an equilibrium system, results in the

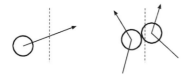

Fig. 2.5 Processes that contribute to the momentum flux. Momentum is transferred through the surface (dashed line). Kinetic transfer (left): a particle crosses the surface, transporting its own momentum. Collisional transfer (right): particles interact through the surface, exchanging momentum between them and, therefore, transferring it from one side to the other.

ideal gas expression, $nk_{\mathrm{B}}T$. The second term is due to the interparticle interactions and is proportional to the force intensity and its range. Attractive forces, where $\mathbf{f}_{ab} \cdot \mathbf{r}_{ab} < 0$, result in a pressure drop, while repulsive forces, on the other hand, increase the pressure.

The kinetic term of the pressure tensor depends on single-particle properties, while the interaction term depends on the positions of sets of two particles. Consequently, its ensemble average is computed using the one- and two-particle reduced distributions.[14]

Similarly to the mass and momentum conservations, we can derive the energy conservation equation. First we have to overcome an ambiguity concerning the energy density. The kinetic energy is naturally allocated at the position of each particle, but we have to decide where the potential energy is located. Many definitions are possible, but simpler expressions are obtained if it is distributed equally on both particles. In this way, the microscopic energy density and energy flux turn out to be

$$\rho_e = \sum_a \frac{p_a^2}{2m}\,\delta(\mathbf{r}-\mathbf{r}_a) + \sum_{a<b} \phi(\mathbf{r}_a - \mathbf{r}_b)\left[\frac{\delta(\mathbf{r}-\mathbf{r}_a) + \delta(\mathbf{r}-\mathbf{r}_b)}{2}\right],$$

$$(2.38)$$

$$\mathbf{J}_e = \sum_a \frac{p_a^2}{2m}\mathbf{v}_a\delta(\mathbf{r}-\mathbf{r}_a) + \sum_{a<b} \phi(\mathbf{r}_a - \mathbf{r}_b)\left[\frac{\mathbf{v}_a\delta(\mathbf{r}-\mathbf{r}_a) + \mathbf{v}_b\delta(\mathbf{r}-\mathbf{r}_b)}{2}\right]$$
$$+ \sum_{a<b} \left(\frac{\mathbf{v}_a\cdot\mathbf{f}_{ab} - \mathbf{v}_b\cdot\mathbf{f}_{ba}}{2}\right)\int_{\mathbf{r}_b}^{\mathbf{r}_a} \delta(\mathbf{r}-\mathbf{s})\,\mathrm{d}\mathbf{s}. \qquad (2.39)$$

The energy flux has three contributions. The first has already been studied in Chapter 1 and corresponds to the transport of kinetic energy by the motion of particles [eqn (1.13)]. Now, when potential energy is considered, the second and third terms appear. The second contribution corresponds to the potential energy that is transported by a pair of interacting particles, similarly to what happens to a hydrogen atom that transports a potential energy of $-13.6\,\mathrm{eV}$ with it as it moves. The third term is associated with the mechanical work done by one particle on another due to the interaction force. These three contributions are represented in Fig. 2.6.

2.6.4   Conservation equations

By construction, the microscopic mass density, momentum density, and energy density are related to the stress tensor and energy flux by the conservation equations,

$$\frac{\partial\rho}{\partial t} = -\nabla\cdot\mathbf{J}, \qquad (2.40)$$

$$\frac{\partial\mathbf{J}}{\partial t} = -\nabla\cdot\mathbb{P}, \qquad (2.41)$$

$$\frac{\partial\rho_e}{\partial t} = -\nabla\cdot\mathbf{J}_e. \qquad (2.42)$$

14 See Exercises 2.4 and 2.5.

Fig. 2.6 Processes that contribute to the energy flux. Energy is transferred through the surface (dashed line). Left: a particle crosses the surface, transporting its own kinetic energy. Middle: a pair of particles cross the surface, transporting their interaction potential energy. Right: energy is transferred from one particle to another due to the interaction force in the form of mechanical work.

One can directly ensemble-average these equations, which being linear in the densities and fluxes, retain their form. They are not closed though, because $\mathbb{P}$ and $\mathbf{J}_e$ are independent fields. To close these equations, they should be complemented by constitutive relations that relate $\mathbb{P}$ and $\mathbf{J}_e$ to the conserved fields and their gradients. If these relations were the viscous Newton law for the stress tensor and the Fourier law for the heat flux, we would obtain the usual hydrodynamic equations of viscous fluids. We should remark here that the conservation equations are a direct consequence of the microscopic conservations and are, therefore, exact. The approximations are made at the level of the constitutive relations, which are hence susceptible to amendment. For example, in the case of the viscous law, a memory effect can be included to take into account viscoelasticity, and noise terms can describe the intrinsic fluctuations in the constitutive relations.[15]

[15]Landau and Lifshitz (1980) developed a theory of fluids where the constitutive relations are indeed fluctuating and the usual expressions are only their average. Based on this hypothesis, they were able to explain the dispersion of light by fluids at thermal equilibrium.

2.7 BBGKY hierarchy

In the previous section, we showed that relevant information regarding a system can be obtained from the first reduced distribution functions.[16] It is then important to derive the equations that describe their temporal evolution. To obtain an explicit form for these equations, we have to specify the Hamiltonian of the system under study. Quite generally, we consider a system of N identical particles that interact with an external potential V and among themselves through a pair potential ϕ, such that the Hamiltonian can be written as a sum of one- and two-particle terms.[17]

[16]In the presented examples only the first two distributions are needed, but other observables could require higher-order distributions, as shown in Exercise 2.7.

[17]Here, we do not contemplate the case of three-particle potentials, which could be relevant in determining the structure of covalent molecules and solids. Also, no magnetic fields are considered; otherwise, we would have to include the vector potential in the one-particle term. This latter case is considered in Exercise 2.13 and will be studied in detail in Chapter 6.

$$H_N = \sum_{a=1}^{N} h_0(a) + \sum_{a<b}^{N} \phi(a,b), \tag{2.43}$$

where

$$h_0(a) = \frac{p_a^2}{2m} + V(\mathbf{r}_a). \tag{2.44}$$

The first step is to multiply the Liouville equation,

$$\frac{\partial F}{\partial t} = -\{H_N, F\}, \tag{2.45}$$

by $N!/(N-n)!$ and integrate the result over the degrees of freedom from $n+1$ to N. This protocol is designed to give directly the time derivative of $F^{(n)}$ on the left-hand side. The right-hand side is more involved and has contributions from h_0 and ϕ, which will be treated separately.

In the contribution of the one-particle terms, the sum is separated between the cases $1 \leq a \leq n$ and $n < a \leq N$, resulting in

$$\frac{N!}{(N-n)!} \sum_{a=1}^{n} \int \{h_0(a), F\} \, \mathrm{d}(n+1) \ldots \mathrm{d}N$$

$$+ \frac{N!}{(N-n)!} \sum_{a=n+1}^{N} \int \{h_0(a), F\} \, \mathrm{d}(n+1) \ldots \mathrm{d}N$$

$$= \sum_{a=1}^{n} \{h_0(a), F^{(n)}\}$$

$$+ \frac{N!}{(N-n)!} \sum_{a=n+1}^{N} \int \left(\frac{\partial h_0}{\partial \mathbf{p}_a} \frac{\partial F}{\partial \mathbf{r}_a} - \frac{\partial h_0}{\partial \mathbf{r}_a} \frac{\partial F}{\partial \mathbf{r}_a} \right) \mathrm{d}(n+1) \ldots \mathrm{d}N.$$

Making the substitution,

$$\frac{\partial h_0}{\partial \mathbf{p}_a} \frac{\partial F}{\partial \mathbf{r}_a} = \frac{\partial}{\partial \mathbf{p}_a} \left(h_0 \frac{\partial F}{\partial \mathbf{r}_a} \right) - \frac{\partial}{\partial \mathbf{r}_a} \left(h_0 \frac{\partial F}{\partial \mathbf{p}_a} \right) + \frac{\partial h_0}{\partial \mathbf{r}_a} \frac{\partial F}{\partial \mathbf{p}_a},$$

the integrand of the second term is rendered as a sum of total derivatives, hence their integral vanishes. Consequently, the contribution from the one-particle terms is simply $\sum_{a=1}^{n} \{h_0(a), F^{(n)}\}$.

For the contributions of the two-particle terms, we separate the sum into three: $\sum_{a<b}^{N} = \sum_{a=1}^{n} \sum_{b=a+1}^{n} + \sum_{a=1}^{n} \sum_{b=n+1}^{N} + \sum_{a=n+1}^{N} \sum_{b=1+1}^{N}$. The first contribution gives

$$\frac{N!}{(N-n)!} \sum_{a<b}^{n} \int \{\phi(a,b), F\} \, \mathrm{d}(n+1) \ldots \mathrm{d}N = \sum_{a<b}^{n} \{\phi(a,b), F^{(n)}\}. \quad (2.46)$$

For the second contribution, the index a does not belong to the integrated variables, while b does. Using the symmetry of the distribution function, we have $N - n$ identical terms, allowing us to take $b = n+1$ and integrate the degrees of freedom from $n+2$ to N, leading to

$$\frac{N!}{(N-n)!} \sum_{a=1}^{n} \sum_{b=n+1}^{N} \int \{\phi(a,b), F\} \, \mathrm{d}(n+1) \ldots \mathrm{d}N$$

$$= \sum_{a=1}^{n} \int \{\phi(a, n+1), F^{(n+1)}\} \, \mathrm{d}(n+1), \quad (2.47)$$

where the factorial term gives exactly the normalisation for $F^{(n+1)}$. Finally, the last term cancels out, similarly to the case of the one-particle terms.

Collecting all terms, we obtain

$$\frac{\partial F^{(n)}}{\partial t} = -\{H_n, F^{(n)}\} - \sum_{a=1}^{n} \int \{\phi(a, n+1), F^{(n+1)}\} \, \mathrm{d}(n+1), \quad (2.48)$$

where we have defined the reduced n-particle Hamiltonian,

$$H_n = \sum_{a=1}^{n} h_0(a) + \sum_{a<b}^{n} \phi(a,b). \quad (2.49)$$

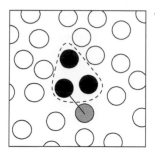

Fig. 2.7 Graphical interpretation of the BBGKY equations. The subsystem (in this case consisting of $n = 3$ particles) evolves according to its internal Hamiltonian [first term of the right-hand side of eqn (2.48)] plus an interaction term for each of the particles inside the subsystem with any (labelled $n+1$) particle outside the subsystem.

[18] The BBGKY hierarchy is a result of a series of articles published between 1935 and 1946 by J. Yvon, N.N. Bogoliubov, J.G. Kirkwood, M. Born, and H.S. Green.

The eqns (2.48) for the reduced distributions with $n = 1, \ldots, N$ are very similar to the Liouville eqn (2.11) for the whole system, except that the Hamiltonian is a reduced one and there is an extra term on the right-hand side. This term indicates that the dynamics of a subsystem of n particles is not closed because any of them can interact with any other outside the subsystem. As the interactions are pairwise additive, only one additional particle is needed, and therefore the configuration of $n+1$ particles must be specified through the corresponding distribution function (Fig. 2.7). This fact implies that the equation for a reduced distribution is not closed and needs information from a higher-order distribution. Only the equation for $n = N$, which reduces to the Liouville equation, is closed. We showed previously that most of the relevant information of a system can be obtained from the first two distributions, but that their equations need information from the third distribution, which can be obtained if we know the fourth distribution, and so on. We thereby obtain a hierarchy of equations, which is usually the case when one deals with reduced systems. The eqns (2.48) are called the Bogoliubov–Born–Green–Kirkwood–Yvon (BBGKY) hierarchy.[18]

2.7.1   Equation for the one-particle distribution

Because of its importance, we will write explicitly the BBGKY equation for $n = 1$. Using the explicit form of the Hamiltonian (2.49), we have

$$\{H_1, F^{(1)}\} = \frac{\mathbf{p}_1}{m} \cdot \frac{\partial F^{(1)}}{\partial \mathbf{r}_1} + \mathbf{F}_1 \cdot \frac{\partial F^{(1)}}{\partial \mathbf{p}_1}, \tag{2.50}$$

where $\mathbf{F} = -\nabla V$. Hence,

$$\frac{\partial F^{(1)}}{\partial t} + \frac{\mathbf{p}_1}{m} \cdot \frac{\partial F^{(1)}}{\partial \mathbf{r}_1} + \mathbf{F}_1 \cdot \frac{\partial F^{(1)}}{\partial \mathbf{p}_1} = -\int \{\phi_{12}, F^{(2)}\} \, \mathrm{d}^3 r_2 \, \mathrm{d}^3 p_2, \tag{2.51}$$

which again is not closed.

This equation can be written in the velocity representation for the distribution function $f(\mathbf{r}, \mathbf{c})$ defined in Chapter 1. The equation will depend on an analogously defined two-particle distribution, $f^{(2)}(\mathbf{r}_1, \mathbf{c}_1, \mathbf{r}_2, \mathbf{c}_2)$. Using that the potential depends only on positions, the Poisson bracket can be simplified, resulting in

$$\frac{\partial f}{\partial t} + \mathbf{c}_1 \cdot \frac{\partial f}{\partial \mathbf{r}_1} + \frac{\mathbf{F}_1}{m} \cdot \frac{\partial f}{\partial \mathbf{c}_1}$$
$$= \int \frac{\partial \phi_{12}}{\partial \mathbf{r}_{12}} \left(\frac{\partial}{\partial \mathbf{c}_1} - \frac{\partial}{\partial \mathbf{c}_2} \right) f^{(2)}(\mathbf{r}_1, \mathbf{c}_1, \mathbf{r}_2, \mathbf{c}_2) \, \mathrm{d}^3 r_2 \, \mathrm{d}^3 c_2. \tag{2.52}$$

The left-hand side of eqn (2.52) can be directly interpreted as the streaming evolution for non-interacting particles. Indeed, without interactions, after a small interval Δt conservation of the number of particles and the Liouville theorem imply, $f(\mathbf{r} + \mathbf{c}\Delta t, \mathbf{c} + \mathbf{F}\Delta t/m, t + \Delta t) = f(\mathbf{r}, \mathbf{c}, t)$. A Taylor expansion gives $\frac{\partial f}{\partial t} + \mathbf{c} \cdot \nabla f + \mathbf{F}/m \cdot \frac{\partial f}{\partial \mathbf{c}} = 0$.

The right-hand side accounts for the interactions and in subsequent chapters, we will approximate either $f^{(2)}$ or directly the integral term for different physical scenarios. The objective will be to find closed equations for f that can later be solved or analysed by different means.

2.8   Generalisation to mixtures

Mixtures are natural extensions of the already described systems. They deal with cases where two or more species are present, for example in plasmas, where we need to treat electrons and ions separately. In the quantum case, we will also be interested in the interaction of electrons and phonons, for example.

To simplify notation, we will treat the case of binary mixtures, although the results can be directly extended to multicomponent systems. Consider a mixture of N_A particles of type A and N_B particles of type B. The phase space point is $\Gamma = (\mathbf{r}_{A,1}, \mathbf{p}_{A,1}, \ldots, \mathbf{r}_{A,N_A},$ $\mathbf{p}_{A,N_A}; \mathbf{r}_{B,1}, \mathbf{p}_{B,1}, \ldots, \mathbf{r}_{B,N_B}, \mathbf{p}_{B,N_B})$. The reduced distribution functions are defined to indicate the state of a subset of n_A and n_B particles, and they are defined—as a natural extension to the monocomponent case—as

$$
F^{(n_A, n_B)}(1, \ldots, n_A; 1, \ldots, n_B; t) = \frac{N_A!}{(N_A - n_B)!} \frac{N_B!}{(N_B - n_B)!}
$$

$$
\times \int \widehat{F}(1, \ldots, N_A; 1, \ldots, N_B; t) \, \mathrm{d}(n_A + 1) \ldots \mathrm{d}N_A \, \mathrm{d}(n_B + 1) \ldots \mathrm{d}N_B,
$$

where $\widehat{F}$ is symmetrised over the separate permutations of the A and B particles.

If the system is described by a general Hamiltonian with two-body interactions,

$$
H = \sum_{a=1}^{N_A} \left(\frac{p_{A,a}^2}{2m_A} + V^A(\mathbf{r}_{A,a}) \right) + \sum_{b=1}^{N_B} \left(\frac{p_{B,b}^2}{2m_B} + V^B(\mathbf{r}_{B,b}) \right)
$$

$$
+ \sum_{a<b}^{N_A} \phi^{AA}(\mathbf{r}_{A,a} - \mathbf{r}_{A,b}) + \sum_{a<b}^{N_B} \phi^{BB}(\mathbf{r}_{B,a} - \mathbf{r}_{B,b}) + \sum_{a=1}^{N_A} \sum_{b=1}^{N_B} \phi^{AB}(\mathbf{r}_{A,a} - \mathbf{r}_{B,b}),
$$

$$
\tag{2.53}
$$

it is possible to construct a BBGKY hierarchy with the same structure as before. That is, the evolution for $F^{(n_A, n_B)}$ is given by the Liouville equation for the reduced subsystem plus the interactions with external particles, which are sampled from an extended distribution with one more particle. As the interactions can be of the form, AA, AB, BA, and BB, this interaction term involves the distributions $F^{(n_A+1, n_B)}$ and $F^{(n_A, n_B+1)}$. Its general expression is rather involved, and it is not necessary to give it here. In the next chapter we will analyse the classical motion of charged particles in a medium made of massive scatterers. There, we will be interested in the one-particle reduced distribution $F^{(1,0)}$ of the moving charges. Also, when studying plasmas in Chapter 6, the motion of electrons and ions is relevant, and we will be interested in the one-particle reduced distributions $F_{\text{electron}} = F^{(1,0)}$ and $F_{\text{ion}} = F^{(0,1)}$. According to the BBGKY hierarchy, the evolution of these distributions will depend on $F^{(2,0)}$, $F^{(1,1)}$, and $F^{(0,2)}$, which we will need to model properly.

2.9   Reduced distributions in equilibrium and the pair distribution function

In thermal equilibrium, the system is described by the Gibbs distribution function, $F_{\text{eq}}(\Gamma) = Z^{-1}e^{-H(\Gamma)/k_{\text{B}}T}$. In general, the reduced distributions are complex to evaluate even in this simple case. Indeed, if they were simple to obtain, the equation of state of an interacting system would be the result of a simple integration over the two-particle distribution function using the virial expression (2.37). Although the problem is still extremely complex, we can obtain general expressions when the system is spatially homogeneous. This homogeneity is achieved by not having external potentials and also requiring that no phase coexistence takes place. Consider then a system interacting with the Hamiltonian

$$H = \sum_{a=1}^{N} \frac{p_a^2}{2m} + \sum_{a<b}^{N} \phi(a,b). \tag{2.54}$$

One can directly verify that the one-particle distribution is the Maxwellian[19]

$$F_{\text{eq}}^{(1)}(\mathbf{p}) = \frac{n}{(2\pi m k_{\text{B}}T)^{3/2}} e^{-p^2/2m k_{\text{B}}T}, \tag{2.55}$$

which does not depend on the position, and where $n = N/\mathcal{V}$ is the number density. Note that the Maxwellian distribution is obtained for any interparticle potential and global densities. It is not a feature of ideal gases only.

For the two-particle reduced distribution, the presence of the interaction terms $\phi(a,b)$ prevents integration of the other degrees of freedom to obtain a closed expression. However, it is possible to show that it has the general form,[20]

$$F_{\text{eq}}^{(2)}(\mathbf{r}_1,\mathbf{p}_1,\mathbf{r}_2,\mathbf{p}_2) = F_{\text{eq}}^{(1)}(\mathbf{p}_1)F_{\text{eq}}^{(1)}(\mathbf{p}_2)g^{(2)}(\mathbf{r}_1 - \mathbf{r}_2), \tag{2.56}$$

in terms of the pair distribution function $g^{(2)}$, which unfortunately cannot be computed explicitly. The pair distribution function measures the spatial correlations between pairs of particles such that, if $g = 0$, it is impossible for two particles to be at a given distance (for example, closer than the hard-core repulsion), whereas $g = 1$ implies that particles are uncorrelated, as occurs when particles are far apart. Experiments and simulations have measured the pair distribution function in simple

[19]See Exercise 2.9.

[20]See Exercise 2.10.

liquids characterised by an interaction potential that has short-range repulsion, as shown in Fig. 2.8. Note that it presents oscillations with a maximum—larger than one—at a distance equal to the hard core, with a height that grows with pressure. This maximum indicates that, at high pressure, particles tend to be in close contact with high probability. In dilute gases, beyond the range of the interaction potential, g equals 1.

2.10   Master equations

In the preceding sections we derived the kinetic equations for systems where the interparticle interactions were known exactly. It was possible to derive the Liouville and BBGKY descriptions starting from Hamilton's equations. It may happen that, in some cases, due to their complexity or because of a lack of detailed information, the interactions are not exactly known, but rather, we are left with a stochastic description.

In the absence of a full description of the interactions, instead of giving the probability distribution for the forces to take some particular values, for example, we adopt a more phenomenological approach. We use a stochastic scheme where we give the probability $P(\Gamma, \Gamma', \Delta t)$[21] that in this time lapse the system transits from the state Γ to a new state Γ'; That is, the stochastic description is of the effect of the interactions rather than of the interactions themselves. Conservation of probability can now be written in the integral form,

$$F(\Gamma, t + \Delta t) - F(\Gamma, t) = -\int F(\Gamma, t) P(\Gamma, \Gamma', \Delta t)\, d\Gamma'$$
$$+ \int F(\Gamma', t) P(\Gamma', \Gamma, \Delta t)\, d\Gamma'. \quad (2.57)$$

The first term takes account of systems that were in the state Γ and transited to any other configuration, while the second one is the opposite, that is, systems that transit to Γ from any other state. We will see that this description in terms of loss and gain terms is quite general in kinetic theory. In particular, we will use this approach for the one-particle distribution function in the next chapter.

Because the interactions have finite intensities, for small Δt, most of the systems end in a state very close to the original, which is equivalent to saying that $P(\Gamma, \Gamma', \Delta t) \approx \delta(\Gamma' - \Gamma)$. Therefore, in the case where the limit exists, we define the transition rates,

$$W(\Gamma, \Gamma') = \lim_{\Delta t \to 0} \frac{P(\Gamma, \Gamma', \Delta t) - \delta(\Gamma' - \Gamma)}{\Delta t}, \quad (2.58)$$

that measure the rate at which a system evolves from one state to a different one. Writing eqn (2.57) in terms of the transition rates, it simplifies to[22]

$$\frac{\partial F(\Gamma, t)}{\partial t} = -\int F(\Gamma, t) W(\Gamma, \Gamma')\, d\Gamma' + \int F(\Gamma', t) W(\Gamma', \Gamma)\, d\Gamma', \quad (2.59)$$

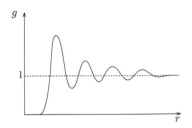

Fig. 2.8 Sketch of the typical form of the radial distribution function $g(r)$ for liquids. At short distances the repulsive potential leads to a vanishing value of $g(r)$. There is a first peak at the minimum of the interatomic potential, and then there is an alternation of minima and maxima that express the short-range order, in the form of layering, that takes place in liquids.

[21] For notational simplicity, we have assumed that the transition probabilities P do not depend explicitly on time, that is, that the dynamics is autonomous.

[22] Note that the effects of the Dirac δ functions cancel as expected because transitions to the same state do not change the distribution function.

which is called the master equation. It simply reflects that probability (the distribution function) is conserved and that the evolution is local in time; in the context of stochastic processes, this is what it is called a Markovian process.

Master equations will be used in the description of Brownian particles in Chapter 5 and in the description of quantum systems in Chapters 7 and 8. In the case of Brownian particles, we will use the formulation based on the equation for a finite time step (2.57), while for quantum systems the differential formulation (2.59) will be used. In the latter cases, the formal integration over Γ' will correspond to an integration over wavevectors and a sum over spin; that is, Γ and Γ' will represent both continuum and discrete variables.

One can verify that the master equation comprises the Liouville equation for deterministic Hamiltonian dynamics. Indeed, according to the deterministic equation in phase space $[\dot{\Gamma} = \Phi(\Gamma)]$, after a small time lapse Δt, the new state is $\Gamma' = \Gamma + \Phi(\Gamma)\Delta t$. Therefore, the transition probabilities are $P_{\text{deterministic}}(\Gamma, \Gamma', \Delta t) = \delta(\Gamma' - \Gamma - \Phi \Delta t)$. We can formally Taylor-expand these transition probabilities, resulting in the following transition rates:

$$W_{\text{deterministic}}(\Gamma, \Gamma') = -\delta'(\Gamma' - \Gamma)\Phi(\Gamma), \qquad (2.60)$$

where δ' is the derivative of the Dirac δ function. When we substitute (2.60) into the master equation, we first note that the loss term vanishes identically. The gain term can be transformed using the properties of the Dirac distributions,[23] and finally the master equation for this deterministic system is

$$\frac{\partial F(\Gamma, t)}{\partial t} = -\nabla_\Gamma \cdot [\Phi(\Gamma)F(\Gamma, t)], \qquad (2.61)$$

which is precisely the Liouville equation in the conservative form (2.9). With the help of Liouville's theorem, we recover the Liouville equation in its standard form.

2.11 Application: systems with overdamped dynamics

Consider an ensemble of particles immersed in a fluid. If the particles are larger than $10\,\mu$m in size, they are subject to the viscous friction force of the fluid while not experiencing Brownian motion.[24] The fluid force is of the form $-\gamma\dot{\mathbf{r}}$, where γ is a friction coefficient and $\dot{\mathbf{r}}$ is the particle velocity. Besides, particles interact among themselves with pairwise forces $\mathbf{f}_{ab}$, such that the Newton equations read,

$$m\ddot{\mathbf{r}}_a = -\gamma\dot{\mathbf{r}}_a + \sum_{b \neq a} \mathbf{f}_{ab}. \qquad (2.62)$$

If R is the typical particle radius, the mass scales as $m \sim R^3$, while according to Stoke's law, the friction coefficient scales as $\gamma \sim R$. Therefore, for small particles the viscous drag dominates and the inertia term

[23]We recall that the Dirac δ function and its derivative satisfy

$$\int f(y)\delta(y - x)\,\mathrm{d}y = f(x),$$

$$\int f(y)\delta'(y - x)\,\mathrm{d}y = -f'(x)$$

for any test function f (see Appendix A).

[24]The description of Brownian motion will be worked out in detail in Chapter 5.

can be dropped off, resulting in the following equations for overdamped motion:

$$\dot{\mathbf{r}}_a = \mathbf{v}_a = \gamma^{-1} \sum_{b \neq a} \mathbf{f}_{ab}, \tag{2.63}$$

where we have defined the drift velocities $\mathbf{v}_a$. Then, phase space reduces to coordinate variables only. Here, we can apply the same procedure as used in Section 2.3 to derive an equation for the distribution function $F(\mathbf{x}, t)$, for the configuration $\mathbf{x} = (\mathbf{r}_1, \ldots, \mathbf{r}_N)$,

$$\frac{\partial F}{\partial t} = -\sum_a \nabla_a \cdot (F \mathbf{v}_a) \tag{2.64}$$

that, in this case, cannot be further simplified because there is no equivalent to the Liouville theorem in overdamped dynamics. Note the similarity of this equation to the mass conservation, $\frac{\partial \rho}{\partial t} = -\nabla \cdot (\rho \mathbf{v})$; the drift velocities generate a flow field in the multidimensional configuration space.

If the forces derive from a potential, then $\mathbf{v}_a = -\gamma^{-1} \nabla_a U$, and the flow field does not have any closed loops in the $3N$-dimensional space.[25] Consequently, in this case, the flow must escape to infinity or end at fixed points, where the potential is minimum (Fig. 2.9).

We are interested in finding the stationary solution for the distribution function when forces derive from a potential. To simplify the analysis, we will assume that there are no fixed points at infinity. The stationary solution corresponds to imposing that $F\mathbf{v}_a = 0$, $\forall a$. This implies that whenever $\mathbf{v}_a \neq 0$, the distribution function must vanish, but at the points where the drift velocities vanish, i.e. at the fixed points, it can take any arbitrary value. Formally, because the fixed points are discrete, the distribution function is a sum of Dirac delta functions,[26]

$$F_0(\mathbf{x}) = \sum_{\mathbf{x}_n \text{ fixed points}} A_n \delta(\mathbf{x} - \mathbf{x}_n), \tag{2.65}$$

where the relative amplitudes A_n depend on the initial conditions. We have found that, if the forces derive from a potential, the final state of an overdamped system is necessarily at a fixed point or to escape to infinity. By performing a stability analysis, it can be shown that only the stable fixed points survive in the stationary distribution (2.65).[27]

The case where the forces do not derive from a potential is quite different. This can be the case for time-dependent interactions or with external forces acting on the system. The general case is extremely complex because here it is possible that, beside the fixed points, the system can flow to limit cycles or strange attractors that can have fractal dimension. For simplicity, we will analyse a simple case with just one degree of freedom. Consider a two-dimensional rotor, described by a single angle θ, subject to a torque that derives from a potential. Additionally, an external torque is added that pushes the system to rotate at a constant angular velocity Ω.[28] The equation of motion is

$$\dot{\theta} = \omega(\theta) + \Omega, \tag{2.66}$$

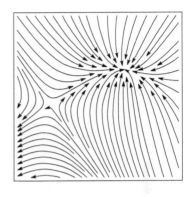

Fig. 2.9 Example of a flow field for forces that derive from a potential. The flow either escapes to infinity or ends at stable fixed points.

[25]This result is a consequence of the Stokes theorem. It states that, if $\nabla \times \mathbf{v}_a = 0$, as occurs when the forces derive from a potential, then $\oint \mathbf{v} \cdot d\mathbf{l} = 0$.

[26]Note that the solution for the equation $(x - x_0) f(x) = 0$ is $f(x) = A\delta(x - x_0)$ (see Appendix A).

[27]See Exercise 2.12.

[28]This external torque does not derive from a potential because the total work over one turn does not vanish.

where $\omega = -\,\mathrm{d}U/\,\mathrm{d}\theta$. The stationary solution of eqn (2.64) is

$$F(\theta)\left[\omega(\theta) + \Omega\right] = J, \tag{2.67}$$

where the flux J is constant. Three cases are possible here. First, if Ω is large enough such that $\omega(\theta) + \Omega > 0, \forall \theta$, then

$$F_0(\theta) = \frac{J}{\omega(\theta) + \Omega}. \tag{2.68}$$

There is a uniform flux driven by the external torque, but the stationary distribution is not uniform: the occupation is larger at those angles where the speed is lower. Here, J plays the role of a normalisation constant that is determined by imposing that F is normalised. The second case is analogous to the first one, when $\omega(\theta) + \Omega < 0, \forall \theta$, where the same solution is found, except that here J is negative to guarantee that F is positive. If, on the contrary, $\omega(\theta) + \Omega$ changes sign in the interval, it is not possible to find a unique J that produces a positive distribution function. The only possibility then is to have $J = 0$, as in the case where all the forces derive from a potential, and the solution (2.65) is obtained. The rotor stops at the stable fixed points, solution of $\omega(\theta) = -\Omega$. In Chapter 5, we will consider this problem again in the presence of small noise to show that the abrupt transitions in terms of the forcing angular velocity are smoothed out (see Exercise 5.10).

Further reading

The principles of classical mechanics in its Hamiltonian formulation can be found in Landau and Lifshitz (1976) or Goldstein *et al.* (2013). The concepts of equilibrium statistical mechanics used in this chapter (entropy, canonical and microcanonical distributions) are presented in all texts in the subject. For example, Huang (1987), Landau and Lifshitz (1980), and Greiner *et al.* (2012).

The Liouville formalism is described in detail in many kinetic theory books; for example, in Ferziger and Kaper (1972), Cercignani (2013), or Liboff (2003). The Liouville equation is also presented in Evans and Morris (1990), Zwanzig (2001), and Kubo *et al.* (1998), this time focused in other non-equilibrium statistical mechanics techniques where, for example, the linear response theory and Green–Kubo formulas are derived. In a more formal approach, Dorfman (1999) describes the relation of the reversibility in Hamiltonian systems and the non-equilibrium behaviour, paying attention to the role of chaos.

The properties of gases and liquids in equilibrium are presented in Hansen and McDonald (1990), where a comprehensive description of the hard sphere mode and the general properties of the pair distribution function are given.

Finally, master equations, in the context of probability theory are analysed in van Kampen (2007) and the addition of fluctuations in hydrodynamics is presented in detail in Lifshitz and Pitaevskii (1980).

Exercises

(2.1) **Solution for the free case.** Consider a system of non-interacting particles described by the Hamiltonian

$$H = \sum_{a=1}^{N} \frac{p_a^2}{2m}.$$

The Liouville equation is

$$\frac{\partial F}{\partial t} = -\{H, F\} = -LF,$$

where we have defined the Liouville operator $L = \{H, \cdot\}$. The time-dependent solution can be written formally as

$$F(\Gamma, t) = e^{-Lt} F(\Gamma, 0).$$

(a) Find the Liouville operator for the non-interacting Hamiltonian.

(b) Using the following property of the translation operator:

$$e^{a\frac{\partial}{\partial x}} f(x) = f(x + a),$$

find the solution for $F(\Gamma, t)$ as a function of the initial condition. This solution can be directly interpreted in terms of the streaming of free particles.

(2.2) **Properties of the Poisson bracket.** Using the Definition (2.2), show that:

(a) If $g(q_1, p_1, q_2, p_2, \ldots)$ and $h(q_1, p_1, q_2, p_2, \ldots)$ are two real functions on the phase space, then

$$\{g, h\} = -\{h, g\}.$$

(b) If $f(q_1, p_1, q_2, p_2, \ldots)$ is any real function on the phase space and s any real function of real variables, both functions at least one time differentiable, then

$$\{f, s(f)\} = 0.$$

(2.3) **BBGKY-like hierarchy for overdamped dynamics.** Derive the equivalent of the BBGKY hierarchy for a system of particles that follow overdamped dynamics with general pairwise interactions plus one-particle forces,

$$\frac{dx_a}{dt} = f_{0a} + \sum_{b \neq a} f_{ab}.$$

(2.4) **Ensemble average of global quantities.** Find expressions for the ensemble average of the following observables in terms of the one- and two-particle distributions:

(a) Total momentum $\mathbf{P} = \sum_a \mathbf{P}_a$,

(b) Electric current density $\mathbf{J} = \mathcal{V}^{-1} \sum_a q_a \mathbf{c}_a$,

(c) Virial expression for the pressure (2.37).

(2.5) **Equation of state in equilibrium.** Using the expression for the two-particle distribution function in equilibrium (2.56), obtain an integral expression for the average pressure in thermal equilibrium using the virial expression (2.37). Consider for simplicity that the forces are central.

(2.6) **Stress tensor and heat flux in equilibrium.** Show that in equilibrium the average stress tensor is diagonal and that the heat flux vanishes. Use the two-particle distribution (2.56).

(2.7) **Average angle in a molecule.** In a molecule made of three atoms, the opening angle can be computed from the relation $\cos\theta = \mathbf{r}_{12} \cdot \mathbf{r}_{13} / r_{12} r_{13}$. Show that the average angle can be expressed in terms of the three-particle distribution function.

(2.8) **Energy flux.** Derive the expression (2.39) for the energy flux starting from the energy density (2.38).

(2.9) **Maxwellian distribution for interacting systems.** Consider a system composed of N particles in a volume at thermal equilibrium with temperature T. The particles interact pairwise but there are no external forces, such that the Hamiltonian is given by eqn (2.54). The objective is to show that the one-particle distribution function is the Maxwellian (2.55). Starting from the equilibrium distribution function, $F_0(\Gamma) = Z^{-1} e^{-H(\Gamma)/k_B T}$, formally integrate the other degrees of freedom to obtain $F^{(1)}(\mathbf{r}, \mathbf{p})$. Show first that it does not depend on the position $\mathbf{r}$ and that the dependence on momentum has the correct form. Note: it is not necessary to perform the integrations in the other positions to show that it does not depend on $\mathbf{r}$.

(2.10) **Pair distribution function.** Considering the same conditions as in the previous problem, the objective here is to obtain the pair distribution function. First, formally integrate the other degrees of freedom to obtain $F^{(2)}(\mathbf{r}_1, \mathbf{p}_1, \mathbf{r}_2, \mathbf{p}_2)$. Do not attempt to perform the integration in positions or to compute the partition function Z; it is impossible to do so in general. Show that $F^{(2)}$ has the correct dependence on momentum and can be written as $F^{(2)}(\mathbf{r}_1, \mathbf{p}_1, \mathbf{r}_2, \mathbf{p}_2) = F^{(1)}(\mathbf{p})F^{(1)}(\mathbf{p})g^{(2)}(\mathbf{r}_1, \mathbf{r}_2)$. Then, show that $g^{(2)}$ only depends on the relative distance between the particles: $g^{(2)}(\mathbf{r}_1, \mathbf{r}_2) = g^{(2)}(\mathbf{r}_1 - \mathbf{r}_2)$.

(2.11) **Forced rotor.** Consider the forced rotor described in Section 2.11 in the specific case where $\omega(\theta) = \sin\theta$. Determine the critical value of the forcing angular velocity Ω_c that separates the cases where the rotor reaches a fixed point from the case where the flux is finite. Write down the stationary distribution function in both cases. Show that the average angular velocity is

$$\langle \dot{\theta} \rangle = \begin{cases} 0 & \text{if } |\Omega| < \Omega_c \\ \mathrm{sgn}(\Omega)\sqrt{\Omega^2 - \Omega_c^2} & \text{if } |\Omega| > \Omega_c, \end{cases}$$

where sgn is the sign function: 1 for positive arguments, -1 for negative arguments, and zero when the argument is null. Draw and interpret the result.

(2.12) **Stability of fixed points in overdamped dynamics.** Consider a system with overdamped dynamics where the drift velocities derive from a potential U. The minima, maxima, and saddle points of U correspond to fixed points of the dynamics, but the objective here is to show that only the minima (stable fixed points) survive in the long-term distribution function and give rise to Dirac delta functions. For this purpose, consider a Gaussian initial condition close to a fixed point $\mathbf{x}_0$. Locally, close to $\mathbf{x}_0$, the potential can be written as $U = U_0 + A_{ik}y_iy_k$, where $\mathbf{y} = \mathbf{x} - \mathbf{x}_0$ and $\mathbb{A}$ is a symmetric matrix. Show that the solution of eqn (2.64) converges to a Dirac delta function if $\mathbb{A}$ is positive definitive (i.e. $\mathbf{x}_0$ is a minimum of U), but separates from $\mathbf{x}_0$ under other conditions of $\mathbb{A}$, associated with saddle points or maxima of U.

(2.13) **Magnetic fields.** Charged particles in an external magnetic field $\mathbf{B}$ with no interactions among particles are described by the Hamiltonian,

$$H = \sum_a \frac{1}{2m}\left(\mathbf{p}_a - q\mathbf{A}\right)^2,$$

where $\mathbf{A}$ is the vector potential, such that $\mathbf{B} = \nabla \times \mathbf{A}$.

(a) Derive the Liouville equation and the first BBGKY equation.

(b) Recalling that, when magnetic fields are present, the momentum is not simply the mass times the velocity, but rather $\mathbf{p}_a = m\mathbf{c}_a + q\mathbf{A}$), rewrite the first BBGKY equation in terms of the velocity, that is, find the equation for $f(\mathbf{r}, \mathbf{c}, t)$.

(2.14) **BBGKY for binary mixtures.** Consider a binary mixture described by the Hamiltonian (2.53). Derive the BBGKY equations for the reduced distributions $F^{(n_A, n_B)}$ and write in detail the equations for the cases $F^{(1,0)}$ and $F^{(0,1)}$.

(2.15) **Simple pendulum.** Consider a simple pendulum with mass m and radius l, described by the angle ϕ.

(a) Draw the phase space portrait, distinguishing between oscillatory and non-oscillatory trajectories.

(b) Consider the initial condition with uncertainties in energy and the angle: $E = E_0 + \Delta E$ and $\phi = \phi_0 + \Delta\phi$. Represent this initial condition in the phase space.

(c) Recalling that for large amplitudes the oscillation period is an increasing function of the energy, represent the evolution of the initial condition.

The Lorentz model for the classical transport of charges

3

3.1 Hypothesis of the model

The first problem we will model using a kinetic equation is the electrical conductivity due to classical charges. This problem was studied first by Drude[1] and later by Lorentz,[2] before quantum mechanics was established. It models electrons as classical particles, and hence, when applied to the transport of electrons in metals and semiconductors, it gives wrong quantitative predictions. However, it is an excellent starting point, as it provides a language, the formalism, and the usual approximations that we will use in the rest of the book, while being relatively simple and intuitive, helping to provide insight into transport phenomena. Also, when we describe the quantum transport of electrons in Chapter 8, the derivation and analysis of the resulting kinetic equation will follow methods similar to those used in this chapter. Finally, the Lorentz model has proved useful to describe transport in porous media or ionic transport in solutions.

In a solid, charge is transported by free electrons that move in a matrix of fixed ions, which are much heavier than the electrons. Therefore, their interaction is asymmetric: in a collision, the electrons will be deflected strongly, while the ions will be only slightly perturbed. This allows us to make the first approximation of the model, namely that the distribution function of the ions will be assumed to be constant, independent of the interaction with the electrons. The effect of this approximation is that, instead of having to deal with the distribution functions of electrons and ions, only the former is relevant while the second is a given function that will depend on how we model the ions. To emphasise this distinction, we will denote by f the electron distribution function and by F that of the ions.

The second approximation is to neglect electron–electron interactions. Several aspects come into play to justify this approximation. First, due to the mass contrast, these interactions produce softer scattering compared with those with ions. Second, as the electron–electron interaction is mediated by the long-range Coulomb force, each electron experiences an almost vanishing force as it is being pushed simultaneously by a huge

[1]P. Drude proposed the model for electrical conductivity in 1900.

[2]H. Lorentz proposed the kinetic model for the transport of charge in metals in 1905.

number of electrons in different incoherent directions (see Chapter 6 for a detailed analysis of the effect of long-range interactions). Note that the ion–electron interaction is not only of Coulomb origin, but also Pauli repulsion with the core electrons comes into play. This latter interaction is short range and is responsible for the large deflections. Finally, if we consider the right-hand side of the first BBGKY equation (2.52) for the electron distribution function, the electron–electron and electron–ion interaction terms scale as n_e^2 and $n_e n_i$, respectively, where n_e is the density of electrons and n_i the ion density. Then, we can assume that the electron density is small enough such that the electron–electron interaction is negligible with respect to the electron–ion interaction.

At this stage we could write down the BBGKY equations for the electron distribution functions. In these, the right-hand side term accounts for the evolution of the distribution functions due to electron–ion interactions. Here, we will make our third approximation, which will simplify the BBGKY equations, specifically the first of these equations. The approximation consists of considering that the electron–ion interaction is short range. This implies that the scattering process takes a short time and, importantly, that electrons interact with ions one by one and that no two ions interact simultaneously with an electron. The trajectory of an electron is then a sequence of free flights (or parabolic ones in the presence of an electric field) and scattering events, which to emphasise their short duration, will be called collisions in what follows. In a collision, given the incoming ion and electron precollisional velocities and their relative positions, it is possible to compute the outgoing postcollisional velocities using scattering theory (see Appendix C).

However, even though we have neglected electron–electron interactions, the hierarchy is still not closed. Indeed, for example, if we consider the first equation of the hierarchy, the right-hand side will depend on $f^{(1,1)}$, the pair distribution of having one electron and one ion with given positions and velocities. Even if we have been given an a priori distribution for the ions, this function is not known and must be computed from the hierarchy itself. The equation for $f^{(1,1)}$ will depend on $f^{(1,2)}$—the distribution of having one electron and two ions—and so on. The origin of the correlations can be easily understood by looking at the trajectory of an electron, dispersed in a distribution of hard-core ions, as presented in Fig. 3.1. At low concentration of ions and if the external field is weak enough, we can safely assume that, after an encounter with an ion, the electron will not meet this ion again and will arrive uncorrelated to the next encounter. Therefore, and this is the final approximation of the model, we will assume that in the precollisional state there are no electron–ion correlations. Note that this and the previous hypothesis become valid if the size of the ions is much smaller than the electron mean free path.

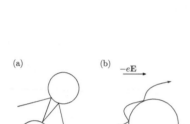

Fig. 3.1 Examples showing the development of correlations between the electrons and the ions. (a) When ions are close, this collision sequence can take place. The second time the electron collides with the top ion, they are already correlated by the first collision. (b) When a high external field **E** is applied, an electron can have several collisions with the same ion, generating correlations. The displayed trajectory corresponds to a case in which the collisions are partially inelastic, increasing the recollision probability.

3.2 Lorentz kinetic equation

The right-hand side of the BBGKY equation can then be written as the difference of two terms: the loss and gain terms. The former represents the number of electrons per unit time with velocity $\mathbf{c}$ that are lost due to collisions; that is, after the collision they end up with another velocity. The latter is the gain term, which corresponds to the number of electrons with some velocity $\mathbf{c}^*$ that, after a collision, end up with velocity $\mathbf{c}$. These collision processes, direct and inverse, are represented in Fig. 3.2.

To compute the loss and gain terms we will use the assumption that there are no correlations in the precollisional states in both the direct and inverse collisions. Their rate of occurrence is hence proportional to $f(\mathbf{c})F(\mathbf{c}_1)$ and $f(\mathbf{c}^*)F(\mathbf{c}_1^*)$, respectively, with some kinematical factors that we analyse below. If we had kept the eventual correlations between electrons and ions, the number of collisions would be proportional to $f^{(1,1)}$. This idea will be put forward when describing dense gases in Section 4.8.1.

Let us consider first the loss term. In classical mechanics, if the interaction potential is spherically symmetric, the geometry of the collision is described by means of the impact parameter b and the azimuthal angle ψ (see Fig. 3.3 and Appendix C for details). For an electron with velocity in the vicinity d^3c of $\mathbf{c}$ to collide with an ion of velocity $\mathbf{c}_1$ with an impact parameter b in a time interval Δt, it must be in the collision cylinder, depicted in Fig. 3.4. It is parallel to the relative velocity and has a volume $\Delta \mathcal{V} = |\mathbf{c} - \mathbf{c}_1|\Delta t\, b\, \mathrm{d}b\, \mathrm{d}\psi$. All electrons in the cylinder will collide with the ion, and electrons outside it will miss the ion or will not reach it in the considered time interval. Recalling that the number of particles in a specific configuration is given by the distribution function times the space and velocity volumes, the average number of collisions with the described characteristics equals the number of electrons in the cylinder volume, given by $f(\mathbf{r},\mathbf{c},t)|\mathbf{c} - \mathbf{c}_1|\Delta t\, b\, \mathrm{d}b\, \mathrm{d}\psi\, \mathrm{d}^3c$, times the number of ions ready to be collided with, given by $F(\mathbf{r},\mathbf{c}_1,t)\,\mathrm{d}^3c_1\,\mathrm{d}^3r$.

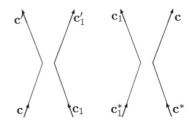

Fig. 3.2 Schematic representation of the direct (left) and inverse (right) collisions.

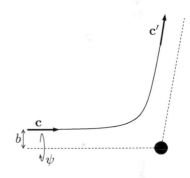

Fig. 3.3 Parameterisation of the scattering process. The incoming velocity is $\mathbf{c}$, and the postcollisional one is $\mathbf{c}'$. The impact parameter b is the minimal distance of the initial straight trajectory to the target. For spherically symmetric potentials, the scattering geometry is planar and the rotation around the dashed axis is quantified by the azimuthal angle ψ.

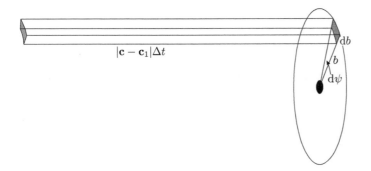

Fig. 3.4 Electrons inside the collision cylinder of length $|\mathbf{c} - \mathbf{c}_1|\Delta t$ and bounded by the circle sector of sides $\mathrm{d}b$ and $b\,\mathrm{d}\psi$ can collide with the ion in Δt. Indeed, it is a right prism but it is customary to call it a cylinder.

In the gain term, the number of relevant collisions in the same time interval Δt is $f(\mathbf{r},\mathbf{c}^*,t)F(\mathbf{r},\mathbf{c}_1^*,t)|\mathbf{c}^* - \mathbf{c}_1^*|\Delta t\, b^*\, \mathrm{d}b^*\, \mathrm{d}\psi^*\, \mathrm{d}^3c^*\, \mathrm{d}^3c_1^*$. This expression, however, can be simplified. Energy and angular momentum conservation imply that the inverse collision is produced by simply reversing the direct collision, while keeping the impact parameter and azimuthal angle. That is, $\mathbf{c}^* = \mathbf{c}'$, $\mathbf{c}_1^* = \mathbf{c}_1'$, and $\mathrm{d}^3 c^* \, \mathrm{d}^3 c_1^* \, b^* \, \mathrm{d}b^* \, \mathrm{d}\psi^* = \mathrm{d}^3 c \, \mathrm{d}^3 c_1 \, b \, \mathrm{d}b \, \mathrm{d}\psi$, where primed velocities are the postcollisional ones of the direct collision (Fig. 3.2-left). This property is derived in Section 4.1.2 using the principles of time reversibility.

With these approximations,[3] we finally arrive at the Lorentz model of classical charge transport, which corresponds to the first equation of the BBGKY hierarchy (2.52), where the interaction term has been written explicitly in terms of the direct and inverse collisions:

$$\frac{\partial f}{\partial t} + \mathbf{c} \cdot \nabla f + \frac{q\mathbf{E}}{m} \cdot \frac{\partial}{\partial \mathbf{c}} f = \int \left[f' F_1' - f F_1 \right] |\mathbf{c} - \mathbf{c}_1| \, b \, \mathrm{d}b \, \mathrm{d}\psi \, \mathrm{d}^3 c_1. \quad (3.1)$$

Here we have adopted the short-hand notation, $f = f(\mathbf{c})$, $f' = f(\mathbf{c}')$, $F_1 = F(\mathbf{c}_1)$, and $F_1' = F(\mathbf{c}_1')$, and all distribution functions are evaluated at the same position and time. The right-hand side is usually called the collision operator and is denoted by $J[f]$. In the Lorentz model, it is a linear integral operator. The collision operator, we recall, is built as the subtraction of a gain term J_+ and a loss term J_-. The factor $|\mathbf{c} - \mathbf{c}_1|$ present in the collision operator reflects the fact that the interaction rate of electrons with ions depends on the relative velocity. Collisions are more frequent if they have a larger relative velocity.

To complete the description given by the Lorentz equation (3.1), we must give the distribution function for the ions and the electron–ion interaction potential. The latter allows us to compute the postcollisional velocities in terms of the incoming ones and the impact parameter. However, the Lorentz equation has many properties that are generic, independent of the particular collision rule, and we postpone discussion on how to model the electron–ion interaction to Section 3.6.

In this chapter we will study unbounded regimes, but boundary conditions can easily be incorporated into the kinetic equation. The procedure described in Section 4.6 for the Boltzmann equation is applied directly to the Lorentz case.

[3] We recall the approximations that lead to the Lorentz model: (i) the ion distribution function is fixed, (ii) no electron–electron interactions, (iii) short-range interactions, and (iv) no precollisional correlations.

3.3   Ion distribution function

The most natural election is to consider ions at thermal equilibrium at temperature T,

$$F(\mathbf{c}_1) = n_i \left(\frac{M}{2\pi k_\mathrm{B} T} \right)^{3/2} e^{-M c_1^2 / 2 k_\mathrm{B} T}, \quad (3.2)$$

where n_i is the ion density and M is their mass.

If the mass contrast M/m is large, the ion distribution becomes very narrow compared with that of the electrons. In the limit of infinite contrast, the ions have vanishing velocities compared with the electrons and their distribution can be approximated by $F(\mathbf{c}_1) = n_i \delta(\mathbf{c}_1)$.

3.4   Equilibrium solution

In the absence of an external field, it can easily be verified that, if the ions are at thermal equilibrium, the Maxwell–Boltzmann distribution is the solution of the Lorentz equation (3.1), and that the temperature of the electrons equals that of the ions. Indeed, energy conservation at collisions, i.e. $mc^2/2 + Mc_1^2/2 = mc'^2/2 + Mc_1'^2/2$, implies directly that $f_{\mathrm{MB}}(\mathbf{c}_1')F_{\mathrm{MB}}(\mathbf{c}_2') = f_{\mathrm{MB}}(\mathbf{c}_1)F_{\mathrm{MB}}(\mathbf{c}_2)$, and the right-hand side of the equation vanishes identically. In Section 3.8 it will be shown that any spatial variations relax with time and the equilibrium solution is homogeneous.

3.5   Conservation laws and the collisional invariants

The Lorentz equation, which is based on the motion of individual electrons, must preserve the basic properties of the underlying dynamics. Here, we study how the microscopic conservation laws are expressed at the kinetic level.

Consider the Lorentz equation and multiply it by any function φ of the velocity and integrate with respect to d^3c. The result, after performing some integration by parts, is

$$\frac{\partial \rho_\varphi}{\partial t} + \nabla \cdot \mathbf{J}_\varphi - \frac{q\mathbf{E}}{m} \cdot \rho_{\mathbf{g}} = \int \varphi(\mathbf{c})\left[f'F_1' - fF_1\right]|\mathbf{c} - \mathbf{c}_1|\, b\, \mathrm{d}b\, \mathrm{d}\psi\, \mathrm{d}^3c\, \mathrm{d}^3c_1,$$

(3.3)

where we have used the definitions of the density ρ_φ and the flux $\mathbf{J}_\varphi$ given in Chapter 1,

$$\rho_\varphi = \int \varphi f\, \mathrm{d}^3c, \qquad\qquad \mathbf{J}_\varphi = \int \varphi f\mathbf{c}\, \mathrm{d}^3c, \qquad (3.4)$$

and $\rho_{\mathbf{g}}$ is the density associated with $\mathbf{g} = \frac{\partial \varphi}{\partial \mathbf{c}}$, which is interpreted below.

Equation (3.3) describes the time evolution of the density field associated with φ. First, the term $\nabla \cdot \mathbf{J}_\varphi$ represents the flux described in Chapter 1. The third term is a source term, analogous to the mechanical work by the field.[4] Finally, the right-hand side term is related to the change of φ at collisions. To give an interpretation, we proceed to rewrite it in a more convenient way. Consider the term with primed velocities:

$$\int \varphi(\mathbf{c})f'F_1'|\mathbf{c} - \mathbf{c}_1|\, b\, \mathrm{d}b\, \mathrm{d}\psi\, \mathrm{d}^3c\, \mathrm{d}^3c_1. \qquad (3.5)$$

As discussed in Section 3.2, the differential factor can be changed, as $|\mathbf{c} - \mathbf{c}_1|\, b\, \mathrm{d}b\, \mathrm{d}\psi\, \mathrm{d}^3c\, \mathrm{d}^3c_1 = |\mathbf{c}^* - \mathbf{c}_1^*|\, b^*\, \mathrm{d}b^*\, \mathrm{d}\psi^*\, \mathrm{d}^3c^*\, \mathrm{d}^3c_1^*$. Here, using the notation of the right panel of Fig. 3.2, we read this term as integrated over precollisional parameters, φ is evaluated at postcollisional velocities, and the distribution function on precollisional velocities. As we have dummy integration variables, we can rename them according to the left

[4]See Exercise 3.9.

panel of Fig. 3.2; that is, precollisional velocities are unprimed whereas postcollisional velocities are primed. Under this renaming of variables, this term can be written as

$$\int \varphi(\mathbf{c}') f F_1 |\mathbf{c} - \mathbf{c}_1| \, b \, db \, d\psi \, d^3c \, d^3c_1. \tag{3.6}$$

Considering the transformation of eqn (3.5) into (3.6), eqn (3.3) reads,

$$\frac{\partial \rho_\varphi}{\partial t} + \nabla \cdot \mathbf{J}_\varphi - \frac{q\mathbf{E}}{m} \cdot \rho_\phi = \int [\varphi(\mathbf{c}') - \varphi(\mathbf{c})] f F_1 |\mathbf{c} - \mathbf{c}_1| \, b \, db \, d\psi \, d^3c \, d^3c_1. \tag{3.7}$$

Now, the right-hand side term has a direct interpretation: it gives the change in ρ_φ as being proportional to the collision frequency times the change of φ at every collision. As a consequence, if φ corresponds to a quantity that is conserved at every collision, the right-hand side term vanishes identically and the transport equation (3.7) simplifies considerably. These functions are called collisional invariants, and they will play a central role in Chapter 4.

The first collisional invariant is $\varphi = 1$, or any constant proportional to it. It should be noted that the cancellation of the right-hand side term is not a coincidence, but rather a direct consequence of the conservation of the number of electrons. Indeed, they are not created or eliminated by the interactions with the ions; only their velocities change. Take $\varphi = q$, where q is the electron charge. In this case, $\mathbf{g} = \frac{\partial \varphi}{\partial \mathbf{c}} = 0$, and the transport equation is

$$\frac{\partial \rho}{\partial t} + \nabla \cdot \mathbf{J} = 0, \tag{3.8}$$

where the electric charge density is defined as

$$\rho(\mathbf{r}, t) = q \int f(\mathbf{r}, \mathbf{c}, t) \, d^3c \tag{3.9}$$

and the electric current is

$$\mathbf{J} = q \int \mathbf{c} f(\mathbf{r}, \mathbf{c}, t) \, d^3c. \tag{3.10}$$

Equation (3.8) is the macroscopic representation of charge conservation. It is the differential form of the Kirchhoff law of electromagnetism.

In general ion models, the charge is the only collisional invariant; that is, there is no other function of the electron velocity that is preserved in every collision. If we consider the case of rigid hard spheres described below, any function of the speed is also a collisional invariant, because the electron energy is conserved at every collision. The transport equations that result from these functions of the speed are, however, trivial and do not carry additional information.

3.6   Kinetic collision models

The Lorentz equation is completed by giving the scattering laws, which permit one to obtain the postcollisional velocities in the collision operator. Bearing this in mind, we will consider two models that simplify the collision operator notably, and study them in detail. To describe specific experimental situations it may be necessary to consider other models that use the scattering properties of the interaction potential. Then, besides analytic methods, numerical tools can be applied, such as those presented in Chapter 10.

3.6.1   Rigid hard spheres

The simplest model, which we will use throughout this chapter, is to consider the ions as rigid hard spheres. In this model, all ions have infinite mass and a hard-core radius R, and are static, with $F(\mathbf{c}_1) = n_i \delta(\mathbf{c}_1)$. The electron–ion collision is parameterised by the normal vector $\hat{\mathbf{n}}$ that forms an angle θ with $\mathbf{c}$, as shown in Fig. 3.5. A collision takes place if the electron is approaching the ion, a condition that is expressed as $\mathbf{c} \cdot \hat{\mathbf{n}} < 0$. Imposing energy conservation and that the instantaneous force points along $\hat{\mathbf{n}}$, the postcollisional velocity is

$$\mathbf{c}' = \mathbf{c} - 2(\mathbf{c} \cdot \hat{\mathbf{n}})\hat{\mathbf{n}}. \tag{3.11}$$

In the hard sphere geometry, the impact parameter is given by $b = R\sin\theta$. Therefore, we can write the differential as $|\mathbf{c} - \mathbf{c}_1|\, b\, db\, d\psi = R^2 \Theta(-\mathbf{c} \cdot \hat{\mathbf{n}})|\mathbf{c} \cdot \hat{\mathbf{n}}|\, d^2\hat{\mathbf{n}}$, where the Heaviside function Θ selects precollisional configurations and we have used that $\mathbf{c}_1 = 0$.

The loss term of the collision operator is then simply given by

$$J_- = n_i R^2 f(\mathbf{c}) \int \Theta(-\mathbf{c} \cdot \hat{\mathbf{n}})|\mathbf{c} \cdot \hat{\mathbf{n}}|\, d^2\hat{\mathbf{n}} = n_i \pi R^2 |\mathbf{c}| f(\mathbf{c}), \tag{3.12}$$

where the tensorial relations described in Appendix B have been used to compute the integral.

In the rigid hard sphere model, the energy of the electrons is preserved at every collision: $|\mathbf{c}'| = |\mathbf{c}|$. Therefore, collisions only change the direction of the velocity. It can easily be shown that collisions with rigid hard spheres with a uniformly distributed normal director $\hat{\mathbf{n}}$ produce an isotropic distribution of postcollisional velocities (see Appendix C). Hence, the gain term of the collision operator can be written as

$$J_+ = n_i \pi R^2 |\mathbf{c}| \mathbb{P} f, \tag{3.13}$$

where the operator $\mathbb{P}$ averages the distribution over all directions:

$$\mathbb{P} f = \frac{1}{4\pi} \int f(\mathbf{r}, \mathbf{c}, t)\, d^2\hat{\mathbf{c}}. \tag{3.14}$$

Finally, the full collision operator in the rigid hard sphere model is

$$J[f] = n_i \pi R^2 |\mathbf{c}| \left(\mathbb{P} f - f \right). \tag{3.15}$$

We can now interpret this collision operator as removing particles at a rate $n_i \pi R^2 |\mathbf{c}|$ and replacing them by particles with an isotropic distribution.

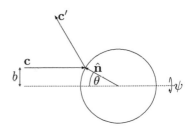

Fig. 3.5 Parameterisation of the collision between an electron and a rigid hard sphere. The incoming velocity is $\mathbf{c}$, and the postcollisional one is $\mathbf{c}'$. The incoming velocity forms an angle θ with the normal vector $\hat{\mathbf{n}}$, ψ is the azimuthal angle, and b is the impact parameter.

In two dimensions, the collisions with rigid hard disks do not produce an isotropic distribution as compared with the three-dimensional case. Consequently, the collision operator is

$$J[f] = n_i R \int f\left(\mathbf{c} - 2(\mathbf{c}\cdot\hat{\mathbf{n}})\hat{\mathbf{n}}\right) |\mathbf{c}\cdot\hat{\mathbf{n}}| \theta(\mathbf{c}\cdot\hat{\mathbf{n}})\, \mathrm{d}\hat{\mathbf{n}} - 2n_i R |\mathbf{c}| f(\mathbf{c}), \quad (3.16)$$

where we have explicitly used the collision rule (3.11) and that in two dimensions, the total cross section is $2R$.

In this model, the magnitudes of the electron velocities are preserved at collisions and only their directions change. Then, any isotropic distribution function is an equilibrium solution, and does not evolve to a Maxwellian.

3.6.2   Thermalising ions: the BGK model

If the ionic mass is not too large compared with that of electrons, they cannot be assumed to be static and a Maxwellian distribution should be used instead. At collisions, electrons exchange energy with the ions. To take into account the effect of the energy exchange, we will consider a model that can be thought of as the opposite limit of the rigid hard sphere model. Here, after each collision, electrons emerge with velocities selected from a Maxwellian distribution, completely forgetting their incoming velocity. For simplicity we will model the ions as hard spheres; having a larger mass than the electrons, their velocity is small, and therefore the relative velocity can be approximated as $|\mathbf{c} - \mathbf{c}_1| \approx |\mathbf{c}|$. The loss term of the collision operator is hence the same as that for hard spheres, but the gain term changes to now model the outcome of Maxwellian velocities. To write the gain term, we note that it should be proportional to the Maxwellian flux distribution $|\mathbf{c}|\widehat{f}_{\mathrm{MB}}(\mathbf{c})$ (see Section 1.3.3 for an interpretation of the flux distribution). The proportionality constant is not arbitrary, though. It must guarantee that charge is conserved, i.e. that any constant function is a collisional invariant. Under these considerations, the collision operator reads,[5]

$$J[f] = n_i \pi R^2 c \left[\widehat{f}_{\mathrm{MB}}(\mathbf{c}) \frac{\int |\mathbf{c}'| f(\mathbf{c}')\, \mathrm{d}^3 c'}{\int |\mathbf{c}'| \widehat{f}_{\mathrm{MB}}(\mathbf{c}')\, \mathrm{d}^3 c'} - f(\mathbf{c}) \right]. \quad (3.17)$$

3.7   Electrical conduction

3.7.1   Conservation equation

To investigate the electrical conductivity, we consider the response of the system to an applied electric field $\mathbf{E}$. The electrical conductivity is produced by the motion of the electrons, so it is natural to study the dynamical evolution of the charge density. The conservation equation (3.8) is not closed, as $\mathbf{J}$ is unknown. Our objectives are to derive Ohm's law, $\mathbf{J} = \sigma\mathbf{E}$, to understand under which conditions it is valid, and to give an expression for the conductivity σ.

[5]In the context of the Boltzmann equation, this is called the BGK model, after P. Bhatnagar, E. Gross, and M. Krook; it was introduced in 1954. It is also called the relaxation time approximation.

3.7.2 Linear response

We consider the case of small electric fields. To do so, we introduce a formal small dimensionless parameter ϵ and write the electric field as $\mathbf{E} = \epsilon \mathbf{E}_0$, which we assume to be constant and uniform. In the absence of an electric field, the electrons will reach an equilibrium Maxwellian distribution. Then, in the presence of a small field, it is expected that the distribution function can be written as[6]

$$f(\mathbf{c}) = f_{\mathrm{MB}}(\mathbf{c}) \left[1 + \epsilon \Phi(\mathbf{c}) \right]. \tag{3.18}$$

The function Φ encodes the response of the system to the external field. This ansatz can be inserted back into the Lorentz equation, resulting in an integral equation for Φ. However, this analysis can be simplified by using some symmetry arguments. By assumption, $\Phi(\mathbf{c})$ is proportional to $\mathbf{E}_0$, but the former is a scalar while the latter is a vector. If the system is isotropic, the electric field must be multiplied by another vector, and the only possibility is $\mathbf{c}$ (see Appendix B). So, the most general form in which this proportionality can be written that respects the tensorial symmetry is

$$\Phi(\mathbf{c}) = \phi(c)\mathbf{c} \cdot \mathbf{E}_0. \tag{3.19}$$

Note that the problem has been greatly simplified, as the new unknown $\phi(c)$ is only a function of the modulus of $\mathbf{c}$, as compared with Φ, which depends on the whole vector.

3.7.3 Ohm's law

The electric current can be computed using (3.18) and (3.19) as

$$\mathbf{J} = q \int \mathbf{c} f_{\mathrm{MB}}(c) \left[1 + \epsilon \phi(c)\mathbf{c} \cdot \mathbf{E}_0 \right] \, \mathrm{d}^3 c = q \int f_{\mathrm{MB}}(c)\phi(c)\mathbf{c}\mathbf{c} \cdot \mathbf{E} \, \mathrm{d}^3 c = \sigma \mathbf{E}, \tag{3.20}$$

recovering Ohm's law with a conductivity tensor σ that depends on ϕ. If the system is isotropic, the conductivity tensor is diagonal and the scalar conductivity is obtained by computing the trace $\sigma = \mathrm{Tr}\,\sigma/3$ as[7]

$$\sigma = \frac{q}{3} \int f_{\mathrm{MB}}(c)\phi(c)c^2 \, \mathrm{d}^3 c. \tag{3.21}$$

We recapitulate that Ohm's law was obtained in the linear regime, appealing only to symmetry arguments. If the field were stronger, the response would be nonlinear.

3.7.4 Electrical conductivity

To compute the electrical conductivity we need the function ϕ. It is found by inserting the perturbation expansion given in (3.18) and (3.19) into the Lorentz equation (3.1) and retaining terms only up to order ϵ. The left-hand side has contributions only from the third term, which up

[6]Here we have made the hypothesis that there is a stationary state in the presence of a field. In the case of the rigid hard sphere model, this is indeed not true as the electrons gain energy by the field and do not exchange it with the ions. However, the energy gain corresponds to the Joule effect and therefore is of order ϵ^2. So, at the linear level, it can be neglected, and the assumption of a stationary linear regime is valid.

[7]By components we have $J_i = \sigma_{ik} E_k$. Under isotropy, $\sigma_{ik} = \sigma \delta_{ik}$, resulting in $J_i = \sigma E_i$ ($\mathbf{J} = \sigma \mathbf{E}$).

to linear order in ϵ reads,

$$\frac{q\mathbf{E}}{m} \cdot \frac{\partial f}{\partial \mathbf{c}} = \frac{q\epsilon\mathbf{E}_0}{m} \cdot \frac{\partial f_{\mathrm{MB}}}{\partial \mathbf{c}} = -\frac{q\epsilon}{k_{\mathrm{B}}T} f_{\mathrm{MB}}\mathbf{c} \cdot \mathbf{E}_0. \qquad (3.22)$$

When substituting the distribution function into the right-hand side of the Lorentz equation, the order-zero term cancels, and the first order in ϵ is

$$J[f] = \epsilon \int f_{\mathrm{MB}}(\mathbf{c}) F_{\mathrm{MB}}(\mathbf{c}_1) \left[\Phi(\mathbf{c}') - \Phi(\mathbf{c})\right] |\mathbf{c} - \mathbf{c}_1|\, b\, \mathrm{d}b\, \mathrm{d}\psi\, \mathrm{d}^3 c_1. \qquad (3.23)$$

We define the linear operator I acting on a function of the velocity as

$$I[\Psi] = \int \widehat{f}_{\mathrm{MB}}(\mathbf{c}) \widehat{F}_{\mathrm{MB}}(\mathbf{c}_1) \left[\Psi(\mathbf{c}) - \Psi(\mathbf{c}')\right] |\mathbf{c} - \mathbf{c}_1|\, b\, \mathrm{d}b\, \mathrm{d}\psi\, \mathrm{d}^3 c_1, \qquad (3.24)$$

where $\widehat{f}_{\mathrm{MB}}$ and $\widehat{F}_{\mathrm{MB}}$ are the Maxwellian distributions of electrons and ions without the density dependence, i.e. normalised to one. Note that, for later convenience, the primed and unprimed terms have been reversed in (3.24) compared with (3.23). In terms of this operator,

$$I[\Phi] = \frac{q}{n_i k_{\mathrm{B}}T} \widehat{f}_{\mathrm{MB}}(c)\mathbf{c} \cdot \mathbf{E}_0. \qquad (3.25)$$

This is a linear integral equation for Φ. If there is no preferred direction in the system, the linear operator is isotropic. Acting on scalars, vectors, and tensors, it produces scalars, vectors, and tensors, respectively. Therefore, inserting (3.19) into (3.25), we obtain an equation for ϕ. Again, we can exploit the linearity to write $\phi = (q/n_i)\hat{\phi}$, which satisfies the equation

$$I[\hat{\phi}(c)\mathbf{c}] = \frac{1}{k_{\mathrm{B}}T} \widehat{f}_{\mathrm{MB}}(c)\mathbf{c}, \qquad (3.26)$$

where it is evident that $\hat{\phi}$ does not depend on the ion or electron density nor on the carrier charge. Coming back to the conductivity, one obtains

$$\sigma = \frac{q^2 n_e}{3 n_i} \int \widehat{f}_{\mathrm{MB}}(c)\hat{\phi}(c)c^2\, \mathrm{d}^3 c. \qquad (3.27)$$

In summary, applying a linear response analysis and using the isotropy of space, we were able to deduce that the electron conductivity is proportional to the carrier density, inversely proportional to the ion density, and independent of the sign of the charge. These results, obtained using this classical model, will be shown to be valid for the quantum transport of electrons after making some reinterpretations. For example, instead of considering the ion density, we use the scatterer density in a more general approach.[8] The linear dependence on the electron density is a direct consequence of considering that electrons do not interact among themselves, therefore the more electrons in the system, the larger the conductivity. The independence from the sign of the charge carrier is due to the double use of q, first in the definition of the electrical current

[8] We will see that electrons are scattered by impurities and lattice vibrations but not by the ions themselves.

and second to obtain their response to the electric field. Changing the sign of the electron charge would make them move in the opposite direction, therefore changing $\langle \mathbf{c} \rangle$ and leaving the charge current unchanged. Of course, there could be a dependence in the scattering process and the linear operator could depend on the sign of q, but we know that, in the case of the rigid hard sphere model or Coulomb scattering, this is not the case (see Appendix C). Finally, the inverse proportionality on the ion density is related to the decrease of the electron mean free path with increasing ion density. As we saw in Chapter 1, the transport coefficients are normally linear in the mean free path.

Rigid hard spheres

To show how the integral equation can be solved and to give a numerical value for the electrical conductivity, we model the ions as rigid hard spheres. In this case, the linear operator is

$$I[\Psi] = \pi R^2 \widehat{f}_{\mathrm{MB}} |\mathbf{c}| (\Psi - \mathbb{P}\Psi). \tag{3.28}$$

We note that, in the conductivity problem, we need to evaluate $\mathbb{P}\hat{\phi}\mathbf{c}$, which vanishes by isotropy. We can therefore proceed directly to solve eqn (3.26), obtaining $\hat{\phi}(c) = (1/\pi R^2 k_{\mathrm{B}} T)|\mathbf{c}|^{-1}$. Replacing in (3.27), the electrical conductivity is

$$\sigma = \frac{q^2 n_e}{3 n_i k_{\mathrm{B}} T \pi R^2} \sqrt{\frac{8 k_{\mathrm{B}} T}{\pi m_e}} = \frac{q^2 n_e \ell}{3 k_{\mathrm{B}} T} \langle |\mathbf{c}| \rangle, \tag{3.29}$$

where in the second expression we have used the mean free path value $\ell = 1/(n_i \pi R^2)$. Note that the Drude formula coincides with this expression. As expected, this result does not give good quantitative predictions when applied to metals, when electrons present quantum behaviour, but it correctly describes the transport of classical charges, as for example in ionic solutions or electrolytes. In these cases, it is more appropriate to use the second formula, in terms of the mean free path, which can be calculated or measured without the need for the assumptions made in the present derivation.

BGK model

The linear operator in this case reads,

$$I[\Psi] = \pi R^2 \widehat{f}_{\mathrm{MB}} |\mathbf{c}| \left(\Psi - \frac{\int \mathrm{d}^3 c' |\mathbf{c}'| \widehat{f}_{\mathrm{MB}}(\mathbf{c}') \Psi(\mathbf{c}')}{\int \mathrm{d}^3 c' |\mathbf{c}'| \widehat{f}_{\mathrm{MB}}(\mathbf{c}')} \right). \tag{3.30}$$

As in the rigid hard sphere case, the loss term evaluated on $\Psi = \hat{\phi}\mathbf{c}$ vanishes. The solution of the linear equation is hence the same as found previously in the rigid hard sphere case, and the conductivity is given by (3.29).

   The agreement of the two models is not a coincidence but, rather, an effect of assuming that the electrons are already thermalised with the same temperature as the ions. Therefore, the only action of the thermal BGK collision operator is to make the distribution isotropic.

Other ion models

If the full scattering model is used, with a given collision rule, the solution of the linear equation (3.26) is not simple. Approximate or numerical solutions must be found. In practice, as the equation is linear, this is done by expanding the unknown function $\hat{\phi}$ in a polynomial basis to later project the equation into this basis. The right-hand side produces terms that are simple Gaussian integrals, while the left-hand side depends on collisional integrals that can be computed using the specific ion model. The matrix problem can be inverted, and the precision of the results increases when using a larger number of elements in the basis.[9]

[9]The associated Laguerre polynomials (Sonine polynomials) are a particularly useful basis for which the calculations can simplify notably.

3.7.5   Frequency response

Using the Lorentz equation one can also compute the response of the system to electric fields that vary temporally. Again we are going to use the linear response, and therefore we can use Fourier analysis and consider electric fields that have a well-defined frequency, $\mathbf{E}(t) = \epsilon\mathbf{E}_\omega e^{-i\omega t}$. The linear response will have the same temporal dependence, and we can write $f(\mathbf{c}, t) = f_{\mathrm{MB}}(\mathbf{c})\left[1 + \epsilon\Phi_\omega(\mathbf{c})e^{-i\omega t}\right]$. Once inserted into the Lorentz equation, we find up to linear order in ϵ that

$$I[\Phi_\omega] - \frac{i\omega}{n_i}\widehat{f}_{\mathrm{MB}}\Phi_\omega = \frac{q}{k_{\mathrm{B}}Tn_i}\widehat{f}_{\mathrm{MB}}\mathbf{E}_\omega \cdot \mathbf{c}. \tag{3.31}$$

By linearity and isotropy we can write $\Phi_\omega(\mathbf{c}) = (q/n_i)\widehat{\phi}_\omega\mathbf{c}\cdot\mathbf{E}_\omega$, resulting in

$$I[\widehat{\phi}_\omega\mathbf{c}] - \frac{i\omega}{n_i}\widehat{f}_{\mathrm{MB}}\widehat{\phi}_\omega\mathbf{c} = \frac{1}{k_{\mathrm{B}}T}\widehat{f}_{\mathrm{MB}}\mathbf{c}. \tag{3.32}$$

The perturbation distribution function becomes complex, and so does the electrical conductivity, $\sigma = \sigma_0 + i\sigma_1$. Solving the equation, we obtain that the sign of the imaginary part is positive. Therefore, the electric current can be written as

$$\mathbf{J} = (\sigma_0 + i\sigma_1)\mathbf{E}_\omega e^{-i\omega t} = \sigma_\omega\mathbf{E}_\omega e^{-i(\omega t - \alpha)}. \tag{3.33}$$

Here, $\sigma_\omega = \sqrt{\sigma_0^2 + \sigma_1^2}$ is the complex conductivity and $\alpha = \arctan\sigma_1/\sigma_0 > 0$ is the phase delay. The sign of the delay indicates that the current is retarded with respect to the field. You will show in Exercise 3.10 that the delay is proportional to the electron mean free path.

3.8   Relaxation dynamics

In Section 3.4 we discussed the equilibrium solution of the Lorentz equation, saying that electrons will reach a stationary Maxwellian distribution. However, we did not discuss how this process takes place. Here, we analyse this problem, considering, for simplicity, that there is no electric field and that initially electrons have a distribution function that is close to, but different from, the Maxwellian.[10] This distribution will be assumed to have a spatial dependence.

We will consider the linear dynamics near equilibrium, and therefore we can use Fourier analysis in the spatial coordinate and consider perturbations that have a unique wavevector $\mathbf{k}$. Initially, the distribution function is $f_0(\mathbf{r}, \mathbf{c}) = f_{\mathrm{MB}}(c)\left[1 + \Phi_{\mathbf{k}}(\mathbf{c})e^{i\mathbf{k}\cdot\mathbf{r}}\right]$, with $\Phi_{\mathbf{k}} \ll 1$, and we assume that, for all times, the deviation from equilibrium will remain small in the form,

$$f(\mathbf{r}, \mathbf{c}, t) = f_{\mathrm{MB}}(c)\left[1 + \Phi_{\mathbf{k}}(\mathbf{c}, t)e^{i\mathbf{k}\cdot\mathbf{r}}\right]. \tag{3.34}$$

Inserting this into the Lorentz equation, each Fourier mode evolves as

$$f_{\mathrm{MB}}\frac{\partial \Phi_{\mathbf{k}}}{\partial t} = -L_{\mathbf{k}}\Phi_{\mathbf{k}}, \tag{3.35}$$

where the operator is $L_{\mathbf{k}}\Phi = n_e n_i I[\Phi] + i\mathbf{k}\cdot\mathbf{c}\,f_{\mathrm{MB}}\Phi$ and the convenience of the minus sign will become evident immediately. This linear equation generates solutions that decay exponentially in time $\Phi_{\mathbf{k}}(\mathbf{c}, t) = \Phi_{\mathbf{k}}(\mathbf{c})e^{-\lambda t}$, which when substituted back give the generalised eigenvalue equation,

$$L_{\mathbf{k}}\Phi_{\mathbf{k}} = \lambda f_{\mathrm{MB}}\Phi_{\mathbf{k}}. \tag{3.36}$$

3.8.1   Properties of the linear operator

Before continuing, it is necessary to describe some of the mathematical properties of the operator $L_{\mathbf{k}}$. First let us consider the homogeneous case, $\mathbf{k} = 0$. It will be shown that L_0 is Hermitian and positive semidefinite. To do so, we define the usual scalar product,

$$(g, h) = \int g^*(\mathbf{c})h(\mathbf{c})\,\mathrm{d}^3 c, \tag{3.37}$$

where g^* is the complex conjugate.

Hermiticity

Consider the scalar product,

$$(g, L_0 h) = \int f_{\mathrm{MB}}(\mathbf{c})F_{\mathrm{MB}}(\mathbf{c}_1)g^*(\mathbf{c})\left[h(\mathbf{c}) - h(\mathbf{c}')\right]|\mathbf{c} - \mathbf{c}_1|\,b\,\mathrm{d}b\,\mathrm{d}\psi\,\mathrm{d}^3 c\,\mathrm{d}^3 c_1. \tag{3.38}$$

The term with the primed velocities can be transformed in a similar way as was done in Section 3.5. We first change the differential to starred variables. Then, we note that the integration is done over precollisional variables, and we change the dummy variables to those of the left panel of Fig. 3.2: the precollisional velocities (starred) are changed to unprimed, and the postcollisional variables (unprimed) are changed to primed variables. Finally, we use that the Maxwellian velocity distributions satisfy $f_{\mathrm{MB}}(\mathbf{c}')F_{\mathrm{MB}}(\mathbf{c}_1') = f_{\mathrm{MB}}(\mathbf{c})F_{\mathrm{MB}}(\mathbf{c}_1)$. Then, the integrand

[10]The case with an electric field is far from trivial. As was discussed, in the long-time limit, Joule heating will take place and electrons will not equilibrate with the ions. See Piasecki (1993) for a detailed analysis of this case.

becomes $f_{\text{MB}}(\mathbf{c})F_{\text{MB}}(\mathbf{c}_1)g^*(\mathbf{c}')h(\mathbf{c})$. With these changes, and summing again the first term of the right-hand side, we get

$$(g, L_0 h) = \int f_{\text{MB}}(\mathbf{c})F_{\text{MB}}(\mathbf{c}_1)\left[g^*(\mathbf{c}) - g^*(\mathbf{c}')\right]h(\mathbf{c})|\mathbf{c}{-}\mathbf{c}_1|\,b\,db\,d\psi\,d^3c\,d^3c_1,$$

$$(3.39)$$

which corresponds to $(L_0 g, h)$, completing the proof that L_0 is Hermitian.

Positivity

Using the same methods as when demonstrating the hermiticity of L_0, it can be shown that

$$(g, L_0 g) = \frac{1}{2}\int \left(g^* - g'^*\right)\left(g - g'\right)f_{\text{MB}}(\mathbf{c})F_{\text{MB}}(\mathbf{c}_1)|\mathbf{c}{-}\mathbf{c}_1|\,b\,db\,d\psi\,d^3c\,d^3c_1,$$

$$(3.40)$$

which is non-negative. Note that it vanishes only if $g' = g$ for all collisions; that is, it only vanishes for a collisional invariant for the electrons. As discussed previously, the only collisional invariant in the Lorentz model is a constant, $g_0 = 1$.[11]

[11]We note that, however, in the case of the rigid hard sphere model, any function of the absolute value of the velocity is a conserved quantity. This case will be analysed in detail in Section 3.8.5.

3.8.2 Kinetic gap

We consider first the case of ion models other than the rigid hard sphere case. As L_0 is Hermitian and positive semidefinite, it turns out that its eigenvalues are real and positive or zero. The zero eigenvalue is non-degenerate with associated eigenfunction $\Psi_0 = 1$, independent of the specific scattering properties with the ions. The other eigenvalues depend on the specific model. It can be proved generically that there is no accumulation point near zero and that the first nonvanishing eigenvalue is at a finite distance Δ from zero, which is of the order of the collision frequency. This property will be shown to be crucial to explain the long-time behaviour of the electron gas. In summary, for homogeneous perturbations, the relaxation dynamics is characterised by the eigenvalues, which are positive. That is, any perturbation to a Maxwellian distribution will decay in time and the equilibrium distribution is reached.

The case when $k \neq 0$ is more complex, because in this case the operator is no longer Hermitian and therefore most of the useful theorems known for this class of operators cannot be applied. Assuming that the eigenvalues behave analytically at $k = 0$, they will evolve in k, as shown in Fig. 3.6. There is then a range of k for which the smallest eigenvalue is the one that is the continuation of the null eigenvalue. If the time evolution of the initial perturbation is expanded in eigenfunctions, the long-time behaviour will be dominated by the smallest eigenvalue. Therefore, as a consequence of this system having a unique collisional invariant, if the initial perturbation is performed with a small enough wavevector, the evolution at long times is only described by a unique eigenvalue. We proceed now to compute this smallest eigenvalue using a perturbative procedure for small wavevectors.

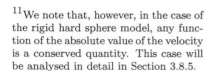
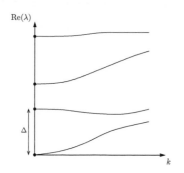

Fig. 3.6 Schematic representation of the spectrum of the linear Lorentz operator as a function of wavevector. Only the real part of the eigenvalues is presented. The kinetic gap Δ is defined as the distance between the first non-vanishing eigenvalue and the null eigenvalue at $k = 0$.

3.8.3 Spectrum of the linear operator

We introduce a formal bookkeeping parameter $\epsilon \ll 1$ for small wavevectors. The linear operator is written as $L_{\mathbf{k}} = n_e n_i I + i\epsilon \mathbf{k} \cdot \mathbf{c} f_{\mathrm{MB}} = L_0 + i\epsilon L_1$. We consider the eigenvalue problem, $L_{\mathbf{k}}\Psi = \lambda f_{\mathrm{MB}}\Psi$, performing a perturbative expansion of the smallest eigenvalue,

$$\lambda = 0 + \epsilon\lambda_1 + \epsilon^2\lambda_2 + \dots, \tag{3.41}$$

$$\Psi = 1 + \epsilon\Psi_1 + \epsilon^2\Psi_2 + \dots. \tag{3.42}$$

The zeroth-order equation in ϵ is immediately satisfied. To linear order in ϵ, the eigenvalue equation is

$$L_0\Psi_1 = (\lambda_1 f_{\mathrm{MB}} - iL_1), \tag{3.43}$$

and we note that there are two unknowns: λ_1 and Ψ_1. Here, we can apply the standard theory of perturbative expansion for the eigenvalue problem, which is explained in Appendix A. First, the linear equation has a solution if the right-hand side is orthogonal to the Ker of $L_0^{\dagger}$.[12] In the present case, L_0 is Hermitian and its Ker is simply $\Psi_0 = 1$. Therefore, in order to have a solution, the condition

$$\int (\lambda_1 f_{\mathrm{MB}} - iL_1)\,\mathrm{d}^3c = 0 \tag{3.44}$$

must be satisfied. By parity (L_1 is odd in $\mathbf{c}$), the second integral vanishes and therefore $\lambda_1 = 0$. With this result, eqn (3.43) simplifies to

$$I[\Psi_1] = -\frac{i\mathbf{c} \cdot \mathbf{k} f_{\mathrm{MB}}}{n_i}, \tag{3.45}$$

which, except for constant factors, has the same structure as (3.25). Therefore, its solution is

$$\Psi_1 = -\frac{i k_{\mathrm{B}} T}{n_i}(\mathbf{c} \cdot \mathbf{k})\widehat{\phi}, \tag{3.46}$$

with $\widehat{\phi}$ being the solution of eqn (3.26).

Now, we consider the eigenvalue equation to second order in ϵ. Using the previous results, it reads,

$$L_0\Psi_2 = (\lambda_2 f_{\mathrm{MB}} - iL_1\Psi_1). \tag{3.47}$$

Again, for this equation to have a solution, the right-hand side must be orthogonal to 1. This gives an equation for λ_2:

$$\lambda_2 = i\frac{\int L_1\Psi_1\,\mathrm{d}^3c}{\int f_{\mathrm{MB}}\,\mathrm{d}^3c} = i\frac{\int L_1\Psi_1\,\mathrm{d}^3c}{n_e}. \tag{3.48}$$

The integral can easily be reduced using the isotropy of the distribution function, resulting in an integral that depends on $\widehat{\phi}$ that is, up to a constant, the same as that which gives the electrical conductivity (3.27). Then, up to second order in ϵ, the smallest eigenvalue associated with the Lorentz equation is

$$\lambda = \left(\frac{k_{\mathrm{B}}T\sigma}{n_e q^2}\right)k^2, \tag{3.49}$$

which is positive.

[12] The Ker of an operator, also called its nucleus, is the set of functions that result in zero when the operator acts on them.

3.8.4   Diffusive behaviour

We have found that, for small wavevectors, all eigenvalues are positive, and that the smallest one goes as k^2. Recalling that the initial perturbation can be written as an expansion in eigenfunctions, this result implies that at large times ($t \gg \Delta^{-1}$) all kinetic modes have relaxed and the evolution is reduced to the smallest wavevector. As this eigenvalue is also positive, the system relaxes finally to a global Maxwellian equilibrium that is homogeneous and stationary. The k^2 dependence of the eigenvalue suggests a diffusive behaviour. Indeed, this is the case, corresponding to the diffusion of the electron density. Integrating the distribution function (3.34) over the velocity and using the expansion in eigenfunctions, it is obtained directly that

$$\frac{\partial \rho}{\partial t} = D \nabla^2 \rho, \tag{3.50}$$

where the diffusion coefficient D is the prefactor of the k^2 term in λ, i.e.

$$D = \left(\frac{k_{\mathrm{B}} T \sigma}{q^2 n_e} \right). \tag{3.51}$$

The previous expression is quite extraordinary. It relates the diffusion coefficient, associated with an initially inhomogeneous distribution, to the electrical conductivity, which gives the response of the electron gas to imposed external fields. This relation, found by Einstein in 1905, is an example of a fluctuation–dissipation relation. Found in the context of the Lorentz model, it does not depend on the particular collisional model and, indeed, is more general than the Lorentz model itself. Its origin is that both transport processes (diffusion and conductivity) are due to the motion of electrons and limited by the scattering with the ions, and therefore they are related. In other chapters we will find other examples of fluctuation–dissipation relations.

Recalling the expression for σ in the rigid hard sphere model, we get $D = \ell \langle |\mathbf{c}| \rangle / 3$, equal to the expression found in Chapter 1 using mean free path arguments.

3.8.5   Rigid hard spheres

In the rigid hard sphere model the energy of the electrons is preserved at every collision. Therefore, any distribution function that depends only on the absolute value of the velocity is an equilibrium solution and will not show temporal evolution. There is no relaxation toward the Maxwellian distribution as in the case of ions with thermal velocities. The dynamics in this model lies only in the angular distribution, with the equilibrium distribution being isotropic.

Again, it can be proved that, in the subspace of functions that depend only on the direction of $\mathbf{c}$, the linear operator (3.28) is Hermitian and positive semidefinite. The relaxation properties are preserved, and the long-time dynamics shows the same diffusive behaviour.

3.8.6 Time scales

At small wavevectors, as the eigenvalues are continuous functions of $\mathbf{k}$ ($\lambda_\mathbf{k} \approx \lambda_0$), the eigenvalues order as $0 < \lambda_{\mathbf{k},0} \ll \lambda_{\mathbf{k},1} < \ldots$, where $\lambda_{\mathbf{k},0}$ is the eigenvalue (3.49) associated to the conserved charge and $\lambda_{\mathbf{k},j}$ are eigenvalues of the kinetic modes. In the absence of spatial dependence, the time scale for the eigenvalues is that of the collision frequency, $\lambda_1 = \nu \widehat{\lambda}_1$, where $\widehat{\lambda}_1$ is a dimensionless number that depends on the ion model.[13] The dynamics of the electron system is then characterised by well-separated time scales (Fig. 3.7). First, at short times, $t \ll \nu^{-1}$, electrons move freely before having collisions with the ions. On this time scale, the distribution function is unchanged from the initial one. This is called the ballistic regime. Then, when $t \sim \nu^{-1}$, a second regime is established. In this kinetic regime the kinetic modes relax and the distribution function approaches a local Maxwellian. Later on, when $t \sim (Dk^2)^{-1} \gg \nu^{-1}$, a third regime is established. Here, the distribution function is a local Maxwellian, but the density is still inhomogeneous and we say that the system is in local thermodynamic equilibrium. It is on this time scale that the hydrodynamic equations[14] describe the system dynamics (in the present case, the diffusion equation). Finally, when $t \gg (Dk^2)^{-1}$, the global thermodynamic equilibrium is reached and the distribution function is a Maxwellian with a uniform density.

[13] In Exercise 3.6 you will estimate this value for the rigid hard sphere model.

[14] Here, we will call hydrodynamic equations those that derive from conservation properties. In the present case, charge is conserved at collisions and the associated hydrodynamic equation is the diffusion equation.

$$t_K = \nu^{-1} \qquad t_H = (Dk^2)^{-1}$$

Free flight   Kinetic regime   Hydrodynamic regime   Equilibrium

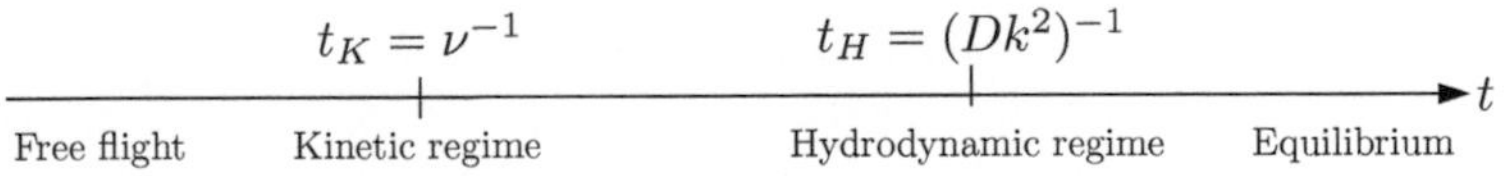

Fig. 3.7 Different regimes in the temporal evolution of the Lorentz gas in the absence of an external field.

The condition for having this time scale separation is that k is small enough such that $Dk^2 \ll \nu$. Taking $k = 2\pi/L$, where L is the characteristic length on which the density varies, the time scale separation is reached if $L \gg 2\pi\ell$, where we have used the rigid hard sphere model to evaluate the collision frequency and electrical conductivity. Then, we find the remarkable result that, whenever the density varies on length scales much larger than the mean free path, the time evolution simplifies to that given by the diffusion equation. As mean free paths are of the order of microns or even smaller, this explains the extraordinary success of the hydrodynamic equations in describing the evolution of macroscopic systems. The condition to have this time scale separation was the existence of a gap in the spectrum due to the presence of collisional invariants. When studying the dynamics of gases in Chapter 4, we will find that, associated with the conservation of mass, vectorial momentum, and energy, there are five null eigenvalues that, for small wavevectors, lead to the appearance of the Navier–Stokes hydrodynamic equations.

3.9   The Chapman–Enskog method

[15]The Chapman–Enskog method is part of a class of methods that systematically analyse the different spatiotemporal scales that can be present in the dynamics of a system. Depending on the context, those methods are called multiple-scale analysis, singular perturbation, or asymptotic analysis.

Chapman and Enskog developed a method to systematically obtain the hydrodynamic equations, at increasing orders in the inhomogeneities, using the key concept of time scale separation.[15] The existence of a unique conservation law in this case, compared with the five conservations in the case of gases, makes the derivation of the method simpler and adequate for this introductory book. The ideas, nevertheless, are general and can be applied to other kinetic models. In Section 4.7.4 we will sketch the method in the case of gases.

The first step in the Chapman–Enskog method is to note that, in the hydrodynamic regime, the distribution function does not depend separately on the velocity, time, and space, but rather that its spatiotemporal dependence is enslaved by that of the density field. Then, the crucial assumption is that the distribution function depends on space and time through the density as

$$f(\mathbf{r}, \mathbf{c}, t) = h(\mathbf{c}; n(\mathbf{r}, t)). \tag{3.52}$$

We now introduce a formal small parameter ϵ that multiplies the gradient term and the electric field in the left-hand side of the Lorentz equation; that is, we assume that the spatial inhomogeneities and the electric field are small. We expand the distribution function as a power series in ϵ as

$$h = h_0 + \epsilon h_1 + \epsilon^2 h_2 + \dots. \tag{3.53}$$

Finally, we make explicit the existence of several time scales. This is done by introducing several time variables, $t_0 = t$, $t_1 = \epsilon t$, $t_2 = \epsilon^2 t$, and so on. The election is done such that each time variable reflects the dynamics on a specific time scale. For example, t_1 will be small whenever $t \ll \epsilon^{-1}$, and therefore we expect that, at this time scale, the distribution function should not depend on t_1, but only on t_0. Also, when $t \gg \epsilon^{-1}$, the time variable t_1 will adopt large values and the system should have reached the stationary regime on this time scale and would be only evolving on the time scale described by t_2. That is, as physical time evolves, the different time variables should turn on and off in sequence. For this to take place in a mathematically well-controlled manner, we should impose that the distribution function should be regular at $t_i \to 0$ and $t_i \to \infty$ for all time scales.[16] We then write $n = n(\mathbf{r}, t_0, t_1, t_2, \dots)$, and using the chain rule, time derivatives are replaced as

[16]If this were not the case and so-called secular divergences developed, it would signal that the multiple time scale scheme is not appropriate because an additional time scale is missing or the expansion is not analytic and fractional powers in ϵ are needed, for example.

$$\frac{\partial n}{\partial t} \to \frac{\partial n}{\partial t_0} + \epsilon \frac{\partial n}{\partial t_1} + \epsilon^2 \frac{\partial n}{\partial t_2} + \dots. \tag{3.54}$$

With all these prescriptions, we can write down the Lorentz equation order by order. At zeroth order in ϵ, the Lorentz equation is

$$\frac{\partial h_0}{\partial n} \frac{\partial n}{\partial t_0} = \int [h_0' F_1' - h_0 F_1] \, |\mathbf{c} - \mathbf{c}_1| \, b \, \mathrm{d}b \, \mathrm{d}\psi \, \mathrm{d}^3 c \, \mathrm{d}^3 c_1. \tag{3.55}$$

When this equation is integrated with respect to d^3c, the right-hand side cancels due to charge conservation. Therefore, we obtain

$$\frac{\partial n}{\partial t_0} = 0. \qquad (3.56)$$

The density, as a consequence of charge conservation, does not evolve on the fast scale, and so we can write $n = n(\mathbf{r}, t_1, t_2, \dots)$. Equation (3.55) admits the solution $h_0 = n\widehat{f}_{\mathrm{MB}}$.

At the first order in ϵ, the Lorentz equation reads,

$$\frac{\partial h_1}{\partial n}\frac{\partial n}{\partial t_0} + \frac{\partial h_0}{\partial n}\frac{\partial n}{\partial t_1} + \mathbf{c} \cdot \nabla h_0 + \frac{q\mathbf{E}}{m} \cdot \frac{\partial}{\partial \mathbf{c}} h_0$$
$$= \int [h_1' F_1' - h_1 F_1] \, |\mathbf{c} - \mathbf{c}_1| \, b \, db \, d\psi \, d^3c \, d^3c_1. \qquad (3.57)$$

Integrating this equation with respect to d^3c, the right-hand side cancels again as well as the last two terms of the left-hand side due to parity in $\mathbf{c}$. This results in

$$\frac{\partial n}{\partial t_1} = 0, \qquad (3.58)$$

and the density only evolves on the slowest time scale $n = n(\mathbf{r}, t_2, \dots)$. We now proceed to solve eqn (3.57) for h_1, noting that now the first two terms of the left-hand side are zero. Fist, we write $h_1 = h_0 \Phi$, and after some manipulations, the equation reads,

$$I[\Phi] = \frac{\widehat{f}_{\mathrm{MB}}(c)}{k_{\mathrm{B}}T}\mathbf{c} \cdot \left[\frac{q\mathbf{E}}{n_i} - \frac{k_{\mathrm{B}}T\nabla n}{nn_i}\right] = \frac{\widehat{f}_{\mathrm{MB}}(c)}{k_{\mathrm{B}}T}\mathbf{c} \cdot \mathbf{G}, \qquad (3.59)$$

where the vector $\mathbf{G}$ has been defined to simplify notation. By linearity and isotropy we look for a solution, $\Phi = \hat{\phi}(c)\mathbf{c} \cdot \mathbf{G}$, resulting in eqn (3.26) previously found in the computation of the electrical conductivity. We can therefore consider this equation solved and proceed to the next order.

Finally, at second order in ϵ, the Lorentz equation can again be integrated with respect to d^3c. The right-hand side cancels again, but in the left-hand side a new term appears:

$$\frac{\partial n}{\partial t_2} + \nabla \cdot \int \mathbf{c} h_1 \, d^3c = 0. \qquad (3.60)$$

We prefer to multiply this equation by q to cast it into the form (3.8). After simple manipulations, the integral associated with $\mathbf{J}$ can be written in terms of σ, defined in (3.27), and D, defined in (3.51), as

$$\mathbf{J} = \sigma\mathbf{E} - D\nabla\rho, \qquad (3.61)$$

and therefore the dynamics at the slow scale is

$$\frac{\partial \rho}{\partial t_2} + \nabla \cdot (\sigma\mathbf{E}) = D\nabla^2\rho. \qquad (3.62)$$

These results are of fundamental importance and are the main outcome of the Chapman–Enskog method. Firstly, we obtain the time scale on which the density evolves and therefore the time scale on which we expect to find hydrodynamic-like behaviour. Secondly, it describes the dynamics on this time scale, namely that the charge current has two possible origins: driven by an electric field according to Ohm's law, and driven by density gradients in a diffusive way. Finally, it gives us the value of the two transport coefficients, the electrical conductivity and diffusion coefficient, which are related by the fluctuation–dissipation relation.

If this procedure is continued to higher orders, nonlinear transport laws are obtained. At large electric fields, the current is nonlinear in the field. Also, nonlinear diffusive behaviour is obtained. These laws are normally subdominant compared with the linear transport law obtained above.

3.10   Application: bacterial suspensions, run-and-tumble motion

Bacteria and other microorganisms live in aqueous solutions, suspended in fluids with mechanical properties very similar to water. The motion of a single microorganism results from the combination of the micromechanical deformation of its body and the viscous response of the surrounding fluid. In the microscopic world[17] it was noted by Purcell in 1977 that a microorganism with a reciprocal motion with only one degree of freedom that performs a back-and-forth motion cannot swim. This result, called the *scallop theorem*, imposes a large restriction on the possible swimming strategies of microorganisms. Among the possible solutions, *Escherichia coli* (*E. coli*) represents a good example of a self-propelled swimmer. It is a cylindrical cell with hemispherical endcaps, of $1\,\mu$m in diameter and $2\,\mu$m in length. The motion is the consequence of the rotation of several helicoidal flagella, where each flagellum is linked to the cell membrane by a nanoscale motor. When all the motors turn in a counterclockwise direction, the flagella rotate in a bundle and this pushes the cell steadily forwards (called a 'run'). When one motor switches to clockwise rotation, the cell can change direction (known as a 'tumble' process). The mean run interval is typically $1\,$s, whereas the mean tumble time is around $0.1\,$s. Alternation of the run and tumble processes generates a trajectory, shown in Fig. 3.8, that at the long scale looks like a diffusive process.

The run-and-tumble motion of *E. coli* can be described using the Lorentz model. Indeed, each swimmer moves in the run phase at a roughly constant speed V in a direction $\hat{\mathbf{n}}$ that remains fixed during the run phase and changes abruptly at the tumble phase, which we will model as an instantaneous process. The run phase does not have a constant duration; rather, the durations are Poisson distributed. This means that there is a constant tumbling rate μ, such that the tumbling probability in a short interval is $\mu\,dt$. Under these assumptions, the

[17]Technically, in fluid dynamics, it is relevant to compute the Reynolds number. It is defined as Re $= VL/(\eta/\rho)$, where V is the swimmer speed, L is its size, and η and ρ are the fluid viscosity and mass density, respectively. The values for these parameters for the *Escherichia coli* bacterium swimming in water are $V \sim 10\,\mu$m/s, $L \sim 2\,\mu$m, $\eta = 8.9 \times 10^{-4}\,$Pa s, and $\rho = 1 \times 10^3\,$kg/m^3, resulting in Re $\sim 2 \times 10^{-5}$. Therefore, we call this the low Reynolds number world.

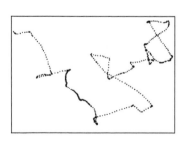

Fig. 3.8 Run-and-tumble motion of *E. coli* observed under the microscope (Berg, 2004). With kind permission from Springer Science and Business Media.

swimmer distribution function $f(\mathbf{r}, \hat{\mathbf{n}}, t)$ satisfies the Lorentz equation,

$$\frac{\partial f}{\partial t} + V\hat{\mathbf{n}} \cdot \nabla f = \mu \int W(\hat{\mathbf{n}}', \hat{\mathbf{n}}) f(\mathbf{r}, \hat{\mathbf{n}}', t) \, \mathrm{d}^2 \hat{\mathbf{n}}' - \mu f(\mathbf{r}, \hat{\mathbf{n}}, t). \qquad (3.63)$$

As for the Lorentz model studied earlier in this chapter, the first term on the right-hand side is the gain term, accounting for the swimmers that were moving in a different direction and performed a tumble ending pointing to $\hat{\mathbf{n}}$. The second term is the loss term that gives the number of swimmers that were moving in the direction $\hat{\mathbf{n}}$ and made a tumble. The kernel W gives the probability that a swimmer moving in a given direction ends in another specified direction after tumble, and we adopt the master equation formalism described in Section 2.10. As for any incoming angle there should be an outcome direction, we have the normalisation,

$$\int W(\hat{\mathbf{n}}', \hat{\mathbf{n}}) \, \mathrm{d}^2 \hat{\mathbf{n}} = 1. \qquad (3.64)$$

In an isotropic medium, W only depends on the angle between the two directions. In the case of *E. coli*, it is biased in the forward direction. If, for simplicity, we consider a flat distribution of outgoing directions, $W(\hat{\mathbf{n}}', \hat{\mathbf{n}}) = 1/4\pi$, the kinetic equation is identical to the Lorentz model for hard spheres (3.15) with the identification $n_i \pi R^2 |\mathbf{c}| = \mu$. Therefore, following the same analysis made in this chapter, we end up with the conclusion that, at long times (longer than μ^{-1}), the swimmers present diffusive motion with a diffusion coefficient given by

$$D = \frac{V^2}{3\mu}. \qquad (3.65)$$

It is possible to carry out this analysis for a more realistic kernel of the form, $W(\hat{\mathbf{n}}', \hat{\mathbf{n}}) = w(\hat{\mathbf{n}}' \cdot \hat{\mathbf{n}})$. We can proceed by either the computation of the eigenvalues in an expansion in small wavevectors or using the Chapman–Enskog procedure.[18] Again a diffusive process is obtained, with a diffusion coefficient that is now given by

[18]See Exercises 3.11 to 3.13.

$$D = \frac{V^2}{3\mu(1 - \alpha)}, \qquad (3.66)$$

where α is a measure of the persistence after tumbling,

$$\alpha = 2\pi \int_0^\pi \sin\theta \cos\theta \, w(\cos\theta) \, \mathrm{d}\theta. \qquad (3.67)$$

For isotropic tumbling $\alpha = 0$ and we recover (3.65). In the other extreme case of very large persistence, with w peaked around $\theta = 0$, α turns out to be very close to one resulting in large diffusion coefficients. *E. coli* has a moderate persistence with $\alpha \approx 0.33$.

It is natural to ask whether bacteria present also a conductivity-like process. In fact they do. It is called chemotaxis, where swimmers, in the presence of a chemical attractant or repellent, move in the direction of the gradient, forward or backward, respectively. To be concrete, we

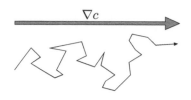
∇c

Fig. 3.9 Schematic representation of the chemotaxis process of a bacterium that performs a run-and-tumble motion. The tumbles reorient the swimmer to a random direction independent of the direction of the chemical gradient, but the run phases are longer if the swimmer is moving along the gradient. The result is that, besides the diffusive motion, there is a net current in the direction of the gradient.

will consider the case of attractants, for which the swimmers must follow the gradient ∇c (Fig. 3.9). To do so, bacteria should sense the chemical gradient, but they are too small to do so with precision. Instead, what they do is to integrate the total chemical signal in a run: if this integral turns out to be large, they continue swimming in this direction for a longer period. This effect is introduced in our model by indicating that the tumble rate depends on the direction of motion relative to the chemical gradient as $\mu \to \mu(\hat{\mathbf{n}} \cdot \nabla c)$, with $\mu(x)$ a monotonically decreasing function. The kinetic model now reads,

$$\frac{\partial f}{\partial t} + V\hat{\mathbf{n}} \cdot \nabla f = \int \mu(\hat{\mathbf{n}}' \cdot \nabla c) w(\hat{\mathbf{n}}' \cdot \hat{\mathbf{n}}) f(\mathbf{r}, \hat{\mathbf{n}}', t) \, \mathrm{d}^2\hat{\mathbf{n}}' - \mu(\hat{\mathbf{n}} \cdot \nabla c) f(\mathbf{r}, \hat{\mathbf{n}}, t). \tag{3.68}$$

This model can be analysed by imposing a fixed constant gradient and looking for the stationary homogeneous solution. If the gradient is small, linear response theory can be applied and the resulting distribution function is

$$f(\hat{\mathbf{n}}) = f_0 \left[1 + \frac{\mu_1 \hat{\mathbf{n}} \cdot \nabla c}{\mu_0} \right], \tag{3.69}$$

where $f_0 = n_0/4\pi$ is the reference distribution for a bacterial density n_0, and the tumble rate has been expanded as $\mu(\hat{\mathbf{n}} \cdot \nabla c) = \mu_0 - \mu_1 \hat{\mathbf{n}} \cdot \nabla c$. The stationary distribution allows us to compute the bacterial flux,

$$\mathbf{J} = \int V\hat{\mathbf{n}} f(\hat{\mathbf{n}}) \, \mathrm{d}^2\hat{\mathbf{n}} = \frac{4\pi\mu_1 V}{3\mu_0} \nabla c, \tag{3.70}$$

which is the sought chemotactic effect.

Further reading

A detailed analysis of the multiple time scales in the Lorentz gas subjected to an external field is presented in (Piasecki, 1993). It is found that, in the rigid hard sphere model, the linear response is only valid in a time window (albeit very wide when the field is small), whereas later, due to the heating by the field, the response is nonlinear.

A system that is related to the Lorentz model is that of billiards. Here, the electrons collide with fixed ions as in the rigid hard sphere model. However, contrary to the Lorentz model, the positions of the ions are known exactly and therefore the collisions are not treated statistically but in an exact way. The billiards system shows chaotic behaviour and has been studied in detail to understand the foundations of statistical mechanics and the

origin of irreversibility from purely reversible equations of motion. One of the most famous examples is Sinai billiards. The reader is invited to refer to the introductory book (Dorfman, 1999) or the more technical one (Szász, 2013).

An interesting introduction to the motion of microorganisms is given in (Purcell, 1977), where the scallop theorem was introduced. There the author discusses the differences that can be found between the fluid dynamics at large scales and in the microscopic world. The diffusive motion of bacteria, considering the run-and-tumble process as well as other mechanisms for diffusion of microorganisms, is described in detail in (Berg, 1993).

Exercises

(3.1) **Mean free path approximation.** Using a mean free path approximation, estimate the electrical conductivity and self-diffusion coefficient in the Lorentz model. Model ions as rigid hard spheres.

(3.2) **Charge conservation.** Show explicitly that $\varphi = 1$ is a collisional invariant of the collision integrals for the hard sphere and BGK models.

(3.3) **Diffusive eigenvalue.** Perform the perturbation analysis and derive eqn (3.49) for the smallest eigenvalue.

(3.4) **Diffusion equation.** Derive the diffusion equation (3.50) by integrating the perturbed distribution.

(3.5) **Thermal conductivity.** Consider a system in which, by some means, the ions are kept at an inhomogeneous temperature. The inhomogeneity is weak, and the temperature varies slightly over one electron mean free path. Locally it is described by a linear function, $T = T_0(1 + \epsilon T_1 z)$, with $\epsilon T_1 \ell \ll 1$, where the z axis is aligned with the temperature gradient. The electrons will tend to reach thermal equilibrium with the ions, with the same temperature. Note, however, that this would imply that their distribution is no longer homogeneous. Solve then the stationary Lorentz equation for the electrons to first order in ϵ, considering a general ion model. Using the solution, compute the heat current,

$$\mathbf{q} = \frac{m}{2} \int c^2 \mathbf{c} f(\mathbf{c}) \, \mathrm{d}^3 c$$

and show that Fourier's law, $\mathbf{q} = -\kappa \nabla T$, is recovered. Using the BGK model, compute the thermal conductivity κ.

(3.6) **Estimation of the kinetic gap.** The aim is to make an estimation of the separation between the first nonvanishing eigenvalue and the null eigenvalue for the Lorentz model. The estimation is based on the variational method for Hermitian operators. Being L_0 Hermitian, the eigenfunctions and eigenvalues $L_0 \Psi_i = \lambda_i \Psi_i$ satisfy the following properties:

- The eigenfunctions are orthogonal: $(\Psi_i, \Psi_k) = \delta_{ik}$;
- The eigenfunctions span a basis of the functional space;

- The eigenvalues are real: $\lambda \in \Re$;
- The eigenvalues can be ordered: $\lambda_0 \leq \lambda_1 \leq \lambda_2 \leq \dots$.

(a) Show that, if Φ is any function with finite norm, then

$$\frac{(\Phi, L_0 \Phi)}{(\Phi, \Phi)} \geq \lambda_0,$$

where λ_0 is the smallest eigenvalue.

(b) In a similar manner, if Φ is orthogonal to the first n eigenfunctions $(\Phi, \Psi_i) = 0$, $i = 0, \dots, n-1$, show that

$$\frac{(\Phi, L_0 \Phi)}{(\Phi, \Phi)} \geq \lambda_n.$$

(c) Knowing that L_0 is Hermitian in the Lorentz model, find a lower bound for the first nonvanishing eigenvalue. Consider the thermal ion model and the rigid hard sphere model. To do this try simple functions (e.g. polynomials) that are orthogonal to Ψ_0. Note that, in the rigid hard sphere model, perturbations that do not change the energy must be considered. Show that λ_1 is related to the collision frequency.

(3.7) **Isokinetic model.** The rigid hard sphere model simplifies the analysis enormously because, in collisions, the absolute value of the velocity is conserved and only its orientation changes. However, when an electric field is applied, it does work and therefore the velocity also changes in magnitude. To retain the simplicity of the model and include the effects of small electric fields, we can devise the so-called isokinetic model, in which the electric field is replaced by its transverse part, $\mathbf{E} \to \mathbf{E}_t = \mathbf{E} - (\mathbf{E} \cdot \mathbf{c}/c^2)\mathbf{c}$, such that it does not perform work but only orients the velocities parallel to the field. Under these conditions, the only relevant variable is the orientation $\hat{\mathbf{c}}$ of the velocity, and therefore $f = f(\mathbf{r}, \hat{\mathbf{c}}, t)$.

(a) Write the associated Lorentz equation.

(b) Assuming that the initial distribution is isotropic, $f_0(\hat{\mathbf{c}}) = 1/4\pi$, compute the electrical conductivity in this model.

(3.8) **Anisotropic diffusion.** Consider a medium composed of hard ellipsoids that have their principal axis pointing in the same direction, say z (Fig. 3.10). When a particle collides with them, it will be dispersed anisotropically. We aim to investigate the effect of the anisotropy on the long-time dynamics. To do so, consider a thermalising model similar to (3.17). The anisotropy can be modelled by modifying the collision rate such that it depends on the direction. Specifically, we take the collision operator,

$$J[f] = n_i \pi R^2 c \left(1 + \alpha(\hat{\mathbf{c}} \cdot \hat{\mathbf{z}})^2\right) \left[\mu \widehat{f}_{MB}(\mathbf{c}) - f(\mathbf{c})\right],$$

where

$$\mu = \frac{\int c' \left(1 + \alpha(\hat{\mathbf{c}}' \cdot \hat{\mathbf{z}})^2\right) f(\mathbf{c}') \, d^3c'}{\int c' \left(1 + \alpha(\hat{\mathbf{c}}' \cdot \hat{\mathbf{z}})^2\right) \widehat{f}_{MB}(\mathbf{c}') \, d^3c'},$$

such that, if $\alpha > 0$, particles moving along the $\pm\hat{\mathbf{z}}$ directions collide more frequently, and the opposite happens if $\alpha < 0$.

Compute the second-order correction in $\mathbf{k}$ to the null eigenvalue. Show that it is given by

$$\lambda = D_1(k_x^2 + k_y^2) + D_2 k_z^2$$

and compute $D_{1,2}$.

Now, when deriving the charge conservation equation, show that it is of the form of an anisotropic diffusion equation,

$$\frac{\partial \rho}{\partial t} = D_1 \left(\frac{\partial^2 \rho}{\partial x^2} + \frac{\partial^2 \rho}{\partial y^2}\right) + D_2 \frac{\partial^2 \rho}{\partial z^2}.$$

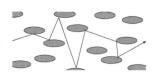

Fig. 3.10 Diffusion of a particle by anisotropic obstacles.

(3.9) **Energy equation.** Consider the kinetic energy transported by electrons $\psi = mc^2/2$. Write down the transport equation for the energy density, $e =$ $\int (mc^2/2) f(\mathbf{c}) \, d^3c$, to show that the source term can be interpreted as describing the Joule heating effect,

$$\frac{\partial e}{\partial t} + \nabla \cdot \mathbf{J}_e = \mathbf{E} \cdot \mathbf{J},$$

where $\mathbf{J}$ is the electric current and $\mathbf{J}_e$ is the energy flux. Use the rigid hard sphere model.

(3.10) **Frequency response.** Consider the rigid hard sphere model subject to an oscillatory electric field. Solve eqn (3.32) and find integral expressions for the real and imaginary parts of the electrical conductivity. Compute the integrals to first order in ω and show that the phase delay is proportional to $\omega\ell/\langle|\mathbf{c}|\rangle$.

(3.11) **Diffusion in the run-and-tumble process.** Compute, using the expansion of the vanishing eigenvalue in powers of the wavevector, the diffusion coefficient in the run-and-tumble process. Consider a general kernel $w(\hat{\mathbf{n}}' \cdot \hat{\mathbf{n}})$.

(3.12) **Chemotaxis in the run-and-tumble process.** Compute, using linear response theory, the bacterial current in the presence of a chemical gradient in the run-and-tumble process. Consider a general kernel $w(\hat{\mathbf{n}}' \cdot \hat{\mathbf{n}})$.

(3.13) **Hydrodynamic description of a bacterial suspension.** Consider the run-and-tumble kinetic equation (3.68) with a general kernel $w(\hat{\mathbf{n}}' \cdot \hat{\mathbf{n}})$. Apply the Chapman–Enskog method to obtain that the hydrodynamic equation for the bacterial density in the long-time regime is

$$\frac{\partial n}{\partial t} = D\nabla^2 n - \nabla \cdot \mathbf{J},$$

where D and $\mathbf{J}$ are given by (3.66) and (3.70).

(3.14) **Radiative transfer in the Sun.** The matter inside the Sun is highly ionised. Therefore, the photons that are produced by the nuclear reactions and other processes suffer many scattering processes with the free charges (electrons and ions). As a result, the mean free path for the photons is extremely small, with the average value: $\ell_{photon} \approx 0.9 \times 10^{-3}$ m. This means that photons do not propagate ballistically but rather in a diffusive manner. Compute the photon diffusion coefficient. Knowing that the Sun radius is $R_\odot \approx 6.96 \times 10^8$ m, estimate the time it takes for a photon to escape from the Sun.

The Boltzmann equation for dilute gases

4

4.1 Formulation of the Boltzmann model

4.1.1 Hypothesis

The Boltzmann equation is by far the most famous kinetic equation. It describes the evolution of classical gases, and since its formulation in 1872, it has served as inspiration for the kinetic modelling of all kinds of systems. Two of the properties of this equation are most remarkable. First, it explains the origin of the irreversible behaviour of macroscopic systems, and second, it succeeds in relating the macroscopic coefficients that describe the irreversible dynamics (viscosity, thermal conductivity, and diffusion coefficient) to the interatomic interactions. The Boltzmann equation gives the conceptual foundations of gas dynamics and also constitutes an extraordinarily accurate description of gases.

With the purpose of giving a simple presentation, in this chapter we will derive and analyse the Boltzmann equation for monocomponent classical gases. In many aspects we will follow a procedure similar to that used in Chapter 3 to derive the Lorentz equation, although with an important difference: the atoms collide among themselves and not with fixed obstacles. In Chapter 2 we analysed the case of monoatomic gases and showed that the one-particle distribution function $f(\mathbf{r}, \mathbf{c}, t)$ obeys the BBGKY equation (2.52), which can be written compactly as

$$\frac{\partial f}{\partial t} + \mathbf{c} \cdot \frac{\partial f}{\partial \mathbf{r}} + \frac{\mathbf{F}}{m} \cdot \frac{\partial f}{\partial \mathbf{c}} = \left(\frac{\partial f}{\partial t} \right)_{\text{coll}}. \tag{4.1}$$

The left-hand side expresses the streaming of non-interacting atoms moving under an external force $\mathbf{F}$. On the right-hand side, the collision or interaction term quantifies the rate of change of the distribution function due to two-particle interactions, which we represent compactly by $\left(\frac{\partial f}{\partial t} \right)_{\text{coll}}$. The collision partner, with velocity $\mathbf{c}_1$, is not isolated though. Rather, it is immersed in a gas and the statistical properties of the pair are given by the function $f^{(2)}(\mathbf{r}, \mathbf{c}, \mathbf{r}_1, \mathbf{c}_1)$. The purpose here is to make an adequate approximation for the interaction integral, valid for gases.

The Boltzmann equation will be valid for dilute gases. Consider a gas composed of N particles in a volume $\mathcal{V}$. The molecules interact via a potential with range r_0, implying a cross section that scales as $\sigma \sim r_0^2$ in three dimensions. The particle density, $n = N/\mathcal{V}$, is a dimensional

Kinetic Theory and Transport Phenomena, First Edition, Rodrigo Soto.
© Rodrigo Soto 2016. Published in 2016 by Oxford University Press.

quantity, and the proper way to define a dilute gas is via the volume fraction occupied by the atoms, $\phi = 4\pi n r_0^3/3$ (modelled as spheres of radius r_0). A dilute gas is characterised by having $\phi \ll 1$, and we will proceed to take the formal limit $\phi \to 0$. In this limiting procedure we aim to respect the dynamical scales that apply in the gas. In Chapter 1 we learnt that the transport coefficients are proportional to the mean free path, which scales as $\ell \sim 1/(n r_0^2)$. Hence, in this limiting process, we will require that ℓ remains finite. Finally, we will demand that the number of particles in a mean free path volume, $N_\ell = n\ell^3$, is large such that we can safely work with the distribution function as a statistical object. These conditions can be achieved simultaneously if $r_0 \to 0$ and $n \to \infty$, while $\mathcal{V}$ is finite.[1] These conditions define the so-called Boltzmann–Grad limit and characterise the regime in which the Boltzmann equation is valid.

In a gas, the mean free path ℓ is much larger than the atom size, and the picture one must adopt is that of atoms travelling unperturbed for long times compared with the duration of a collision, until they meet another atom. Collisions are therefore binary to an excellent approximation. Indeed, three-particle collisions are extremely rare, proportional to ϕ, vanishing in the Boltzmann–Grad limit.[2]

Particles that meet in a collision come from regions far apart, having collided with many other particles before. Here, Boltzmann made the approximation that, under these conditions, it is reasonable to assume that particles that collide are statistically independent (uncorrelated). In our language, that is to say that, for the precollisional state, $f^{(2)}(\mathbf{c}_1, \mathbf{c}_2)_{\text{precoll}} \approx f(\mathbf{c}_1) f(\mathbf{c}_2)$. This approximation is usually termed the *molecular chaos* hypothesis.[3] To be more quantitative, we can estimate the importance of the correlations between colliding pairs, by noting that particles can become correlated via recollision events, such as those shown in Fig. 4.1. In Section 4.8.2, we will see that the fraction of collisions that are a result of a recollision turns out to be proportional to ϕ, where again we are led to the conclusion that, in the Boltzmann–Grad limit, recollisions and precollisional correlations are irrelevant.

As a final consideration, we will derive a kinetic equation valid on the mesoscale, that is, for time and length scales on the order of or larger than the mean free time and mean free path, respectively. Atoms are much smaller than this length scale, allowing us to neglect their size and to consider collisions as taking place at a point, with both particles sharing one position. Equally, the duration of the collision is vanishingly small compared with the other time scales.[4] Hence, collisions will be treated as instantaneous events, and particles will not experience the effect of the external forces during the collision process.

[1] In the atmosphere, under normal conditions, $\ell = 1/(\sqrt{2}\pi r_0^2 n) = 8.4 \times 10^{-8}$ m, $\phi = 4\pi n r_0^3/3 = 1.1 \times 10^{-3}$, and $N_\ell = 1.6 \times 10^5$, where $n = 2.7 \times 10^{26}$ m^{-3} and $r_0 \sim 1$ Å.

[2] For an estimation of their frequency see Exercise 4.2.

[3] This hypothesis was first introduced by Maxwell in 1867 and later used by Boltzmann, who coined the German word *Stosszahlansatz*, collision number hypothesis.

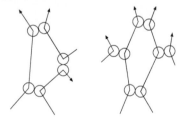

Fig. 4.1 Examples of recollisions that generate dynamic correlations. Left: three-particle recollision with particle exchange. Right: a four-particle recollision event.

[4] In the atmosphere, under normal conditions, the relevant distances are $r_0 \approx 10^{-10}$ m while $\ell \approx 8.4 \times 10^{-8}$ m. The collision time is $t_{\text{coll}} = r_0/c_{\text{th}} \approx 2.4 \times 10^{-13}$ s while the mean free time is $\tau \approx 1.8 \times 10^{-10}$ s.

4.1.2 Kinematics of binary collisions

To derive the Boltzmann kinetic equation, we will model the effect of the collisions. Two collision processes induce changes in the distribution function $f(\mathbf{r}, \mathbf{c}, t)$. First, a particle with velocity $\mathbf{c}$ can collide with another, moving with arbitrary velocity $\mathbf{c}_1$ (the direct collision, shown in the left panel of Fig. 4.2). The atom changes its velocity $\mathbf{c}$ to $\mathbf{c}'$, hence reducing $f(\mathbf{c})$. Conversely, after the collision of two particles, one of them can emerge with velocity $\mathbf{c}$, increasing the value of the distribution function. We know that this second process (the inverse collision, shown in the right panel of Fig. 4.2) must exist due to the microscopic reversibility of classical mechanics for elastic interactions. Indeed, we can use this principle to relate the kinematics of the direct and inverse collision. In a direct collision between particles with velocities $\mathbf{c}$ and $\mathbf{c}_1$, they will reach a point of maximum approach, and we designate by $\mathbf{n}$ the vector that joins the two particles at that instant. After the collision, they emerge with velocities $\mathbf{c}'$ and $\mathbf{c}'_1$. The microscopic dynamics are reversible in time, and we assume that the interaction respects parity.[5] Then, for any process that takes place, its reverse in time and space can also take place (that is, playing the movie reflected in a mirror and backward in time depicts a perfectly possible process that respects all conservations). Figure 4.2 (right) shows precisely such a time- and space-reflected process of the direct collision (on the left). The parity operation changes the sign of $\mathbf{n}$, but the simultaneous time and parity operations keep velocities unchanged. Therefore, the output velocities are $\mathbf{c}$ and $\mathbf{c}_1$, which is precisely the definition of the inverse collision. In summary, we have not only shown that the inverse collision always exists, but also found how to build it: one simply takes the outcome velocities of the direct collision and reverses the collision vector $\mathbf{n}$.

To parameterise three-dimensional collisions, we will use the concepts of classical scattering theory, described in Appendix C. First, we change to the centre of mass, $\mathbf{G} = (\mathbf{c} + \mathbf{c}_1)/2$ and relative velocity $\mathbf{g} = \mathbf{c} - \mathbf{c}_1$, where for notational simplicity we assume a monocomponent gas with all particles having the same mass m. In the Boltzmann–Grad limit, the collision time is vanishingly small and the external forces do not act, implying that momentum is conserved, $\mathbf{G}' = \mathbf{G}$. Also energy $E = mG^2 + mg^2/4$ is conserved, which together with the previous result gives $g' = g$. Hence, in a collision only the direction of $\mathbf{g}$ changes. Placed in the centre of mass, the collision is parameterised by the impact parameter b and the azimuthal angle ψ (shown in Fig. 4.3). Angular momentum conservation implies that the origin, $\mathbf{g}$, and $\mathbf{g}'$ lie in the same plane and also that the impact parameter is preserved ($b' = b$). Hence the inverse collision is characterised by the same value of b, while $\psi' = \psi + \pi$.

The rotation of $\mathbf{g}$ to $\mathbf{g}'$ depends on the impact parameter and the magnitude of the relative velocity, being a function of the interatomic potential. Its calculation, needed for example to obtain numerical values of the transport coefficients, uses the formalism of the differential cross section in classical mechanics (see Appendix C). To obtain explicit results, we will apply the general expressions found in this chapter to the hard sphere model. For other interaction models, numerical simulations can be used. In particular, the direct simulation Monte Carlo (DSMC) method described in Chapter 10 is particularly well suited, where the deflection produced by the interatomic potential is used as an input. Also,

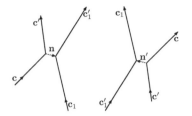

Fig. 4.2 Direct (left) and inverse (right) collisions. The vector $\mathbf{n}$ joins the points of closest approach in the collision.

[5]Here, we are not considering magnetic interactions and assume that the interparticle potential is spherically symmetric. For the analysis of the Boltzmann equation for anisotropic interactions, see Exercise 4.4.

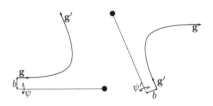

Fig. 4.3 Direct (left) and inverse (right) collisions in the centre-of-mass frame (shown as a black dot). The relative velocity $\mathbf{g}$ changes direction to $\mathbf{g}'$, preserving its magnitude. The geometry of the collision is described by the impact parameter b and the azimuthal angle ψ.

analytical methods for specific potentials have been developed (Ferziger and Kaper, 1972).

4.2 Boltzmann kinetic equation

4.2.1 General case

We now have all the ingredients to model the right-hand side of eqn (4.1), which we recall represents the rate of change of the distribution function at $\mathbf{c}$. For a particle with velocity $\mathbf{c}$ to collide with another having velocity $\mathbf{c}_1$ and geometrical parameters b and ψ, the partner must be located in the collision cylinder, depicted in Fig. 4.4, which has volume $|\mathbf{g}|\Delta t\, b\, \mathrm{d}b\, \mathrm{d}\psi$. Therefore, the number of direct collisions with fixed

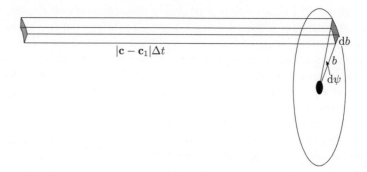

Fig. 4.4 Collision cylinder in which a particle must be located to collide with the partner in a time interval Δt. Indeed, it is a right prism but it is customary to call it cylinder.

geometrical parameters in that time interval between particles with velocity $\mathbf{c}$, within a range of size d^3c, and partners having velocities in the vicinity of size d^3c_1 of $\mathbf{c}_1$ and position in the vicinity of size d^3r of $\mathbf{r}$ is

$$N_{\mathrm{dir.coll}} = f(\mathbf{r}, \mathbf{c})f(\mathbf{r}, \mathbf{c}_1)|\mathbf{g}|\Delta t\, b\, \mathrm{d}b\, \mathrm{d}\psi\, \mathrm{d}^3r\, \mathrm{d}^3c\, \mathrm{d}^3c_1, \qquad (4.2)$$

where we have used the different hypotheses presented in Section 4.1, namely that (i) the two particles are uncorrelated, (ii) the collision is instantaneous, (iii) the two particles are located at the same position, and (iv) the collision is binary. Similarly, the number of inverse collisions with the specified parameters is

$$N_{\mathrm{inv.coll}} = f(\mathbf{r}, \mathbf{c}')f(\mathbf{r}, \mathbf{c}_1')|\mathbf{g}'|\Delta t\, b'\, \mathrm{d}b'\, \mathrm{d}\psi'\, \mathrm{d}^3r\, \mathrm{d}^3c'\, \mathrm{d}^3c_1'. \qquad (4.3)$$

For the inverse term, we first use that $|\mathbf{g}'| = |\mathbf{g}|$ and $b'\, \mathrm{d}b'\, \mathrm{d}\psi' = b\, \mathrm{d}b\, \mathrm{d}\psi$, as discussed in Section 4.1.2. Second, microscopic reversibility implies that $\mathrm{d}^3c'\, \mathrm{d}^3c_1' = \mathrm{d}^3c\, \mathrm{d}^3c_1$.[6] Hence

$$N_{\mathrm{inv.coll}} = f(\mathbf{r}, \mathbf{c}')f(\mathbf{r}, \mathbf{c}_1')|\mathbf{g}|\Delta t\, b\, \mathrm{d}b\, \mathrm{d}\psi\, \mathrm{d}^3r\, \mathrm{d}^3c\, \mathrm{d}^3c_1. \qquad (4.4)$$

Every direct collision provokes the loss of a particle with velocity $\mathbf{c}$, hence decreasing the distribution function. Conversely, for every inverse collision, there is a gain in the number of particles with the specified velocity, therefore increasing the distribution function. The total rate of

[6]There are several ways to show this property; for example, in the case of hard spheres, one can directly show that the Jacobian of the collision rule (4.7) is one. The general demonstration is given in Appendix C.

change for the distribution function $\left(\frac{\partial f}{\partial t}\right)_{\text{coll}}$ is obtained by integrating $(N_{\text{inv.coll}} - N_{\text{dir.coll}})$ over all collision parameters and velocities while keeping $\mathbf{c}$ and $\mathbf{r}$ fixed, i.e.

$$\left(\frac{\partial f}{\partial t}\right)_{\text{coll}} = \int [f(\mathbf{r}, \mathbf{c}')f(\mathbf{r}, \mathbf{c}'_1) - f(\mathbf{r}, \mathbf{c})f(\mathbf{r}, \mathbf{c}_1)] |\mathbf{g}| b \, db \, d\psi \, d^3 c_1. \quad (4.5)$$

In summary, the Boltzmann kinetic equation is

$$\frac{\partial f}{\partial t} + \mathbf{c} \cdot \frac{\partial f}{\partial \mathbf{r}} + \frac{\mathbf{F}}{m} \cdot \frac{\partial f}{\partial \mathbf{c}} = J[f, f] = \int [f'f'_1 - ff_1] |\mathbf{g}| b \, db \, d\psi \, d^3 c_1, \quad (4.6)$$

where we have introduced the compact notation $f = f(\mathbf{r}, \mathbf{c})$, $f' = f(\mathbf{r}, \mathbf{c}')$, $f_1 = f(\mathbf{r}, \mathbf{c}_1)$, and $f'_1 = f(\mathbf{r}, \mathbf{c}'_1)$ and defined the collision operator J. Note that, as a consequence of atoms interacting with themselves, the collision operator and the kinetic equation are nonlinear. This property makes it extremely difficult to find analytic solutions under general conditions. However, we will see that it is possible to extract interesting conclusions from it.

In this section we have followed the standard derivation of the Boltzmann equation using the hypotheses already described. It is also possible to prove formally that the Liouville equation reduces to the Boltzmann equation in the Boltzmann–Grad limit. This would require advanced mathematics which lies beyond the scope and focus of the present book. The interested reader is referred to the books cited in the 'Additional reading' section.

4.2.2 Hard sphere model

Hard spheres are usually employed to obtain explicit expressions to described classical gases. It is a tractable model and also describes in good detail atomic collisions at high energies, where the attractive wells are negligible compared with the kinetic energy. In this model, particles of radius r_0 interact with a central potential that is infinitely hard when particles meet at a distance $D = 2r_0$. The momentum exchange goes along the normal vector $\hat{\mathbf{n}}$ with a magnitude that is obtained by imposing energy conservation, leading to the following collision rule:[7]

$$\mathbf{c}' = \mathbf{c} - [(\mathbf{c} - \mathbf{c}') \cdot \hat{\mathbf{n}}] \hat{\mathbf{n}}, \qquad \mathbf{c}'_1 = \mathbf{c}_1 + [(\mathbf{c} - \mathbf{c}') \cdot \hat{\mathbf{n}}] \hat{\mathbf{n}}. \quad (4.7)$$

The postcollisional relative velocity is $\mathbf{g}' = \mathbf{g} - 2(\mathbf{g} \cdot \hat{\mathbf{n}})\hat{\mathbf{n}}$.

When the interactions are described by the hard sphere model, we can simplify the collision operator. The impact parameter is $b = D \sin \theta$, where the angle θ is defined in Fig. 4.5. This implies that $|\mathbf{g}| b \, db \, d\psi = D^2 |\mathbf{g}| \sin \theta \cos \theta \, d\theta \, d\psi = D^2 (\mathbf{g} \cdot \hat{\mathbf{n}}) \, d^2\hat{\mathbf{n}}$. Hence, the Boltzmann equation reads,

$$\frac{\partial f}{\partial t} + \mathbf{c} \cdot \frac{\partial f}{\partial \mathbf{r}} + \frac{\mathbf{F}}{m} \cdot \frac{\partial f}{\partial \mathbf{c}} = D^2 \int [f'f'_1 - ff_1] (\mathbf{g} \cdot \hat{\mathbf{n}}) \Theta(\mathbf{g} \cdot \hat{\mathbf{n}}) \, d^2\hat{\mathbf{n}} \, d^3 c_1, \quad (4.8)$$

where the postcollisional velocities are given by the rules (4.7). The Heaviside step function Θ limits the normal vector to $\mathbf{g} \cdot \hat{\mathbf{n}} > 0$, corresponding to the precollisional configuration.

[7]See Exercise 4.5.

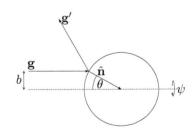

Fig. 4.5 Collision of hard spheres of radius r_0, analysed in the centre-of-mass frame. Particles collide when their distance is $D = 2r_0$. The incoming relative velocity $\mathbf{g}$ and the postcollisional one $\mathbf{g}'$ form an angle θ with the normal vector $\hat{\mathbf{n}}$. ψ is the azimuthal angle, and b is the impact parameter.

4.3   General properties

4.3.1   Balance equations and collisional invariants

From the Boltzmann equation, it is possible to derive balance equations for different relevant fields defined in terms of the distribution function. Take a function $\varphi(\mathbf{c})$ and the associated density field,

$$\rho_\varphi(\mathbf{r}, r) = \int f(\mathbf{r}, \mathbf{c}, t)\varphi(\mathbf{c})\, d^3c. \qquad (4.9)$$

The choices $\varphi = m$, $m\mathbf{c}$, and $mc^2/2$ are particularly relevant because they lead to equations for the mass, momentum, and energy density, which are conserved fields. The balance equations are obtained by multiplying the Boltzmann equation by φ and integrating over the velocity as

$$\frac{\partial \rho_\varphi}{\partial t} + \nabla \cdot \mathbf{J}_\varphi - S_\varphi = \int J[f, f](\mathbf{c})\varphi(\mathbf{c})\, d^3c, \qquad (4.10)$$

where

$$S_\varphi = \int f(\mathbf{r}, \mathbf{c}, t)\frac{\mathbf{F}}{m} \cdot \frac{\partial \varphi}{\partial \mathbf{c}}\, d^3c, \qquad \mathbf{J}_\varphi = \int f(\mathbf{r}, \mathbf{c}, t)\varphi(\mathbf{c})\mathbf{c}\, d^3c \qquad (4.11)$$

are the source due to the external force and the flux. The integral with the collision operator can be transformed into a simpler form. First, we note that $\mathbf{c}$ and $\mathbf{c}_1$ play a symmetric role in the integral. Stating this symmetry explicitly leads to

$$I_\varphi = \frac{1}{2} \int [\varphi(\mathbf{c}) + \varphi(\mathbf{c}_1)] \left[f' f_1' - f f_1\right] |\mathbf{g}| b\, db\, d\psi\, d^3c\, d^3c_1. \qquad (4.12)$$

Now, we relabel the dummy integration variables $\mathbf{c} \to \mathbf{c}'$ and $\mathbf{c}_1 \to \mathbf{c}_1'$ and use that, as shown in Fig. 4.2 (right), the postcollisional velocities of the pair $(\mathbf{c}', \mathbf{c}_1')$ are $(\mathbf{c}, \mathbf{c}_1)$. Then, for consistency, we must relabel $\mathbf{c}' \to \mathbf{c}$ and $\mathbf{c}_1' \to \mathbf{c}_1$ as well, giving

$$I_\varphi = \frac{1}{2} \int [\varphi(\mathbf{c}') + \varphi(\mathbf{c}_1')] \left[f f_1 - f' f_1'\right] |\mathbf{g}'| b'\, db'\, d\psi'\, d^3c'\, d^3c_1'. \qquad (4.13)$$

Using the properties of the inverse collision discussed in Sections 4.1.2 and 4.2, we can combine eqns (4.12) and (4.13) to obtain the following two equivalent expressions:

$$I_\varphi = \frac{1}{4} \int [\varphi + \varphi_1 - \varphi' - \varphi_1'] \left[f' f_1' - f f_1\right] |\mathbf{g}| b\, db\, d\psi\, d^3c\, d^3c_1, \qquad (4.14)$$

$$I_\varphi = \frac{1}{2} \int [\varphi' + \varphi_1' - \varphi - \varphi_1] f f_1 |\mathbf{g}| b\, db\, d\psi\, d^3c\, d^3c_1. \qquad (4.15)$$

The form (4.15) of the collisional integral can be easily interpreted. It indicates that the rate of change of the density ρ_φ is directly the product of the change of φ in each collision between the post- and precollisional states times the collision frequency. The factor of $1/2$ compensates for the double counting produced by the exchange of $\mathbf{c}$ and $\mathbf{c}_1$, where each collision is counted twice. Notably, if φ corresponds to a quantity conserved in collisions (m, $m\mathbf{c}$, or $mc^2/2$), the integral vanishes. These are called collisional invariants and generate conservation equations for the associated fields in (4.10) with a vanishing right-hand side. For general interaction potentials, besides mass, momentum, energy, and their linear combinations, there are no other collisional invariants. In binary collisions described by classical mechanics, angular momentum is also conserved. Here, however, there is no balance equation associated with angular momentum. The reason is that one of the hypotheses used to derive the Boltzmann equation is that collisions take place with both particles at the same point. Hence, no internal angular momentum exists and the total angular momentum is directly given in terms of the total linear momentum.[8]

We now proceed to write explicitly the equations for the mass, velocity, and energy density fields. When $\varphi = m$, one rapidly obtains the mass conservation equation

$$\frac{\partial \rho}{\partial t} + \nabla \cdot (\rho \mathbf{v}) = 0. \tag{4.16}$$

Next, we work out in detail the momentum equation. Using tensorial notation, we consider the Cartesian component $\varphi = mc_i$. The flux J is a tensor which is readily identified as the total momentum flux $\widetilde{P}_{ik}$ [defined in eqn (1.12)], which is related to the stress tensor by $\widetilde{P}_{ik} = P_{ik} + \rho v_i v_k$ [eqn (1.20)]. Finally, the term associated with the external force is $S = \rho F_i/m$, giving

$$\frac{\partial}{\partial t}(\rho v_i) + \frac{\partial}{\partial x_k}(P_{ik} + \rho v_i v_k) - \rho F_i/m = 0. \tag{4.17}$$

With the help of the mass conservation equation, this can be transformed into

$$\rho \left(\frac{\partial \mathbf{v}}{\partial t} + (\mathbf{v} \cdot \nabla)\mathbf{v} \right) = -\nabla \cdot \mathbb{P} + \rho \mathbf{F}/m, \tag{4.18}$$

where $(\mathbf{v} \cdot \nabla) = v_k \frac{\partial}{\partial x_k}$. Applying a similar treatment for the energy equation, we finally obtain[9]

$$\frac{3}{2} k_{\mathrm{B}} \rho \left(\frac{\partial T}{\partial t} + (\mathbf{v} \cdot \nabla)T \right) = -\nabla \cdot \mathbf{q} - \mathbb{P} : \nabla \mathbf{v}, \tag{4.19}$$

with $\mathbb{P} : \nabla \mathbf{v} = P_{ik} \frac{\partial v_i}{\partial x_k}$.

The conservation equations (4.16), (4.18), and (4.19) are a direct consequence of the fact that the Boltzmann equation respects the microscopic conservations. They resemble the hydrodynamic equations, but it is not correct to make this identification because they are not closed. Indeed, there is no specification of the value of the stress tensor $\mathbb{P}$ or the heat flux $\mathbf{q}$. New balance equations could be obtained for these unknowns with the aim of closing the aforementioned equations, but two problems emerge. First, the associated fluxes J are, by construction, moments of higher order in powers of $\mathbf{c}$ (for example $J_{\mathbf{q}} = \rho \langle \mathbf{cc}mc^2/2 \rangle$).

As a consequence, a hierarchy of equations emerges, each time requiring one to go to higher order. Second, the stress tensor and heat flux are not associated with conserved quantities, and hence the collisional integrals do not vanish. Their evaluation requires knowledge of the full distribution function, rendering the balance equations we were building completely useless.

However, in the next sections, we will see that these equations can be closed to obtain hydrodynamic equations, thanks to two properties of the Boltzmann equation: the approach toward equilibrium (Section 4.3.2) and the scale separation between the different modes (Section 4.4.3). These results will allow us to perform a systematic development known as the Chapman–Enskog method. In Chapter 10, a different approach, called the moment expansion, will also use these results to close the conservation equations in Grad's moment scheme.

4.3.2 H-theorem

For a gas described by a distribution function $f(\mathbf{r}, \mathbf{c})$, it is possible to compute the Boltzmann entropy. First, we divide the one-particle phase space $\mathbf{r}$–$\mathbf{c}$ into parcels of size Δ, large enough to have many particles inside but still small enough for the distribution to be homogeneous in each. The conditions for the existence of an appropriate Δ that fulfils both constraints lie at the basis of kinetic theory, and we are not introducing new hypotheses. The parcels are labelled $k = 1, 2, \ldots$, and the number of particles in each is $N_k = \Delta f_k$. The Boltzmann entropy $S = k_{\mathrm{B}} \ln \Omega$ is computed in terms of the total number of configurations Ω compatible with the specification of the values N_k. In the Boltzmann–Grad limit, the atoms are independent, resulting in Ω being simply the combinatorial result of all the forms in which the atoms can be arranged in the parcels:

$$\Omega = \frac{N!}{N_1! N_2! \ldots N_k! \ldots}, \tag{4.20}$$

where $N = \sum_k N_k$ is the total number of particles. Then

$$S = k_{\mathrm{B}} \ln N! - k_{\mathrm{B}} \sum_k \ln N_k! = -k_{\mathrm{B}} \sum_k N_k \ln(N_k/N), \tag{4.21}$$

where in the last expression we have used the Stirling approximation, $\ln N! \approx N \ln N - N$, valid for large N. Transforming the sum into an integral, we obtain finally

$$S = -\int f(\mathbf{r}, \mathbf{c}) \ln(f(\mathbf{r}, \mathbf{c}) \Delta/N) \, \mathrm{d}^3 r \, \mathrm{d}^3 c. \tag{4.22}$$

Motivated by the previous expression for the entropy, we define the H functional for a homogeneous gas as[10]

$$H[f] = \int f(\mathbf{c}) \ln[f(\mathbf{c})/f_0] \, \mathrm{d}^3 c, \tag{4.23}$$

where f_0 is an immaterial constant, inserted only to make the argument of the log function dimensionless.

[10]The historical order was the inverse: Boltzmann first proved the H-theorem and then proposed the formula for the entropy as the log of the number of states. Here, for clarity, we choose to reverse the order.

The time derivative of H can be obtained using the chain rule via the evolution of f as

$$\frac{dH}{dt} = \int [\ln(f/f_0) + 1] \frac{\partial f}{\partial t} \, d^3c,$$

$$= \int [\ln(f/f_0) + 1] [f' f_1' - f f_1] \, |\mathbf{g}| b \, db \, d\psi \, d^3c \, d^3c_1,$$

$$= \int [\ln f + \ln f_1 - \ln f' - \ln f_1'] [f' f_1' - f f_1] \, |\mathbf{g}| b \, db \, d\psi \, d^3c \, d^3c_1,$$

$$= -\int [\ln(f f_1) - \ln(f' f_1')] [f f_1 - f' f_1'] \, |\mathbf{g}| b \, db \, d\psi \, d^3c \, d^3c_1, \quad (4.24)$$

where we have used eqn (4.14) to derive the third line. The argument of the integral has the structure $(\ln x - \ln y)(x - y)$, with $x = f f_1$ and $y = f' f_1'$, which is always non-negative because the logarithm is a monotonically increasing function. Hence, it results that

$$\frac{dH}{dt} \leq 0; \quad (4.25)$$

that is, there is an irreversible evolution toward decreasing values of H (increasing values of Boltzmann entropy). Two possibilities emerge: either H decreases forever, or there is a lower bound. We can prove by contradiction that the latter possibility is indeed the case. Imagine that there were no lower bound. Then, there would be at least one distribution $\overline{f}$ for which $H[\overline{f}] = -\infty$. Given that the integrand $f \ln(f/f_0)$ is bounded from below by $-f_0 e^{-1}$, H can diverge only if the integrand remains negative over an infinitely large volume. This is achieved if, for large c, there is a positive number A such that $-\ln(\overline{f}/f_0)\overline{f}c^2 > Ac^{-1}$, where the factor c^2 accounts for the three-dimensional integral. At the same time, energy is finite and conserved, implying that there is another positive constant B such that $\overline{f} < Bc^{-5}$. Combining these two inequalities, one obtains that $\overline{f} < f_0 e^{-Ac^2/B}$. However, under this condition, H turns out to be finite, contradicting the original assumption.

Therefore, H has a lower steady value. It is obtained if $f f_1 = f' f_1'$ or, equivalently, if $\ln f + \ln f_1 = \ln f' + \ln f_1'$ for all collision parameters and velocities in (4.24); that is, the lowest value of H is achieved when $\ln f$ is a collisional invariant. In Section 4.3.1 it was shown that mass, momentum, energy, and their linear combinations are the only collisional invariants. Then, for the steady distribution f_0,

$$\ln f_0(\mathbf{c}) = \alpha m + \boldsymbol{\beta} \cdot m\mathbf{c} + \gamma mc^2/2, \quad (4.26)$$

where α, $\boldsymbol{\beta}$, and γ are free parameters. Exponentiating this expression gives the Maxwell–Boltzmann distribution,

$$f_0(\mathbf{c}) = n \left(\frac{m}{2\pi k_B T} \right)^{3/2} e^{-m(\mathbf{c}-\mathbf{v})^2/2k_B T}, \quad (4.27)$$

where the density n, average velocity $\mathbf{v}$, and temperature T are expressed in terms of the linear combination coefficients.

Summarising, we have proved the H-theorem, which states that the Boltzmann equation prescribes an irreversible evolution characterised by the monotonic decrease of H. This irreversible dynamics stops when the system reaches thermal equilibrium, which is described by a Maxwell–Boltzmann distribution, where the density, mean velocity, and temperature are determined by the initial mass, momentum, and energy, which remain constant during the evolution. Our proof was achieved for a homogeneous gas with no external forces, but this demonstration can be extended to inhomogeneous systems if the external forces are conservative.

There is something extraordinarily peculiar in the H-theorem that motivated wide debate when it was formulated. The Boltzmann equation was derived based on the kinematics of classical particles, explicitly using the microscopic reversibility of collisions to derive the inverse collision term. How is it then possible that the resulting equation shows a preferred arrow of time? Where did we introduce this irreversibility? Which hypothesis imposed a difference between past and future?

The latter question gives us a clue to understanding the apparent contradiction. When we derived the Boltzmann equation, we imposed that the two-particle distribution function could be factorised into one-particle distributions, because particles that were approaching a collision had never met before, arriving fresh and without correlations. However, we only demanded the precollisional state to be uncorrelated whereas no assumption was made regarding the postcollisional state. Indeed, in Exercise 4.3 the postcollisional correlations are explicitly computed and shown to be nonzero. Herein lies the difference between past and future introduced in the Boltzmann equation, explaining why macroscopically the evolution is irreversible: the precollisional state is always uncorrelated, while the postcollisional state can be correlated.

Going back to the derivation of the Boltzmann equation, we note that it was not simply a matter of replacing $f^{(2)}(\mathbf{c}_1, \mathbf{c}_2) = f(\mathbf{c}_1)f(\mathbf{c}_2)$ in (4.1), because both pre- and postcollisional states are present in the collision term. It was only after writing the BBGKY collision integral in terms of the direct and inverse collisions that we could implement the factorisation hypothesis. Replacing $f^{(2)}$ by the factorised product directly in the BBGKY equation is only justified for long-range interactions, as we will see in the study of plasmas in Chapter 6. In that case, the Vlasov equation is obtained, which consequently presents reversible dynamics.

We can now consider again the irreversibility in the Lorentz model, studied in Chapter 3, as manifested in the positivity of the collision operator. In that case, it was also assumed that the electrons and ions showed no correlations prior to a collision. However, besides this, in the Lorentz model the scatterers take out momentum and energy. The fact that the distribution function of the scatterers remains unperturbed can be interpreted as their having infinite heat capacity and therefore acting as a thermal or momentum bath. It may then be conceivable that the origin of the irreversibility is that the electrons were in contact with a bath. However, in the case of the Boltzmann equation, the atoms collide among themselves and there is no external bath, while irreversibility is still observed. We therefore need a more profound understanding of the origin of the irreversible behaviour in classical gases.

4.3.3   On the irreversibility problem

The emergence of irreversibility in macroscopic systems lies at the origin of a long-running debate that is still not closed. The problem can be divided into the following three aspects. First, how can we reconcile the reversibility that takes place in microscopic dynamics with the observed irreversibility in the macroscopic world? Second, how does irreversibility emerge in the kinetic models when explicit reversible rules have been added? Finally, how well justified are the hypotheses that are responsible for the irreversible behaviour?

Formally, a classical system obeying microscopic reversible dynamics (e.g. Hamiltonian dynamics) satisfies the Poincaré recurrence theorem. This states that, after a long but finite time, the system will return close to any initial condition. The recurrence time, however, grows extremely fast with the system size (roughly, $\tau_{\text{recurr}} \sim 2^N$, where N is the number of degrees of freedom).

To see how irreversibility appears despite the existence of recurrence, we will consider an oscillator chain, which is a simple system that can be fully solved analytically. Take N particles of mass m, placed in a one-dimensional ring with periodic conditions (Fig. 4.6). The particles are joined to their neighbours by springs of constant k. Initially, all particles are in their equilibrium positions at rest and one particle is given a velocity V_0. The equations of motion are linear and can be solved using normal mode decomposition, resulting in the velocity of the kicked particle,

$$V(t) = V_0 \sum_{n=0}^{N-1} \cos(\omega_n t)/N, \qquad (4.28)$$

where $\omega_n = 2\sqrt{\frac{k}{m}} \sin\left(\frac{\pi n}{N}\right)$ are the frequencies of the normal modes. The structure is quasiperiodic, consistent with Poincaré recurrence. Indeed, instants that are approximate common multiples of the periods result in $V \approx V_0$. However, as the frequencies are irrationals, the least common multiple grows rapidly with the number of modes N, as does the recurrence time (Fig. 4.7). In the thermodynamic limit $N \to \infty$, the sum transforms into an integral, which can be evaluated in terms of a Bessel function,

$$V(t) = V_0 J_0(2t\sqrt{k/m}). \qquad (4.29)$$

Remarkably, this expression presents damped oscillations that never come back close to the initial value. We have obtained that, in the thermodynamic limit, the recurrence time is pushed to the infinite future, and irreversible behaviour is observed. In this case, irreversibility takes place by distributing the energy, originally located in one particle,

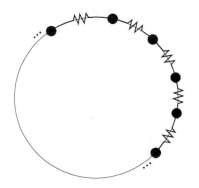

Fig. 4.6 N particles of equal mass m are placed on a ring. The particles are joined by springs of constant k. Initially, all particles are at rest in their equilibrium position and one particle receives a kick with velocity V_0.

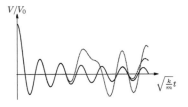

Fig. 4.7 Velocity of the kicked particle as a function of time for different numbers of particles: $N = 10$ (dotted line), $N = 20$ (dashed line), and $N \to \infty$ (solid line).

to the rest of the system. For large but finite N, it is shown that the recurrence time is $\tau_{\text{recurr}} \approx \sqrt{2}\pi e^N$ (Zwanzig, 2001). This example shows that recurrence is lost in the thermodynamic limit because information flows to infinity and cannot come back. Of course, the information does not disappear immediately, being characterised in the oscillator chain example by the decay of the Bessel function.

Returning to the Boltzmann model, after a collision in a gas, the pair of particles explore a large volume of phase space (with many other collisions), before coming again to a new collision. The volume of the explored phase space volume increases on reducing ϕ and diverges in the Boltzmann–Grad limit, resulting in the observed irreversibility. That is, irreversibility appears because the information is carried and transferred by the multiple collisions (tending to infinity in the limit), being finally lost. In other statistical mechanics systems, the situation is similar. Irreversibility is obtained for long times only after the system size has been made large. Formally, the limits do not commute and the correct order to observe irreversible behaviour is $\lim_{t\to\infty} \lim_{N,\mathcal{V}\to\infty}$.

We need to make a distinction here between the thermodynamic limit and the Boltzmann–Grad limit, because it is only the latter that is needed in the context of Boltzmann theory to generate irreversibility. Indeed, on many occasions, gases will be in a finite volume but there is still a definitive arrow of time. The Boltzmann–Grad limit states that $N_\ell \gg 1$, implying that many degrees of freedom are involved in the dynamics, diluting the information and giving rise to a divergent recurrence time.

The discussion presented above have focused on dilute gases, which is the subject of study of this chapter. However, in dense gases and liquids, the macroscopic dynamics is also irreversible, for which an explanation is needed. Now we cannot use the hypothesis that recollisions do not take place, and correlations are indeed present in the precollisional state. However, again, irreversibility appears here in the thermodynamic limit. When two particles collide and come back for a recollision, they necessarily collided with others in between which take momentum and energy out. If the system is infinitely large, energy and momentum will flow to infinity, without coming back. Then, information (correlations) flows away, a process that is responsible for the irreversible behaviour. When these fluxes are analysed in detail, it is found that correlations do decay irreversibly but, contrary to expectation, instead of an exponential decay, power-law tails are observed (Pomeau and Resibois, 1975).

4.4   Dynamics close to equilibrium

4.4.1   Linear Boltzmann operator

The H-theorem indicates that a gas will approach equilibrium. To study this dynamics in detail, we will place ourselves in the later phases of the evolution,[11] when the system is already close to equilibrium. We

---

[11] In Section 4.4.3, we will describe in detail the different time scales involved in the evolution.

linearise, therefore, the distribution

$$f(\mathbf{c}) = f_{\mathrm{MB}}(\mathbf{c})\left[1 + \Phi(\mathbf{c})\right], \tag{4.30}$$

which is substituted into the collision integral. Recalling that the collision integral vanishes for the Maxwell–Boltzmann distribution and using the notation

$$J[f, f] = -n^2 I[\Phi], \tag{4.31}$$

to first order in Φ one obtains the linear Boltzmann operator,

$$I[\Phi](\mathbf{c}) = \int \widehat{f}_{\mathrm{MB}} \widehat{f}_{\mathrm{MB1}} \left[\Phi + \Phi_1 - \Phi' - \Phi_1'\right] |\mathbf{g}| b\, db\, d\psi\, d^3 c_1. \tag{4.32}$$

For the last expression, we have used that $f_{\mathrm{MB}}' f_{\mathrm{MB1}}' = f_{\mathrm{MB}} f_{\mathrm{MB1}}$ and defined the normalised distribution $\widehat{f}_{\mathrm{MB}} = f_{\mathrm{MB}}/n$, where n is the density of the reference state.

Defining the bracket product,

$$[\Psi, \Phi] = \int \Psi^*(\mathbf{c}) I[\Phi](\mathbf{c})\, d^3 c, \tag{4.33}$$

one can directly prove that, $[\Psi, \Phi] = [\Phi, \Psi]^*$, $[\Phi, \Phi] \geq 0$, and $[\Psi, \Phi] = 0$ if either Φ or Ψ is a collisional invariant. Therefore, for the usual scalar product,

$$(g, h) = \int g^*(\mathbf{c}) h(\mathbf{c})\, d^3 c, \tag{4.34}$$

the linear Boltzmann operator is positive semidefinite. Also, using the same transformations that led to eqn (4.15), it can be proved that I is an Hermitian operator.[12]

[12] See Exercise 4.7.

4.4.2   Spectrum of the linear Boltzmann equation

Consider a gas close to equilibrium without external forces. The Boltzmann equation is linearised, and we can therefore consider Fourier modes for the spatial dependence,

$$f(\mathbf{r}, \mathbf{c}, t) = f_{\mathrm{MB}}(\mathbf{c})\left[1 + \Phi_{\mathbf{k}}(\mathbf{c}, t) e^{i\mathbf{k}\cdot\mathbf{r}}\right], \tag{4.35}$$

resulting in the following equation:

$$f_{\mathrm{MB}}\frac{\partial \Phi_{\mathbf{k}}}{\partial t} + i\mathbf{k}\cdot\mathbf{c} f_{\mathrm{MB}}\Phi_{\mathbf{k}} = -n^2 I[\Phi_{\mathbf{k}}], \tag{4.36}$$

which can be analysed in terms of the eigenvalue problem $L_{\mathbf{k}}\Phi_{j,\mathbf{k}} = -\lambda_{j,\mathbf{k}} f_{\mathrm{MB}}\Phi_{j,\mathbf{k}}$ for the linear operator $L_{\mathbf{k}} = n^2 I + i\mathbf{k}\cdot\mathbf{c} f_{\mathrm{MB}}$. The evolution of the distribution is finally

$$f(\mathbf{r}, \mathbf{c}, t) = f_{\mathrm{MB}}(\mathbf{c})\left[1 + \sum_j A_{j,\mathbf{k}}\Phi_{j,\mathbf{k}}(\mathbf{c}) e^{i\mathbf{k}\cdot\mathbf{r} - \lambda_{j,\mathbf{k}} t}\right], \tag{4.37}$$

where j labels the modes and the amplitudes $A_{j,\mathbf{k}}$ are obtained from the initial condition.

We proceed similarly as we did for the Lorentz model. For $\mathbf{k} = 0$, the operator I is Hermitian and positive semidefinite with five null eigenvalues (the conserved modes), while all the others are positive and real (the kinetic modes). Therefore, there is an irreversible evolution to equilibrium, respecting the conservations, consistent with the H-theorem. A detailed mathematical analysis shows that, for usual interatomic potentials, the positive eigenvalues are at a finite distance from zero. There is a kinetic gap, of the order of the collision frequency ν, between the conserved and nonconserved modes.

Assuming that for small $|\mathbf{k}|$ no accidents occur, the eigenvalues will be continuous in $\mathbf{k}$. Hence, the kinetic modes will continue to have eigenvalues with positive real parts and will relax on the scale of the mean free time. At long times, only the conserved modes survive and dominate the dynamics, justifying their detailed study. The perturbation method, described in Appendix A, can be applied to compute the corrections to the eigenvalues. The calculations are lengthy, and only a schematic presentation is given here. It follows the same method used in Section 3.8.3 for the Lorentz model, with the additional complexity that for $\mathbf{k} = 0$ the eigenvalues are degenerate, with eigenfunctions $\Phi = \{1, c_x, c_y, c_z, c^2\}$. To show how the degeneration is broken, we choose an explicit direction for $\mathbf{k} = k\hat{\mathbf{x}}$. It results the transverse modes c_y and c_z not coupling with the others and with the perturbation analysis being straightforward for them. These are called the shear or transverse ($\perp$) modes, because they correspond to perturbations in the velocity field $\mathbf{v} = (V_y\hat{\mathbf{y}} + V_z\hat{\mathbf{z}})e^{ikx}$ perpendicular to $\mathbf{k}$, shown in Fig. 4.8. The other three modes couple into what is called a heat (H) mode and two sound ($\pm$) modes. Without doing the explicit calculation, it can be shown by symmetry arguments that the eigenvalues present the following structure at the dominant orders:[13]

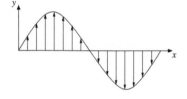

Fig. 4.8 Transverse or shear mode, where the velocity field is perpendicular to the wavevector.

[13]Exercise 4.8 considers the detailed calculation of the shear mode.

$$\lambda_\perp^{y,z} = \nu k^2 \qquad \Phi = c_{y,z} + \mathcal{O}(k), \qquad (4.38)$$

$$\lambda_\pm = \pm i c_s k + \Gamma k^2 \qquad \Phi = c_x + \mathcal{O}(k), \qquad (4.39)$$

$$\lambda_H = D_T k^2 \qquad \Phi = (mc^2/2 - 3k_B T/2) + \mathcal{O}(k). \qquad (4.40)$$

The values of the coefficients ν, D_T, c_s, and Γ will be given in Section 4.7.4. Nevertheless, this result can be analysed already at the qualitative level. The real parts of all of them are positive, implying that the modes relax to a homogeneous equilibrium state. The sound modes have an imaginary part, being therefore associated to waves $e^{ik(x\mp c_s t)}$ with sound speed c_s. These are compressive waves of longitudinal velocity ($\mathbf{v} \parallel \mathbf{k}$), which damp at a rate Γk^2, being the only oscillatory modes. All relaxation rates are proportional to k^2, therefore corresponding to slow, diffusive modes. These are the heat diffusion, described by the heat equation, and the momentum diffusion, associated with the viscosity. The latter has two rates (νk^2 and Γk^2), depending on whether the momentum is transverse or parallel to the wavevector.

4.4.3 Time scales

At the macroscopic level (small k), there is a clear separation of time scales between the fast kinetic modes, which relax on the time scale of the mean free time, and the conserved modes. The latter present two time scales, one associated with the sound propagation $\tau_\pm = 1/(c_s k)$ and a much longer one related to the diffusive relaxation (Fig. 4.9). Note that sound corresponds to reversible dynamics while only the kinetic and diffusive modes are irreversible, driving the system to equilibrium.

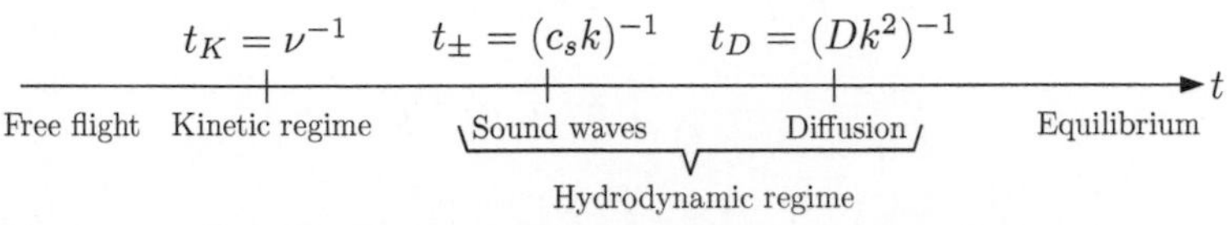

Fig. 4.9 Different regimes in the temporal evolution of a gas described by the Boltzmann equation. D represents any of the diffusive relaxation coefficients ν, D_T, or Γ.

In contrast to the Lorentz gas, the analysis of the linear modes does not provide full information on the gas dynamics. Indeed, the Boltzmann equation is nonlinear, and we have only obtained the evolution close to equilibrium. There is no complete quantitative picture of the full nonlinear evolution that, for example, includes processes as complex as turbulence or thermal convection. The qualitative picture, however, is well understood. In the kinetic regime, information has not had time to travel between distant zones and all evolution takes place locally. The H-theorem indicates that, locally, collisions in the kinetic regime will drive the system to a Maxwellian distribution. This is a local equilibrium distribution because regions have not equilibrated among themselves. Thereafter, the gas evolves in states of slightly perturbed local equilibrium states (they are not exactly in local equilibrium, because if this were the case, the momentum and heat fluxes would vanish). Different regions exchange mass, momentum, and energy through slow mechanisms. Besides the sound and diffusive modes present in the linear cases, there is also nonlinear convective transport [terms $\nabla \cdot (\rho \mathbf{v})$ in (4.16) and $(\mathbf{v} \cdot \nabla)$ in (4.18) and (4.19)]. The time scale of convective transport is $t_{\text{conv}} \sim 1/(vk)$, also being a slow process. Note that all transport processes are slow because they involve homogenisation of the system. The conserved quantities, which cannot be created or destroyed, must be transported in a process that takes longer and longer times as the relevant length scales grow. This homogenisation regime is governed by hydrodynamics, which will finally lead the system to thermodynamic equilibrium.

4.5 BGK model

The Boltzmann equation, with the collisional operator based on the direct and inverse collisions, gives an accurate description of gases. However, because of its complexity, it is challenging to make detailed calculations using it. Any simplified model should maintain the irreversible evolution toward the equilibrium Maxwell–Boltzmann distribution, while respecting the collisional invariants of the original model. This objective is achieved by the Bhatnagar–Gross–Krook (BGK) model, which replaces the collision operator by

$$J_{\mathrm{BGK}}[f] = \nu \left\{ f_{\mathrm{MB}}(\mathbf{c}; n[f], \mathbf{v}[f], T[f]) - f(\mathbf{r}, \mathbf{c}) \right\}, \qquad (4.41)$$

where ν is the collision frequency and the density, mean velocity, and temperature are obtained from the distribution as (see Section 1.3)

$$n[f] = \int f(\mathbf{r}, \mathbf{c}) \, \mathrm{d}^3 c, \qquad (4.42)$$

$$\mathbf{v}[f] = \frac{1}{n(\mathbf{r}, t)} \int \mathbf{c} f(\mathbf{r}, \mathbf{c}) \, \mathrm{d}^3 c, \qquad (4.43)$$

$$\frac{3}{2} k_{\mathrm{B}} T[f] = \frac{1}{n(\mathbf{r}, t)} \int \frac{m(\mathbf{c} - \mathbf{v})^2}{2} f(\mathbf{r}, \mathbf{c}) \, \mathrm{d}^3 c. \qquad (4.44)$$

One can directly prove that mass, momentum, and energy are collisional invariants of this model. Also, it is possible to prove the H-theorem for it.[14]

The BGK model, which can be interpreted as a relaxation time approximation,[15] is highly nonlinear and indeed is very complex to use in full detail as it stands. However, the linear operator is notably simpler. Considering the perturbation $f = f_{\mathrm{MB}}(1 + \Phi)$, to linear order in Φ, one has

$$n[f] = n_0 \left[1 + \int \widehat{f}_{\mathrm{MB}}(\mathbf{c}) \Phi(\mathbf{c}) \, \mathrm{d}^3 c \right], \qquad (4.45)$$

$$\mathbf{v}[f] = n_0 \int \widehat{f}_{\mathrm{MB}}(\mathbf{c}) \Phi(\mathbf{c}) \mathbf{c} \, \mathrm{d}^3 c, \qquad (4.46)$$

$$T[f] = T_0 \left[1 + \frac{2}{3} \int \widehat{f}_{\mathrm{MB}}(\mathbf{c}) \Phi(\mathbf{c}) \left(\frac{mc^2}{2k_{\mathrm{B}} T_0} - \frac{3}{2} \right) \mathrm{d}^3 c \right], \qquad (4.47)$$

with $\widehat{f}_{\mathrm{MB}} = f_{\mathrm{MB}}/n_0$. Then,

$$n_0 I_{\mathrm{BGK}}[\phi] = \nu \widehat{f}_{\mathrm{MB}}(\mathbf{c}) \Phi(\mathbf{c}) - \nu \widehat{f}_{\mathrm{MB}}(\mathbf{c}) \left[\int \widehat{f}_{\mathrm{MB}}(\mathbf{c}') \Phi(\mathbf{c}') \, \mathrm{d}^3 c' \right.$$

$$+ \frac{m\mathbf{c}}{k_{\mathrm{B}} T_0} \cdot \int \widehat{f}_{\mathrm{MB}}(\mathbf{c}') \Phi(\mathbf{c}') \mathbf{c}' \, \mathrm{d}^3 c'$$

$$\left. + \frac{2}{3} \left(\frac{mc^2}{2k_{\mathrm{B}} T_0} - \frac{3}{2} \right) \int \widehat{f}_{\mathrm{MB}}(\mathbf{c}') \Phi(\mathbf{c}') \left(\frac{mc'^2}{2k_{\mathrm{B}} T_0} - \frac{3}{2} \right) \mathrm{d}^3 c' \right]. \qquad (4.48)$$

This collisional operator relaxes all perturbations [term $\nu \Phi(\mathbf{c})$], but it also creates terms proportional to 1, $\mathbf{c}$, and $(mc^2/2k_{\mathrm{B}} T_0 - 3/2)$. The first one simply adjusts the density, compensating the relaxation already described in order to keep the total mass conserved. The second and third terms are more subtle and should be understood in the context of a linear perturbation. Indeed, a function, $f(\mathbf{c}) = f_{\mathrm{MB}}(\mathbf{c}) \left[1 + \epsilon \mathbf{a} \cdot \mathbf{c} + \epsilon b \left(\frac{mc^2}{2k_{\mathrm{B}} T_0} - \frac{3}{2} \right) \right]$ for $\epsilon \ll 1$ is identical to

[14]See Exercise 4.11.

[15]The BGK model should not be confused with the simplest relaxation time approximation, $J[f] = \nu \{ f_{\mathrm{MB}}(\mathbf{c}; n_0, \mathbf{v}_0 = 0, T_0) - f(\mathbf{r}, \mathbf{c}) \}$, which does not conserve mass, momentum, or energy and hence is simply wrong. However, at the linear level, if the symmetry of the perturbations is such that they do not generate modifications to the conserved fields, the relaxation time approximation can be used as, for example, in the study of the electron conductivity in Chapter 8.

a Maxwellian, where the velocity and temperature have changed by a factor proportional to ϵ. In summary, the linear BGK collision term relaxes all modes except those associated with conservations,[16] which are adjusted consistently. For example, momentum that could be in other kinetic modes is moved to the mode $\mathbf{c}$, which to linear order gives rise to a displaced Maxwellian.

The spectrum of the linear BGK model is trivial. It has five vanishing eigenvalues, associated with the conservations, while all the others are equal to ν. To prove this, we can build a basis of orthogonal polynomials with weight function $\widehat{f}_{\mathrm{MB}}$. The first polynomials are precisely 1, $\mathbf{c}$, and $(mc^2/2k_{\mathrm{B}}T - 3/2)$. For these, the action of I vanishes (see Exercise 4.12), while for the others, due to orthogonality, the action of I is simply to multiply by ν.

[16]See Exercise 4.12 for an analysis of the conservations in the linear BGK model.

4.6 Boundary conditions

The boundary conditions for the distribution function reflect the microscopic dynamics of the atoms colliding with the walls. Different models are possible depending on the atom–wall interaction. The action of a wall oriented along $\hat{\mathbf{n}}$ is characterised by the transition probability $W(\mathbf{c}, \mathbf{c}')$, normalised as $\int W(\mathbf{c}, \mathbf{c}')\,\mathrm{d}^3 c' = 1$, that an incoming velocity $\mathbf{c}$ changes to $\mathbf{c}'$ (Fig. 4.10). The normalisation of the outgoing distribution is obtained by imposing that the outgoing flux equals the incoming one. Then, a general boundary condition can be written as

$$\mathbf{c}' \cdot \hat{\mathbf{n}} f(\mathbf{c}') = \int W(\mathbf{c}, \mathbf{c}') f(\mathbf{c}) |\mathbf{c} \cdot \hat{\mathbf{n}}| \Theta(-\mathbf{c} \cdot \hat{\mathbf{n}})\,\mathrm{d}^3 c; \quad \mathbf{c}' \cdot \hat{\mathbf{n}} > 0. \quad (4.49)$$

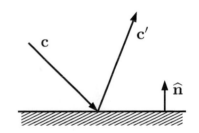

Two limiting cases are normally used. First, elastic walls reflect atoms specularly, where the normal component of the velocity is reversed, i.e. $\mathbf{c}' = \mathbf{c} - 2(\mathbf{c} \cdot \hat{\mathbf{n}})\hat{\mathbf{n}}$, giving $W(\mathbf{c}, \mathbf{c}') = \delta[\mathbf{c}' - \mathbf{c} + 2(\mathbf{c} \cdot \hat{\mathbf{n}})\hat{\mathbf{n}}]$. In the other limit, when there is a strong interaction, atoms thermalise with the walls and emerge with a flux Maxwellian distribution (1.24) at the wall temperature; hence $W(\mathbf{c}, \mathbf{c}') = \mathbf{c}' \cdot \hat{\mathbf{n}} \widehat{f}_{\mathrm{MB}}(\mathbf{c}') / \int \mathbf{c}' \cdot \hat{\mathbf{n}} \widehat{f}_{\mathrm{MB}}(\mathbf{c}')\,\mathrm{d}^3 c'$, irrespective of the incoming velocity.

Fig. 4.10 When a particle collides with a wall oriented along $\hat{\mathbf{n}}$, the incoming velocity $\mathbf{c}$ transforms into $\mathbf{c}'$, satisfying $\mathbf{c} \cdot \hat{\mathbf{n}} < 0$ and $\mathbf{c}' \cdot \hat{\mathbf{n}} > 0$.

4.7 Hydrodynamic regime

4.7.1 The hydrodynamic equations

At long times, the dynamics is governed by the transport of the conserved quantities, which we have shown to be a slow process. In this regime, the stress tensor and heat flux adopt simple forms, being proportional to the gradients of the thermodynamic fields. Specifically, we will show that the heat flux is given by the Fourier law,

$$\mathbf{q} = -\kappa \nabla T, \quad (4.50)$$

where κ is the thermal conductivity, and the stress tensor is given by Newton's law,

$$\mathbb{P} = p\mathbb{I} - \eta \left[(\nabla \mathbf{v}) + (\nabla \mathbf{v})^T - \frac{2}{3}(\nabla \cdot \mathbf{v})\mathbb{I} \right]. \tag{4.51}$$

Here $p = nk_\mathrm{B}T$ is the ideal gas pressure and η is the shear viscosity. These constitutive relations correspond to the linear response expressions, and at higher inhomogeneities, corrections appear.

Once substituted into the conservation equations [eqns (4.16), (4.18), and (4.19)], one recovers the Navier–Stokes hydrodynamic equations for gases. The transport coefficients, as was anticipated in Chapter 1 and we will see in the next sections, depend on temperature. Therefore, they cannot be taken out of the divergences of the fluxes in (4.18) and (4.19), rendering the final form for the hydrodynamic equations rather involved, which we will not write down explicitly.

The hydrodynamic equations can be analysed for their long-time dynamics to compare with the kinetic analysis performed in Section 4.4.2. Consider a homogeneous reference state with vanishing velocity, being slightly perturbed as: $n = n_0 + \epsilon n_1(\mathbf{r}, t)$, $\mathbf{v} = \epsilon \mathbf{v}_1(\mathbf{r}, t)$, and $T = T_0 + \epsilon T_1(\mathbf{r}, t)$, with $\epsilon \ll 1$. Expanding the hydrodynamic equations in ϵ gives

$$\frac{\partial n_1}{\partial t} = -n_0 \nabla \cdot \mathbf{v}_1, \tag{4.52}$$

$$mn_0 \frac{\partial \mathbf{v}_1}{\partial t} = -k_\mathrm{B} \nabla(\rho_0 T_1 + T_0 \rho_1) + \eta_0 \nabla^2 \mathbf{v}_1, \tag{4.53}$$

$$\frac{3}{2}k_\mathrm{B} mn_0 \frac{\partial T_1}{\partial t} = -\kappa_0 \nabla^2 T_1 + p_0 \nabla \cdot \mathbf{v}, \tag{4.54}$$

where $\eta_0 = \eta(n_0, T_0)$, $\kappa_0 = \kappa(n_0, T_0)$, and $p_0 = k_\mathrm{B} n_0 T_0$. These linear equations can be studied in spatial Fourier modes, allowing the associated eigenvalues to be determined. Expanding them for small wavevector values, the five eigenvalues are given by the same expressions (4.38) to (4.40) obtained for the kinetic equation, with

$$\nu = \eta_0/(mn_0), \tag{4.55}$$

$$\Gamma = \frac{2\kappa_0}{15 k_\mathrm{B} n_0} + \frac{2\eta_0}{mn_0}, \tag{4.56}$$

$$D_T = \frac{3\kappa_0}{5 k_\mathrm{B} n_0}, \tag{4.57}$$

$$c_s = \sqrt{\frac{5 k_\mathrm{B} T}{3m}}, \tag{4.58}$$

where in the sound attenuation coefficient Γ it is apparent how the heat and longitudinal momentum equations mix.

At long times, the hydrodynamic equations have the same dynamical structure as the Boltzmann equation, presenting the same dynamical modes. Moreover, the identification of the eigenvalues of the linearised Boltzmann equation with the hydrodynamic transport coefficients provides a method to compute the transport coefficients, which are hence related to the microscopic dynamics.[17]

[17]See, for example, Exercise 4.8, where the shear viscosity η_0 is computed.

In the next section, a different method will be presented to compute transport coefficients, which is equivalent to the analysis by eigenvalues.

4.7.2   Linear response

We will apply the linear response theory to the calculation of transport coefficients, in particular the shear viscosity, in a manner similar to how the electrical conductivity was obtained in Chapter 3. Consider a gas with uniform density n_0 and temperature T_0 that is subject to a constant shear flow by walls located far apart, as in Fig. 4.11 (see Section 1.6.2 for an analysis of this geometry). The gas develops a velocity profile $\mathbf{v} = V' y \hat{\mathbf{x}}$, with a shear rate V' that is constant in the stationary regime. The shear rate has units of inverse time and should therefore be compared with the collision frequency ν. We will consider the linear response regime for $|V'| \ll \nu$.[18] The collision frequency is the rate of approach to local thermodynamic equilibrium, while V' is the rate at which the gas is driven away from equilibrium. So, in the linear response regime, the gas will be very close to local equilibrium and we can write for the stationary regime,

$$f(\mathbf{r}, \mathbf{c}) = f_{\mathrm{MB}}(\mathbf{c}; n_0, T_0, \mathbf{v}(\mathbf{r})) \left[1 + \Phi(\mathbf{c})\right], \tag{4.59}$$

where Φ is of order V'/ν. When this distribution is inserted into the Boltzmann equation, the left-hand side gives a term that is due to the spatial derivative of the distribution, whereas the right-hand side reduces to the linear collision operator to first order in the perturbation,

$$V' \frac{m(c_x - v_x)c_y}{k_{\mathrm{B}} T_0} f_{\mathrm{MB}} = -n_0^2 I[\Phi], \tag{4.60}$$

showing that the perturbation was indeed proportional to the shear rate. We exploit the linearity of the equation to write $\Phi(\mathbf{c}) = -V'\widehat{\Phi}(\mathbf{c})/n_0$, and we change to the comoving reference frame. This gives

$$I[\widehat{\Phi}] = \frac{m c_x c_y}{k_{\mathrm{B}} T} \widehat{f}_{\mathrm{MB}}(\mathbf{c}). \tag{4.61}$$

Assuming the solution of (4.61) is known, we can compute the off-diagonal component of the stress tensor,

$$P_{xy} = \int f(\mathbf{c}) m c_x c_y \, \mathrm{d}^3 c = -V' \int \widehat{f}_{\mathrm{MB}}(\mathbf{c})\widehat{\Phi}(\mathbf{c}) m c_x c_y \, \mathrm{d}^3 c, \tag{4.62}$$

where the equilibrium distribution gives no contribution by symmetry. Newton's law for viscosity is obtained as $P_{xy} = -\eta \frac{\partial v_x}{\partial y}$, giving an expression for the viscosity of

$$\eta = \int \widehat{f}_{\mathrm{MB}}(\mathbf{c})\widehat{\Phi}(\mathbf{c}) m c_x c_y \, \mathrm{d}^3 c. \tag{4.63}$$

Note that no density factors appear in the integral equation for $\widehat{\Phi}$, which is therefore independent of density. Consequently, the viscosity is independent of density for ideal gases, regardless of the particular interaction potential which enters into eqn (4.61). This result was anticipated in Chapter 1, in the mean free path approximation.

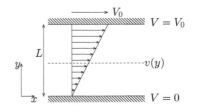

Fig. 4.11 Couette geometry used to study viscous flows. The gas is bounded between two parallel walls: the bottom one is kept fixed, while the top wall moves at a constant speed V_0. The gas develops a velocity profile $\mathbf{v} = \frac{V_0}{L} y \hat{\mathbf{x}}$.

[18] In practice, this is normally the case, because $\nu \sim 5.7 \times 10^9 \, \mathrm{s}^{-1}$, a very large value compared with the shear rate in usual experimental conditions.

4.7.3   Variational principle

Combining eqns (4.61) and (4.63), we obtain

$$\eta = k_{\mathrm{B}}T[\widehat{\Phi}, \widehat{\Phi}], \tag{4.64}$$

where the definition of the bracket product (4.33) has been used, implying that η is always positive.

Imagine that one could find a function $\widehat{\Psi}(\mathbf{c})$ such that $[\widehat{\Psi}, \widehat{\Psi}] = [\widehat{\Psi}, \widehat{\Phi}]$. Fulfilment of this condition does not require one to know $\widehat{\Phi}$ by the definition of the square product, and it indeed reduces to

$$[\widehat{\Psi}, \widehat{\Psi}] = \int \widehat{\Psi}(\mathbf{c})\frac{mc_xc_y}{k_{\mathrm{B}}T}\widehat{f}_{\mathrm{MB}}(\mathbf{c})\,\mathrm{d}^3c. \tag{4.65}$$

Then, the positivity of the square bracket, i.e. $[\widehat{\Psi} - \widehat{\Phi}, \widehat{\Psi} - \widehat{\Phi}] \geq 0$, finally leads to the following inequality:

$$\eta \geq k_{\mathrm{B}}T[\widehat{\Psi}, \widehat{\Psi}], \tag{4.66}$$

which serves as a variational principle by varying the test function $\widehat{\Psi}$. Nontrivial test functions must have the same symmetry as c_xc_y, otherwise the right-hand side of (4.65) would vanish. The polynomial series

$$\widehat{\Psi} = c_xc_y\sum_{n=0}^{N-1} a_nS_n(c^2) \tag{4.67}$$

[19]The Sonine (or associated Laguerre) polynomials are defined as

$$S_p(x) = \sum_{n=0}^{p} \frac{(-1)^n(p+1/2)!}{(n+1/2)!(p-n)!n!}x^n.$$

The first ones are

$$S_0(x) = 1,$$
$$S_1(x) = \frac{3}{2} - x,$$
$$S_2(x) = \frac{15}{8} - \frac{5x}{2} + \frac{x^2}{2}.$$

is proposed, where usually Sonine polynomials are used,[19] for which many integrals simplify. Here, we stop at one polynomial ($N = 1$), which has only one degree of freedom, a_0. Imposing the condition (4.65) gives

$$a_0 = \frac{4m}{k_{\mathrm{B}}T}\frac{\int c_x^2c_y^2\widehat{f}_{\mathrm{MB}}(\mathbf{c})\,\mathrm{d}^3c}{\int(\Delta c_xc_y)^2\widehat{f}_{\mathrm{MB}}(\mathbf{c})\widehat{f}_{\mathrm{MB}}(\mathbf{c}_1)|\mathbf{g}|\,b\,\mathrm{d}b\,\mathrm{d}\psi\,\mathrm{d}^3c\,\mathrm{d}^3c_1}, \tag{4.68}$$

where $\Delta c_xc_y = c_x'c_y' + c_{x1}'c_{y1}' - c_xc_y - c_{x1}c_{y1}$ is the collisional change of c_xc_y. The viscosity is therefore estimated by the bound,

$$\eta_0 = \frac{4m^2}{k_{\mathrm{B}}T}\frac{\left[\int c_x^2c_y^2\widehat{f}_{\mathrm{MB}}(\mathbf{c})\,\mathrm{d}^3c\right]^2}{\int(\Delta c_xc_y)^2\widehat{f}_{\mathrm{MB}}(\mathbf{c})\widehat{f}_{\mathrm{MB}}(\mathbf{c}_1)|\mathbf{g}|\,b\,\mathrm{d}b\,\mathrm{d}\psi\,\mathrm{d}^3c\,\mathrm{d}^3c_1}. \tag{4.69}$$

[20]See Exercise 4.9.

The calculations can be done explicitly for the hard sphere model, and the result is[20]

$$\eta_0^{\mathrm{HS}} = \frac{5}{16D^2}\sqrt{\frac{mk_{\mathrm{B}}T}{\pi}}. \tag{4.70}$$

4.7.4   The Chapman–Enskog method

The Chapman–Enskog method, presented in detail in Chapter 3, allows one to deal with the different time scales present in the kinetic description, providing the relevant dynamics at each time scale. It does not assume linearity and, therefore, will result in the full, nonlinear hydrodynamic equations. However, in contrast to the Lorentz gas, here there are five conserved quantities, rendering the problem much more tedious. The methodology is nevertheless the same, and we proceed here to give only the main steps and results; the details can be found elsewhere (Chapman and Cowling, 1970; Ferziger and Kaper, 1972; Liboff, 2003).

A formal parameter $\epsilon \ll 1$ is placed in the Boltzmann equation, multiplying the spatial gradient and external force terms, to signal that they are going to be considered as small. The H-theorem states that the system tends to relax toward local equilibrium and that the long-term evolution is governed by the slow conserved fields. We then state that the distribution function is enslaved to these fields as[21] $f(\mathbf{r}, \mathbf{c}, t) = h(\mathbf{c}; n(\mathbf{r}, t), \mathbf{v}(\mathbf{r}, t), T(\mathbf{r}, t))$, being moreover expanded in ϵ as $h = h_0 + \epsilon h_1 + \epsilon^2 h_2 + \dots$. Finally, new time variables are introduced that will take account of the different scales: $t_0 = t$ accounts for the fast kinetic regime, $t_1 = \epsilon t$ will describe the slow hydrodynamic regime (sound), and $t_2 = \epsilon^2 t$ will be responsible for describing slower hydrodynamic modes (diffusion). We stop here, but it is possible to go to higher orders that would result in the Burnett and super-Burnett equations. The conserved fields are made to depend on these variables; for example, $n(\mathbf{r}, t) \to n(\mathbf{r}, t_0, t_1, t_2, \dots)$. Hence, the time derivatives are expanded as $\frac{\partial}{\partial t} = \frac{\partial}{\partial t_0} + \epsilon \frac{\partial}{\partial t_1} + \epsilon^2 \frac{\partial}{\partial t_2} + \dots$. Finally, all spatial and temporal derivatives are transformed by the chain rule into derivatives of the fields. To order zero, the Boltzmann equation reduces to

$$\frac{\partial h_0}{\partial n} \frac{\partial n}{\partial t_0} + \frac{\partial h_0}{\partial \mathbf{v}} \cdot \frac{\partial \mathbf{v}}{\partial t_0} + \frac{\partial h_0}{\partial T} \frac{\partial T}{\partial t_0} = J[h_0, h_0]. \tag{4.71}$$

Computing the associated conservation equations (i.e. multiplying eqn (4.71) by 1, $\mathbf{c}$, and c^2 and integrating over the velocity) gives vanishing right-hand sides, implying that none of these fields depend on t_0; that is, in the kinetic regime, collisions do not affect the conserved fields because there has been no time to transport them. With this result, the equation reduces to $J[h_0, h_0] = 0$, whose solution is a local Maxwellian $h_0 = f_{\mathrm{MB}}$.

The conserved fields do not depend on t_0, implying that, by the enslaving hypothesis, the distribution function also does not depend on t_0 at any order. At first order in ϵ, the Boltzmann equation reads,

$$\frac{\partial h_0}{\partial t_1} + \mathbf{c} \cdot \nabla h_0 = -n^2 I[\Phi], \tag{4.72}$$

where we have written $h_1 = h_0 \Phi$. Any linear combination of the collisional invariants is a homogeneous solution of the previous equation. We choose to fix this freedom by imposing that the conserved fields are entirely given by h_0. That is, the perturbation h_1 and the subsequent orders do not contribute to these fields ($\int h_1 \, \mathrm{d}^3 c = \int \mathbf{c} h_1 \, \mathrm{d}^3 c = \int c^2 h_1 \, \mathrm{d}^3 c = 0$).

Now, computing the conservation equations for (4.72) gives a vanishing right-hand side.[22] The first term of the left-hand side gives the time derivatives of the conserved fields. The second term contributes with the

[21] The enslaving is imposed by stating that all the spatiotemporal dependence of f is via the conserved fields.

[22] The procedure presented here to compute the conservation equations at each order is equivalent to applying the solvability condition which is normally used in the derivation of the Chapman–Enskog method. For a symmetric matrix A that is not invertible, the linear equation $Ax = b$ admits solutions only if b is orthogonal to all solutions of the homogeneous equation, which in the case of the Boltzmann equation are the collisional invariants.

divergence of the fluxes, which are computed from $h_0 = f_{MB}$, resulting in $\mathbb{P} = nk_BT\mathbb{I}$ and $\mathbf{q} = 0$. In summary, one obtains

$$\frac{\partial \rho}{\partial t_1} + \nabla \cdot (\rho \mathbf{v}) = 0, \tag{4.73}$$

$$\rho \left(\frac{\partial \mathbf{v}}{\partial t_1} + (\mathbf{v} \cdot \nabla)\mathbf{v} \right) = -\nabla p + \rho \mathbf{F}/m, \tag{4.74}$$

$$\frac{3}{2}k_B\rho \left(\frac{\partial T}{\partial t_1} + (\mathbf{v} \cdot \nabla)T \right) = -p\nabla \cdot \mathbf{v}, \tag{4.75}$$

which are the Euler fluid equations for a compressible gas. Considering that there are no irreversible fluxes, the Euler equations turn out to be time reversible. Now, using the chain rule to compute the temporal and spatial derivatives of h_0 in (4.72) and using the Euler equations, gives

$$n^2 I[\Phi] = - \left[\left(\frac{mC^2}{2k_BT} - \frac{5}{2} \right) \mathbf{C} \cdot \nabla \ln T \right.$$
$$\left. + \frac{m}{k_BT} \left(\mathbf{CC} - \frac{C^2}{3}\mathbb{I} \right) : \nabla \mathbf{v} \right] f_{MB}(c), \tag{4.76}$$

where $\mathbf{C} = \mathbf{c} - \mathbf{v}$ is the peculiar velocity. The correction Φ is linear in $\nabla \ln T$ and $\nabla \mathbf{v}$, allowing separation of the equation into one for each contribution,

$$I[A(C)\mathbf{C}] = - \left(\frac{mC^2}{2k_BT} - \frac{5}{2} \right) \mathbf{C}\widehat{f}_{MB}(C), \tag{4.77}$$

$$I\left[B(C)\left(\mathbf{CC} - \frac{C^2}{3}\mathbb{I} \right) \right] = -\frac{m}{k_BT}\left(\mathbf{CC} - \frac{C^2}{3}\mathbb{I} \right)\widehat{f}_{MB}(C), \tag{4.78}$$

where we have used the isotropy of the linear operator to write,

$$\Phi = -\frac{1}{n}A(C)\mathbf{C} \cdot \nabla \ln T - \frac{1}{n}B(C)\left(\mathbf{CC} - \frac{C^2}{3}\mathbb{I} \right) : \nabla \mathbf{v}. \tag{4.79}$$

The vectorial part is analogous to what we did in Chapter 3, and for the tensorial part we have used that the right-hand side (and thus the unknown) is a symmetric traceless tensor. The linear equations (4.77) and (4.78) depend on the specific interaction potential, and for the next step we can consider them to be solved. The x–y component of the second equation has already been found when computing the viscosity (4.61), where it was shown how the variational principle allows one to obtain the solutions.

Finally, at second order in ϵ, computing the conservation equations and using all previous results gives

$$\frac{\partial \rho}{\partial t_2} = 0, \quad \rho \frac{\partial \mathbf{v}}{\partial t_2} = -\nabla \cdot \mathbb{P}_1, \quad \frac{3}{2}k_B\rho \frac{\partial T}{\partial t_2} = -\nabla \cdot \mathbf{q}_1, \tag{4.80}$$

where the fluxes are computed using the distribution $h_1 = f_{MB}\Phi$. By symmetry, the heat flux will have a contribution coming from ∇T and the stress tensor one from $\nabla \mathbf{v}$. Explicitly,

$$\mathbb{P}_1 = -\eta \left[(\nabla \mathbf{v}) + (\nabla \mathbf{v})^T - \frac{2}{3} (\nabla \cdot \mathbf{v}) \mathbb{I} \right], \qquad \mathbf{q}_1 = -\kappa \nabla T, \qquad (4.81)$$

with the transport coefficients,

$$\eta = \frac{k_B T}{10} \left[B \left(\mathbf{CC} - \frac{C^2}{3} \mathbb{I} \right), B \left(\mathbf{CC} - \frac{C^2}{3} \mathbb{I} \right) \right], \qquad \kappa = \frac{k_B}{3} [A\mathbf{C}, A\mathbf{C}].$$
$$(4.82)$$

It is obtained that, on the slow scales, the dissipative fluxes appear, having the form of Newton's viscous law and the Fourier law for heat flux. The Chapman–Enskog method also gives the viscosity and thermal conductivity, which are computed by solving the linear equations that involve the interatomic interaction potential (4.77) and (4.78). Finally, the total time derivative for the conserved fields is obtained by summing the derivatives for all scales. The result is the Navier–Stokes equations.

For a hard sphere gas, the viscosity to first order in Sonine polynomials is the same as that found using the linear response formalism, while the thermal conductivity to the same order is

$$\kappa_0^{HS} = \frac{75 k_B}{64 D^2} \sqrt{\frac{k_B T}{\pi m}}. \qquad (4.83)$$

Using eqns (4.70) and (4.83) we can compute the Prandlt number for the hard sphere model, $\mathrm{Pr} = c_p \eta / \kappa = 2/3$. This result is in agreement with the experimental values (Table 1.1), improving the value obtained using the mean free path model ($\mathrm{Pr}_{mfp} = 5/3$).

The Chapman–Enskog method shows that, in a gas, the dominant contribution to the distribution function is the local Maxwellian, and that the time evolution depends on the time scales, which are clearly differentiated. On the fast scale, corresponding to molecular collisions, the distribution is unchanged. This is so because collisions conserve mass, momentum, and energy; therefore the parameters that characterise the Maxwellian—density, average velocity, and temperature—do not change. On the intermediate scale, the evolution is dominated by the rapid flux associated with momentum transfer expressed by pressure. These fluxes are reversible, giving rise to the Euler fluid equations. Finally, there is a slower temporal scale on which the irreversible fluxes appear; these are the diffusive transport of momentum and energy, corresponding to viscous stress and heat flux. The resulting dynamic equations for the Maxwellian parameters are the Navier–Stokes fluid equations, which describe the evolution of the gas toward the final thermodynamic equilibrium state with no fluxes.

4.8   Dense gases

4.8.1   The Enskog model for hard sphere gases

The Boltzmann kinetic equation was derived for dilute gases in the Boltzmann–Grad limit. In 1921, Enskog proposed an extension for dense hard sphere gases using hypotheses similar to those of Boltzmann. It is not a systematic derivation, and contrary to the dilute case, it cannot be derived from the Liouville formalism. In spite of being based on phenomenological arguments, the Enskog equation is widely used because of its simplicity and excellent predictive capability.

Enskog considers only the hard sphere model for which, even at high concentrations, collisions always take place as a sequence of binary encounters; that is, with probability zero three particles can collide. It is exact, hence, to describe the evolution of the distribution function in terms of direct and inverse binary collisions (Fig. 4.12). In hard sphere collisions, the centres of the particles are at a finite distance D. Hence, the second particle in the collision terms must be located at $\mathbf{r}_1 = \mathbf{r} \pm \mathbf{n}$, where $\mathbf{n} = D\hat{\mathbf{n}}$ is the vector that joins the sphere centres at collision. The upper sign is for the direct collision and the lower for the inverse one. Under these considerations, the collision integral reads,

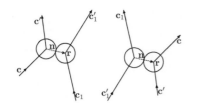

Fig. 4.12 Direct (left) and inverse (right) collisions for hard spheres. The vector $\mathbf{n} = D\hat{\mathbf{n}}$ joins the sphere centres at collision. The position of the centre of the particle with velocity $\mathbf{c}$ is $\mathbf{r}$, and the partner is located at $\mathbf{r}_1 = \mathbf{r} + \mathbf{n}$ for the direct collision and $\mathbf{r}_1 = \mathbf{r} - \mathbf{n}$ for the inverse collision.

$$J[f](\mathbf{c}) = D^2 \int \Big[f^{(2)}(\mathbf{r}, \mathbf{c}', \mathbf{r} - \mathbf{n}, \mathbf{c}'_1)$$
$$-f^{(2)}(\mathbf{r}, \mathbf{c}, \mathbf{r} + \mathbf{n}, \mathbf{c}_1) \Big] (\mathbf{g} \cdot \hat{\mathbf{n}}) \Theta(\mathbf{g} \cdot \hat{\mathbf{n}})\, \mathrm{d}^2\hat{\mathbf{n}}\, \mathrm{d}^3 c_1. \quad (4.84)$$

To close this equation, we need to approximate the two-particle distribution function for the precollisional states present in the integral. In equilibrium, as discussed in Section 2.9, the two-particle distribution function can be factorised as

$$f^{(2)}_{\text{eq}}(\mathbf{r}_1, \mathbf{c}_1, \mathbf{r}_2, \mathbf{c}_2) = f^{(1)}_{\text{eq}}(\mathbf{c}_1) f^{(1)}_{\text{eq}}(\mathbf{c}_2) g^{(2)}(\mathbf{r}_1 - \mathbf{r}_2), \quad (4.85)$$

where the pair distribution function g does not depend on velocities. Defining $\chi = g(r = D)$, the pair distribution at contact, the Enskog approximation consists of extending this result to nonequilibrium states,

$$f^{(2)}(\mathbf{r}_1, \mathbf{c}_1, \mathbf{r}_2, \mathbf{c}_2)|_{|\mathbf{r}_1 - \mathbf{r}_2| = D} = \chi f(\mathbf{r}_1, \mathbf{c}_1) f(\mathbf{r}_2, \mathbf{c}_2). \quad (4.86)$$

The pair distribution function depends on the gas density, and for an inhomogeneous gas, we approximate it as evaluated at the midpoint.[23] The Enskog collision operator then reads,

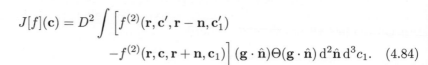

$$J_{\text{Enskog}}[f](\mathbf{c}) = D^2 \int [\chi(\mathbf{r} - \mathbf{n}/2) f(\mathbf{r}, \mathbf{c}') f(\mathbf{r} - \mathbf{n}, \mathbf{c}'_1)$$
$$-\chi(\mathbf{r} + \mathbf{n}/2) f(\mathbf{r}, \mathbf{c}) f(\mathbf{r} + \mathbf{n}, \mathbf{c}_1)] (\mathbf{g} \cdot \hat{\mathbf{n}}) \Theta(\mathbf{g} \cdot \hat{\mathbf{n}})\, \mathrm{d}^2\hat{\mathbf{n}}\, \mathrm{d}^3 c_1. \quad (4.87)$$

The Enskog equation for dense gases shares many properties with the Boltzmann equation. Notably, there is an H-theorem, it has the

[23] The pair distribution function at contact χ is an equilibrium quantity for which there is no exact expression, although some approximate theories such as the Percus–Yevick integral equation are very accurate. The Carnahan–Starling approximation gives a simple but extremely precise formula based on the virial expansion for the pressure,

$$\chi = \frac{1 - \phi/2}{(1 - \phi)^3},$$

where $\phi = \pi n D^3/6$ is the volume fraction occupied by the gas.

same collisional invariants, and the equilibrium distribution is also a Maxwellian.

When particles have a finite size, besides the kinetic transport for momentum and energy, collisional transport shows up as well.[24] In Section 2.6.3, it was shown in the context of the formal Liouville equation how collisional fluxes appear, and here we will derive them from the Enskog equation. We first derive the momentum conservation equation by multiplying it by $m\mathbf{c}$ and integrating over velocity. The left-hand side gives the same result as for the Boltzmann equation, where the kinetic stress tensor $\mathbb{P}_K = n\langle m\mathbf{c}\mathbf{c}\rangle$ appears. The right-hand side, however, does not vanish now. Momentum is conserved at the collision, but it is not locally conserved: at the collision, $\Delta\mathbf{p} = m(\mathbf{g}\cdot\hat{\mathbf{n}})\hat{\mathbf{n}}$ is transferred a finite distance D. The objective is to write this collision integral as a divergence such that the conservation equation can be written as

$$\rho\left(\frac{\partial\mathbf{v}}{\partial t} + (\mathbf{v}\cdot\nabla)\mathbf{v}\right) = -\nabla\cdot(\mathbb{P}_K + \mathbb{P}_C) + \rho\mathbf{F}/m, \qquad (4.88)$$

where $\mathbb{P}_C$ is the collisional contribution to the stress tensor that emerges from the collisional integral. First, the change of variables, $(\mathbf{c},\mathbf{c}_1) \rightarrow (\mathbf{c}',\mathbf{c}_1')$ and $\hat{\mathbf{n}} \rightarrow -\hat{\mathbf{n}}$, is performed in the inverse collision term, then the variables are relabelled back to nonprimed ones. Adding the direct contribution, the collision integral results as

$$I = D^2\int m[\mathbf{c}'-\mathbf{c}]\chi(\mathbf{r}-\mathbf{n}/2)f(\mathbf{r},\mathbf{c})f(\mathbf{r}-\mathbf{n},\mathbf{c}_1)(\mathbf{g}\cdot\hat{\mathbf{n}})\Theta(\mathbf{g}\cdot\hat{\mathbf{n}})\,\mathrm{d}^2\hat{\mathbf{n}}\,\mathrm{d}^3c\,\mathrm{d}^3c_1,$$
$$(4.89)$$

where the collision rule (4.7) gives $\mathbf{c}' - \mathbf{c} = -(\mathbf{g}\cdot\hat{\mathbf{n}})\hat{\mathbf{n}}$. Now, a new change of variables is made: $\mathbf{c}\rightarrow\mathbf{c}_1$, $\mathbf{c}_1\rightarrow\mathbf{c}$, and $\hat{\mathbf{n}}\rightarrow-\hat{\mathbf{n}}$, which allows us to write the integral as the average between the resulting expression and (4.89). Finally, the same strategy as in Section 2.6.3 to make a divergence appear is used in the factor that appears in the collision integral,

$$[\chi(\mathbf{r}-\mathbf{n}/2)f(\mathbf{r}-\mathbf{n},\mathbf{c})f(\mathbf{r},\mathbf{c}_1) - \chi(\mathbf{r}+\mathbf{n}/2)f(\mathbf{r},\mathbf{c})f(\mathbf{r}+\mathbf{n},\mathbf{c}_1)]$$
$$= -\nabla\cdot\hat{\mathbf{n}}\int_0^1\chi(\mathbf{r}-s\mathbf{n}/2)f(\mathbf{r}-s\mathbf{n},\mathbf{c})f(\mathbf{r}+(1-s)\mathbf{n},\mathbf{c}_1)\,\mathrm{d}s. \quad (4.90)$$

This finally gives for $\mathbb{P}_C$

$$\mathbb{P}_C = mD^2\int_0^1\int\chi(\mathbf{r}-s\mathbf{n}/2)f(\mathbf{r}-s\mathbf{n},\mathbf{c})f(\mathbf{r}+(1-s)\mathbf{n},\mathbf{c}_1)$$
$$(\mathbf{g}\cdot\hat{\mathbf{n}})^2\hat{\mathbf{n}}\hat{\mathbf{n}}\Theta(\mathbf{g}\cdot\hat{\mathbf{n}})\,\mathrm{d}^2\hat{\mathbf{n}}\,\mathrm{d}^3c\,\mathrm{d}^3c_1\,\mathrm{d}s. \quad (4.91)$$

This expression has the structure already discussed in Section 2.6.3 where momentum is transferred along the line that joins the two centres (see Fig. 4.13).

In equilibrium, under homogeneous conditions, the integral can be computed directly, giving

$$\mathbb{P}_C = \frac{2\pi n^2 D^3 k_B T\chi}{3}\mathbb{I}. \qquad (4.92)$$

[24]In the collision rule (4.7) for hard spheres, the transferred momentum is along the line that joins the particle centres. That is, no torque is exerted. As a consequence, as for dilute gases, angular momentum is not an independent field for dense gases.

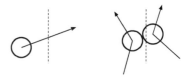

Fig. 4.13 Processes that contribute to the momentum flux. Momentum is transferred through the surface (dashed line). Kinetic transfer (left): a particle crosses the surface, transporting its own momentum. Collisional transfer (right): particles interact through the surface, exchanging momentum between them and, therefore, transferring it from one side to the other.

This gives an isotropic contribution to the pressure, which is finally

$$p = nk_\text{B}T\left(1 + \frac{2\pi nD^3\chi}{3}\right). \qquad (4.93)$$

That is, due to the finite size of the particles and the spatial correlations, the pressure is no longer that of an ideal gas; rather, density effects are present. This expression also corresponds to the virial expression that can be obtained from equilibrium statistical mechanics. A similar approach can be made to extract the collisional contribution to the heat flux,[25] which vanishes in equilibrium.

[25]See Exercise 4.16.

Now, for the nonequilibrium contributions, the Chapman–Enskog, linear response, or eigenvalue methods can be applied to obtain the transport coefficients. The finite density has two effects here. First, the transport coefficients are no longer density independent. As the hard sphere model does not have any intrinsic energy scale, the transport coefficients show the same temperature dependence as for dilute gases, namely being proportional to $\sqrt{T}$. The second effect is that, at finite densities, there is a finite bulk viscosity ζ and the stress tensor is now

$$\mathbb{P} = p\mathbb{I} - \eta\left[(\nabla\mathbf{v}) + (\nabla\mathbf{v})^T - \frac{2}{3}(\nabla\cdot\mathbf{v})\mathbb{I}\right] - \zeta(\nabla\cdot\mathbf{v})\mathbb{I}, \qquad (4.94)$$

as discussed in Appendix B (Exercise B.10). Notably, ζ vanishes for dilute gases. Using the Chapman–Enskog method, the transport coefficients are

$$\eta = \left[1 + \frac{16}{5}\phi\chi + 12.2(\phi\chi)^2\right]\frac{\eta_0}{\chi}, \qquad (4.95)$$

$$\zeta = 16\phi^2\chi\eta_0, \qquad (4.96)$$

$$\kappa = \left[1 + \frac{24}{5}\phi\chi + 12.1(\phi\chi)^2\right]\frac{\kappa_0}{\chi}. \qquad (4.97)$$

4.8.2   Virial expansion

During the second half of the twentieth century, there were serious and systematic attempts to develop a formal theory to extend the Boltzmann equation to higher densities. Based on the BBGKY hierarchy and the Bogoliubov hypothesis, which states that, in the kinetic and hydrodynamic regimes, $f^{(1)}$ is the only slow variable, a series expansion in clusters of interacting particles was performed. This led to the Choh–Uhlenbeck equation, which can be written in pictorial form as

$$\frac{\partial f}{\partial t} + \mathbf{c}\cdot\nabla f = J[f,f] + K[f,f,f] + L[f,f,f,f] + \dots, \qquad (4.98)$$

where J is the usual Boltzmann collision operator and K, L, $\dots$ are collision operators including three, four, and more particles.

Three-particle collisions mean either that the three particles are simultaneously interacting, which for hard spheres cannot take place, or that they lead to a recollision; that is, they do not consider simple sequences of binary collisions, which are already included in J. Figure 4.1 show examples of three- and four-particle collisions. The frequency of three-particle recollisions can be estimated for hard spheres by considering the geometry presented in Fig. 4.14. In a time interval Δt, the total number of collisions is an integral $f(\mathbf{c}_1)f(\mathbf{c}_2)f(\mathbf{c}_3)$ over all possible configurations. In the reference frame of $\mathbf{c}_1'$, shown at the bottom, it is easy to parameterise the collision by the free time t_{23} that particle 2 travels after the first collision before encountering particle 3. The distance travelled is $l_{23} = g_{23}t_{23}$, where g_{23} is the relative velocity. Having fixed l_{23}, the velocity $\mathbf{c}_3$ is constrained so that particle 2 must come back to collide with one. This limits the solid angle of the velocity to be within $\alpha \sim r_0/l_{23}$, where r_0 is the particle radius. Then, the differential of the velocity $\mathbf{c}_3$ is $\mathrm{d}^3 c_3 \approx c_3^{d-1}\alpha^{d-2}\,\mathrm{d}c$, where $d = 2, 3$ is the dimensionality of the space, that we leave free for the moment. The rate of three-particle recollision events is then obtained by integrating over t_{23} as $\mathrm{d}N_3/\mathrm{d}t \sim \phi^3 \int \mathrm{d}t_{23}/t_{23}^{d-1}$, where the factor ϕ^3 comes from the distribution functions.

The integral over t_{23} converges in three dimensions, and we are led to the conclusion that the ratio of the three-particle collision rate to the usual binary collision rate is $(\mathrm{d}N_3/\mathrm{d}t)/(\mathrm{d}N_2/\mathrm{d}t) \sim \phi$. This result justifies the use of the Boltzmann equation in the Boltzmann–Grad limit, and also looks promising in that it seems to give a systematic procedure for higher densities. However, in two dimensions, the integral goes as $\int \mathrm{d}t_{23}/t_{23}$, which diverges logarithmically for long times (at short times, the estimation should be improved and there is no divergence). The temporal divergence is based on the procedure of the Choh–Uhlenbeck equation that expands in small clusters of isolated particles. It was soon realised that the temporal divergence could be fixed by noting that free flights are limited by the mean free time, $\tau \sim 1/(c_{\mathrm{th}}D^{d-1}n)$ (formally this is achieved by resumming the full series). The series is now finite, but $n \log n$ terms appear. In three dimensions, the $\log n$ singularities appear in the four-particle collision term. These contributions are nonanalytic, and the Taylor-like Choh–Uhlenbeck equation was put into question. To date, there is no formal systematic procedure to build kinetic equations to arbitrary order in density.

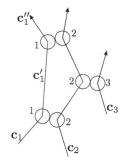

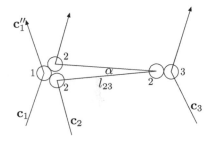

Fig. 4.14 Recollision event involving three particles with precollisional velocities $\mathbf{c}_1$, $\mathbf{c}_2$, and $\mathbf{c}_3$. Particles 1 and 2 collide twice. After the first collision, the velocity of particle 1 is $\mathbf{c}_1'$, and after the second collision $\mathbf{c}_1''$. Representation in the laboratory frame (top) and in a frame moving with velocity $\mathbf{c}_1'$ (bottom), such that particle 1 stays at repose after the first collision. Between collisions, particle 2 travels a distance l_{23}, where the opening angle is $\alpha \sim r_0/l_{23}$ for spheres of radius r_0.

4.9 Application: granular gases

Granular materials are ubiquitous in nature and many industrial processes. Matter composed of macroscopic grains appears in avalanches and other geological flows, in the processing and transport of pharmaceutical or agricultural products, or on the astrophysical scale in protoplanetary disks or the rings of Saturn. In all cases, the dominant feature is that grains are hard particles that lose energy at collisions, which is dissipated in the form of sound, vibrations, fracture, or by surface abrasion. The inelastic hard sphere model has become a prototype to describe granular materials in the rapid flow regime, where the energy injected by some mechanism is large enough to sustain dynamical states. Grains are modelled as spheres of equal diameter D and mass m, where collisions are inelastic and characterised by a restitution coefficient α ($0 \leq \alpha \leq 1$), such that the relative velocity after the collision is α times the incoming one. The collision rule that respects momentum conservation is

$$\mathbf{c}' = \mathbf{c} - \frac{1+\alpha}{2}\left[(\mathbf{c} - \mathbf{c}_1) \cdot \hat{\mathbf{n}}\right]\hat{\mathbf{n}}, \quad \mathbf{c}_1' = \mathbf{c}_1 + \frac{1+\alpha}{2}\left[(\mathbf{c} - \mathbf{c}_1) \cdot \hat{\mathbf{n}}\right]\hat{\mathbf{n}},$$

(4.99)

where $\alpha = 1$ corresponds to the elastic case and $\alpha = 0$ to completely plastic collisions. At a collision, the energy loss is

$$\Delta E = \frac{m}{2}(c'^2 + c_1'^2 - c^2 - c_1^2) = -\frac{m(1-\alpha)^2}{4}\left[(\mathbf{c} - \mathbf{c}_1) \cdot \hat{\mathbf{n}}\right]^2. \quad (4.100)$$

Energy is no longer a collisional invariant, and the collision integral for the energy equation does not vanish. Its evaluation requires the distribution function, but a good estimation can be made by assuming a Maxwellian distribution, which is an excellent approximation for quasielastic gases with $\alpha \approx 1$. In this case, the energy equation reduces to[26]

$$\frac{3}{2}mn\frac{\partial T}{\partial t} = -2\sqrt{\frac{\pi}{m}}(nD)^2(1-\alpha^2)T^{3/2}, \quad (4.101)$$

with solution $T = T_0/(1+t/t_0)^2$, known as the Haff law.[27] It describes how a homogeneous granular gas cools down. Detailed analysis of the hydrodynamic equations shows that the homogeneous cooling state becomes unstable to compressive waves of large wavelength, giving rise to clustering. The mechanism is that dense regions created by fluctuations have a higher dissipation rate and therefore cool faster than less dense regions, decreasing the pressure, leading to progressive clustering and cooling.

One method to sustain fluid states is to uniformly shear a granular medium. Consider a uniform shear flow $\mathbf{v} = V'y\hat{\mathbf{x}}$. The viscous heating term in (4.19), $\mathbb{P} : \nabla\mathbf{v} = \eta V'^2$, can compensate for the collisional dissipation, resulting in the energy balance, $2\sqrt{\pi/m}(nD)^2(1-\alpha^2)T^{3/2} = \eta V'^2$. For quasielastic grains, the viscosity is close to that of elastic hard spheres (4.70), which allows one to compute the resulting granular temperature,

$$T \sim \frac{mV'^2}{(1-\alpha)^2 n^2 D^4}, \quad (4.102)$$

known as the Bagnold scaling. Note that the temperature is not an independent variable but rather is enslaved by the velocity gradient and can be quite small for dissipative grains. As the viscosity is proportional to $\sqrt{T}$, this results in the shear stress now going as $P_{xy} \sim V'^2$ and hence the response becomes nonlinear.

It is also possible to write a Boltzmann or Enskog equation for granular gases in order to perform a more systematic study of them. The

[26]In granular gases, as the particles are macroscopic and the thermodynamic temperature plays no role, it is usual to define the granular temperature as the velocity fluctuations about the mean velocity, without the Boltzmann constant, i.e.

$$\frac{3}{2}T = \left\langle \frac{m(\mathbf{c} - \mathbf{v})^2}{2} \right\rangle.$$

[27]See Exercise 4.17.

procedure to derive it follows the same route as presented in this chapter, with one precaution. Energy dissipation implies that collisions are not reversible and the velocities for the inverse collision $\mathbf{c}^*$ and $\mathbf{c}_1^*$ are not the outgoing velocities of the direct collision (4.99). It is easy to verify that, rather, the inverse velocities are obtained by making the substitution $\alpha \to 1/\alpha$ in the mentioned collision rule. The kinematics give $\mathrm{d}^3 c^* \, \mathrm{d}^3 c_1^* = \mathrm{d}^3 c \, \mathrm{d}^3 c_1 / \alpha$ and $|\mathbf{g}^* \cdot \hat{\mathbf{n}}| = |\mathbf{g} \cdot \hat{\mathbf{n}}|/\alpha$, resulting in the following Boltzmann equation:

$$\frac{\partial f}{\partial t} + \mathbf{c} \cdot \frac{\partial f}{\partial \mathbf{r}} + \frac{\mathbf{F}}{m} \cdot \frac{\partial f}{\partial \mathbf{c}} = D^2 \int \left[\frac{f^* f_1^*}{\alpha^2} - f f_1 \right] (\mathbf{g} \cdot \hat{\mathbf{n}}) \Theta(\mathbf{g} \cdot \hat{\mathbf{n}}) \, \mathrm{d}^2 \hat{\mathbf{n}} \, \mathrm{d}^3 c_1. \tag{4.103}$$

4.10 Application: the expanding universe

In the early universe, kinetic processes took place that involve the production and equilibration of the primordial particles. In Section 7.9 we will apply the tools of kinetic theory to the description of the quark–gluon plasma, which is believed to be present in the early phases of the universe. Besides collisions and reactions, the expansion of the universe played an important role on how particles behave. In the simplest approach, consider the Friedmann–Lemaître–Robertson–Walker model for cosmology, where the geometry of the universe is characterised by a unique scale factor $a(t)$[28]. In cosmology it is usual to assume that at large scales, the space is homogeneous and isotropic. Hence, the particles are described by the four-momentum distribution, $f(p^\mu, t) = f(p^0, |\mathbf{p}|, t)$, where $p^\mu = (p^0, p^i)$ and $p^0 = E = \sqrt{\mathbf{p}^2 c^2 + m^2 c^4}$. Note that as a result of the expansion, the momentum $\mathbf{p}$, which can be interpreted as the inverse of the wavelength, decreases with time. Considering the scale factor, the total number of particles in a unitary comoving volume is $N = a^3 \int f \, \mathrm{d}^3 p$. It can be proved that, in the absence of collisions, the kinetic equation that conserves the number of particles is

$$\frac{\partial f}{\partial t} - \frac{\dot{a}}{a} |\mathbf{p}| \frac{\partial f}{\partial |\mathbf{p}|} = 0. \tag{4.104}$$

Collisions are included as a Boltzmann term resulting in an H-theorem. This implies that in equilibrium the logarithm of the distribution function is a linear combination of the four-momentum components, which are the collisional invariants. By isotropy we obtain

$$f_{\mathrm{eq}} = e^{-\alpha(t) - \beta(t) E(p)}. \tag{4.105}$$

For massless particles $E = cp$, substituting (4.105) into (4.104) gives

$$\frac{\dot{\alpha}}{\beta} = cp \left(1 - \frac{\dot{a}}{a} \frac{\beta}{\beta} \right), \tag{4.106}$$

with solution

$$\alpha(t) = \text{cst.}, \qquad\qquad \beta(t) = \text{cst.} \times a(t). \tag{4.107}$$

[28]The scale factor relates the proper distance d between cosmological objects at any arbitrary time t to their distance d_0 at a reference time t_0 by $d(t) = a(t) d_0$. The Hubble parameter that measures the expansion of the universe is $H \equiv \dot{a}(t)/a(t)$. At the present time $H_0 = 1/(14.4 \text{ billion years})$.

Interpreting β as the temperature inverse we obtain that the particles cool down as the universe expands, in agreement with the Big Bang model and the astronomical observations.

The inhomogeneous case can be studied perturbatively, with predictions that show an excellent agreements with the measurements of the cosmic microwave background and the large-scale structures.

Further reading

The Boltzmann and Enskog equations are presented in many books on statistical mechanics and kinetic theory. In particular, detailed descriptions can be found in Chapman and Cowling (1970), Ferziger and Kaper (1972), Lebowitz and Montroll (1983), Cercignani (2013), Huang (1987), or Liboff (2003).

In Lebowitz (1993) there is an interesting discussion on the irreversibility problem. The oscillator chain was originally proposed by Rubin in 1960 to derive Brownian motion. The recurrence time was computed by Kac in 1939. The model is analysed in detail in Zwanzig (2001) in the context of the origin of irreversibility. The proper way of taking the thermodynamic limit to observe irreversibility is discussed in Evans and Morris (1990) in the context of transport processes; for strongly nonequilibrium conditions, the transport coefficients depend on wavevector and frequency, and the usual coefficients are obtained by taking $\lim_{\omega \to 0} \lim_{\mathbf{k} \to 0}$ in that order, that is, first the thermodynamic limit, and later the limit of long times.

The review articles by Cohen (1993) and Ernst (1998) give a detailed presentation of the Bogoliubov–Choh–Uhlenbeck approach and some perspectives for future developments, as well as the formal derivation of the Boltzmann equation. Analysis of the hydrodynamic modes is found in Hansen and McDonald (1990), as well as the description of the equilibrium properties of the hard sphere gas.

Finally, kinetic theory applied to granular gases can be found in Brilliantov and Pöschel (2010) and Andreotti *et al.* (2013) and the application to cosmological problems is found in Dodelson (2003) and Bernstein (2004).

Exercises

(4.1) **Inverse collision for hard spheres.** Show that, for the hard sphere collision rule (4.7), the Jacobian of the transformation is indeed 1.

(4.2) **Three-particle collisions.** Estimate the number of collisions Λ_3, per unit time, that take place between three particles. Consider a gas composed of N particles in a volume $\mathcal{V}$, where the interaction potential extends to a distance r_0, which is anyway much smaller than the system length. Compare the three-particle collision rate, $\nu_3 = \Lambda_3/N$, with the usual two-particle collision rate ν and show that, in the Boltzmann–Grad limit, it is vanishingly small.

(4.3) **Postcollisional correlations.** The objective is to show that postcollisional correlations appear in a gas. Consider a gas of hard spheres in a steady uniform shear flow, $\mathbf{v} = V'y\hat{\mathbf{x}}$. The gas is close to equilibrium, but the shear flow produces a deviation that is characterised by the Grad distribution, studied in Exercise 1.12,

$$f(\mathbf{c}) = f_0 \left[1 + \frac{mP_{xy}}{nk_{\mathrm{B}}^2 T^2}(c_x - v_x)c_y \right],$$

where

$$f_0 = n \left(\frac{m}{2\pi k_B T} \right)^{3/2} e^{-mC^2/2k_B T}$$

is the local Maxwellian distribution and $\mathbf{C} = \mathbf{c} - \mathbf{v}(\mathbf{r}, t)$ is the peculiar velocity. For simplicity, we will place ourselves in the comoving frame, where $\mathbf{C} = \mathbf{c}$.

When two particles collide, their velocities come from the distribution f, and the outgoing velocities $\mathbf{c}'$ and $\mathbf{c}'_1$ are obtained from the collision rule (4.7).

(a) Show that the joint distribution $F(\mathbf{c}', \mathbf{c}'_1)$ is not factorisable.

(b) To measure the degree of correlation, compute $\langle \mathbf{c}' \cdot \mathbf{c}'_1 \rangle$. Show that it vanishes for $V' = 0$, implying that in equilibrium, collisions do not create correlations.

(4.4) **Anisotropic interactions.** The existence of the inverse collision was justified using symmetries under temporal and spatial inversion, a condition that was satisfied with central conservative forces. As an example, to show that if this condition is violated inverse collisions do not necessarily exist, consider an abstract model where the collision in the centre-of-mass frame is represented by a non-symmetrical hard body (for example, a triangle in Fig. 4.15). Convince yourself that, for the direct collision shown in the figure, there is no associated inverse collision.

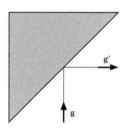

Fig. 4.15 Anisotropic collision in the centre of mass.

Investigate how a Boltzmann equation can be written in this case, when inverse collisions are not guaranteed to exist.

(4.5) **Hard sphere collision rule.** The objective is to derive the collision rule for hard spheres of equal mass (4.7). Consider two spheres that at collision are joined by the vector $D\hat{\mathbf{n}}$, where D is the contact distance. Because for hard spheres the collision is instantaneous, it can be modelled by a momentum transfer, $\Delta\mathbf{p}$. The spheres are smooth, implying that they do not exert tangential forces and the

momentum transfer is parallel to $\hat{\mathbf{n}}$. Imposing energy conservation, derive the collision rule.

(4.6) **Energy conservation equation.** Derive eqn (4.19). For this purpose, multiply the Boltzmann equation by $\psi = mc^2/2$ and integrate over $\mathbf{c}$. Use the mass (4.16) and momentum equations (4.18) to simplify the result.

(4.7) **Properties of the linear collision operator.** Show that the linear collision operator I is Hermitian and semidefinite positive. Follow the same procedure as in Chapter 3.

(4.8) **Eigenvalue for the shear mode.** Perform the perturbation analysis of the eigenvalue associated with the shear mode. Show that, to second order in k, it goes as $\lambda = \nu k^2$ and obtain an equation for ν. Consider that $\mathbf{k} = k\hat{\mathbf{x}}$, and $\Phi_0 = c_y$ as the unperturbed eigenfunction.

(4.9) **Shear viscosity for the hard sphere gas.** Consider a hard sphere gas for particles with mass m and diameter D. Using the collision rule (4.7), compute the integrals in (4.69) needed to obtain the shear viscosity of the gas.

(4.10) **Shear viscosity for the hard disc gas.** Consider a gas in two dimensions, where particles are modelled as hard discs. Obtain the viscosity for this system.

(4.11) **BGK model.** Show that the BGK model has the same collisional invariants as the Boltzmann equation and that an H-theorem exists.

(4.12) **Linear BGK model.** Deduce the expression (4.48) for the linear BGK operator. Show that the linear collision operator in the BGK model conserves mass, momentum, and energy.

(4.13) **Viscosity in the BGK model.** Compute the shear viscosity for a gas described by the BGK model by solving eqn (4.61).

(4.14) **Thermal equilibration of a mixture.** Consider a mixture of two gases, modelled as hard spheres of equal diameter D and mass m. Both gases have the same density n, but initially their temperatures are different: T_{10} and T_{20}. The objective is to show that, in the final state, the temperatures are equal.

(a) Write the Boltzmann equation for the distribution of both gases, using the rule (4.7) for all collisions.

(b) Derive the equation for the temperature of each gas. In this case, energy is not a conserved quantity due to the cross collisions, and a collision integral remains.

(c) Evaluate the collision integrals assuming that both gases are well described at all times by Maxwellian distributions with vanishing mean velocity.

(d) Analyse the resulting equations and show that the system evolves to a state with equal temperatures.

(4.15) **Chemical reaction.** Consider a chemical reaction where each time two atoms collide they react and the product disappears from the system $(A + A \to \emptyset)$. This can easily be achieved on a solid surface where the product desorbs. Model the atoms as hard spheres.

(a) Write the kinetic equation for the distribution function.

(b) Derive the equations for the mass density, velocity, and temperature fields. Note that now these are not conserved quantities and the collision integrals do not vanish.

(c) Evaluate the collision integrals assuming that the distribution is a Maxwellian with vanishing average velocity, closing the hydrodynamic equations. Solve them for an initial homogeneous condition of $n(\mathbf{r}, t = 0) = n_0$ and $T(\mathbf{r}, t = 0) = T_0$.

(4.16) **Collisional heat flux in dense gases.** Obtain the collisional contribution to the heat flux $\mathbf{q}_C$, performing an analysis similar to that applied to obtain the collisional stress tensor (4.91). Compute the collision integral for $\varphi = mc^2/2$ in the Enskog equation and render it as a divergence, $-\nabla \cdot \mathbf{q}$. Show that it has the structure discussed in Section 2.6.3, where the third contribution (potential transport) is not present for hard spheres.

(4.17) **Granular gases.** Derive the equation for the temperature in the inelastic hard sphere model and evaluate the collision integral assuming a Maxwellian distribution. Solve the resulting equation for an initial homogeneous condition of $T(\mathbf{r}, t = 0) = T_0$.

(4.18) **Conservation of the number of particles in the expanding universe.** Show that the kinetic equation (4.104) conserves the total number of particles in a unitary comoving volume, $N = a^3 \int f \, d^3p$.

(4.19) **Massive particles in the expanding universe.** Consider massive particles with $E = mc^2 + |\mathbf{p}|^2/2m$. Show that for large m, eqn (4.104) admits the equilibrium solution (4.105). Find the evolution of α and β. Finally, show that for finite m no equilibrium solution is found.

Brownian motion

5

5.1 The Brownian phenomenon

In 1828, the botanist R. Brown placed pollen grains on top of a water surface. He observed the pollen grains under a microscope and noted that some small particles were ejected from the grains. To his surprise, these particles permanently showed erratic motion. After discarding the first naïve explanations based on the activity of the particles for biological reasons, he was led to the conclusion that this was not an effect of living organisms and that the effect takes place for inert particles as well.

The explanation for the observed motion was given by A. Einstein in 1905. He argued that the liquid that suspends the particles is made of atoms and molecules, which are ultimately discrete. So, the liquid interacts with the particles through discrete collision events. Even though the average force exerted by the molecules on the particle vanishes, it is a fluctuating quantity. This fluctuating force explains the observed erratic motion. This idea, which we will develop in subsequent sections, was revolutionary at the time in its use of the concept of atoms and molecules, being one of the first applications of early kinetic theory. The theory of Einstein successfully explained the experimental observations, constituting one of the most important proofs of the atomic theory and the ideas introduced by Boltzmann, i.e. that the kinetic description of atomic motion is able to explain irreversible phenomena.

The particles emitted from the pollen grains are microscopic, with a typical size of one micron. Therefore, there is a dramatic contrast between the sizes and masses of the particles and the water molecules. This contrast implies that the molecules are fast while the particles exhibit comparatively slow motion. Generally, we call Brownian particles those that are large enough to exhibit this important separation of temporal and spatial scales, while still being small enough to respond to the effects of the fluctuating forces. In the next sections, we will develop a kinetic model for Brownian particles considering this separation of spatial and temporal scales.

In the long term, Brownian particles move away from their original positions as a result of the random kicks they experience. There is no preferred direction of motion, and the particles perform what is called a random walk or Brownian motion (see Fig. 5.1). One peculiar property of this kind of motion is that the mean distance of a particle from its

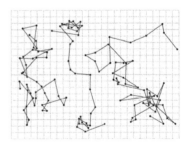

Fig. 5.1 Reproduced from the book of Jean Baptiste Perrin (1991), *Les Atomes*. Tracing of three colloidal particles of radius 0.53 µm suspended on water, as seen under a microscope. Successive positions are recorded every 30 s, joined by straight-line segments for clarity. The mesh size is 3.2 µm.

initial position grows as the square root of time, a characteristic of a diffusion process.

5.2   Derivation of the Fokker–Planck equation

Here we derive a kinetic equation—called the Fokker–Planck equation—that describes the evolution of a collection of Brownian particles. In the model, these particles interact with the suspending fluid, but no interaction among particles will be considered for the moment. To make the derivation simpler, no other external forces will be included at first. Because of their size, the Brownian particles are classical, and therefore, their motion is described by Newton's law, $M\dot{\mathbf{c}} = \mathbf{F}$, where $\mathbf{F}$ is the total force exerted by the fluid on the particle. As a first approximation, we can model the force as a viscous friction in the form, $\mathbf{F} = -\gamma\mathbf{c}$, where γ is the friction coefficient, which for a sphere immersed in a fluid takes the value $\gamma = 6\pi\eta R$, where η is the fluid viscosity and R is the particle radius. The resulting equation is relaxational, and the velocity vanishes on the time scale $\tau_{\mathrm{B}} = M/\gamma$. Obviously, this cannot be the full model for a Brownian particle, because it does not present the observed erratic motion. The key point here is to notice that the viscous drag expression is a phenomenological law. When we consider the individual molecules that constitute the fluid, they will collide with a Brownian particle randomly, and the total force they exert will be given by the drag force only on average. On top of that, there are continuous fluctuations of the force produced by the molecules. We model this by saying that the total force is

$$\mathbf{F} = -\gamma\mathbf{c} + \mathbf{F}_{\mathrm{fluct}}. \tag{5.1}$$

That is, the total force is separated into a deterministic part, which is obtained from the macroscopic laws (the hydrodynamic equations in this case), and a fluctuating part, which originates from the corpuscular nature of the fluid. This part takes into account all the microscopic degrees of freedom that were eliminated from the simple hydrodynamic description that gives the average force.

The fluctuating force takes into account the variations of the total force due to the individual collisions of the molecules with the particle. A collision takes a time that can be estimated as $\tau_{\mathrm{col}} = d/c_{\mathrm{th}}$, where $d \sim 1\,\text{Å}$ is the typical size of a molecule (or similarly the mean free path in a liquid) and $c_{\mathrm{th}} \sim 526\,\text{m/s}$ is the thermal velocity of the molecules (see Section 1.2). This gives $\tau_{\mathrm{col}} \sim 1.9 \times 10^{-13}\,\text{s}$. On the other hand, a Brownian particle is of microscopic size, and the relaxation time can be estimated as $\tau_{\mathrm{B}} \sim 7.5 \times 10^{-7}\,\text{s}$.[1]

As anticipated in Section 5.1, the size contrast between the particles and molecules implies an enormous contrast in time scales ($\tau_{\mathrm{B}}/\tau_{\mathrm{col}} \sim 10^6$). We will not pretend to give a comprehensive description of the

---

[1] To estimate this value we use that typical Brownian particles have mass densities similar to that of the water in which they are suspended. Then, if $R = 1\,\mu\text{m}$, the mass is $M = 1.3 \times 10^{-14}\,\text{kg}$ and the friction coefficient is $\gamma = 1.7 \times 10^{-8}\,\text{kg s}^{-1}$, with the viscosity of water being $\eta = 8.9 \times 10^{-4}\,\text{Pa s}$.

water molecules, because this is impossible in practice and, more importantly, because this level of detail is irrelevant. Rather, here we are interested in modelling the motion of the Brownian particles noting that, in their scale, the fluctuating force evolves on very short time scales. In this context, being called a noise. The Fokker–Planck approach is a probabilistic description where the action of the fluctuating force on the particles is treated statistically.

The starting point is the master equation (2.57) (see Section 2.10) for the evolution of the distribution function $f(\mathbf{r}, \mathbf{c}, t)$ in a time interval Δt,

$$
\begin{aligned}
f(\mathbf{r}, \mathbf{c}, t + \Delta t) = {} & f(\mathbf{r} - \mathbf{c}\Delta t, \mathbf{c}, t) - \int f(\mathbf{c}, t) P(\mathbf{c}, \mathbf{c}', \Delta t)\, \mathrm{d}^3 c' \\
& + \int f(\mathbf{c}', t) P(\mathbf{c}', \mathbf{c}, \Delta t)\, \mathrm{d}^3 c', \quad (5.2)
\end{aligned}
$$

where $P(\mathbf{c}, \mathbf{c}', \Delta t)$ is the probability density that, in Δt, the particle changes its velocity from $\mathbf{c}$ to $\mathbf{c}'$.

In the time step Δt, if we neglect fluctuations, the new velocity would be with full certitude $\mathbf{c}' = \mathbf{c} - \gamma \mathbf{c} \Delta t / M$. In the language of the master equation, this means that the kernel is represented by a Dirac delta function in the deterministic evolution. Fluctuations in the force will imply a broadening of this sharp, delta-peaked kernel. However, note that, due to the separation of mass scales, the effect of the fluctuating forces is small and the kernel is still peaked around the deterministic value. To exploit this feature, we define the jump kernel $K(\mathbf{c}, \Delta \mathbf{c}, \Delta t) = P(\mathbf{c}, \mathbf{c} + \Delta \mathbf{c}, \Delta t)$, which measures the transition probability for a jump of size $\Delta \mathbf{c}$ (see Fig. 5.2). This jump kernel is peaked in its second argument, but it is a smooth function of its first argument, a property that will allow us to make a Taylor expansion of it. In terms of this jump kernel, the master equation reads,

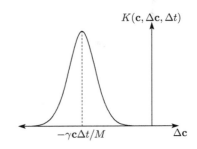

$K(\mathbf{c}, \Delta \mathbf{c}, \Delta t)$

$-\gamma \mathbf{c}\Delta t / M$ $\Delta \mathbf{c}$

Fig. 5.2 Jump kernel $K(\mathbf{c}, \Delta \mathbf{c}, \Delta t)$, which is peaked around the average velocity change, $-\gamma \mathbf{c}\Delta t / M$.

$$
\begin{aligned}
\frac{\partial f(\mathbf{r}, \mathbf{c})}{\partial t} + \mathbf{c} \cdot \nabla f(\mathbf{r}, \mathbf{c}) = {} & \frac{1}{\Delta t} \left[- \int K(\mathbf{c}, \mathbf{u}, \Delta t) f(\mathbf{r}, \mathbf{c})\, \mathrm{d}^3 u \right. \\
& \left. + \int K(\mathbf{c} - \mathbf{u}, \mathbf{u}, \Delta t) f(\mathbf{r}, \mathbf{c} - \mathbf{u})\, \mathrm{d}^3 u \right]. \quad (5.3)
\end{aligned}
$$

The second integral can be expanded for small jumps $\mathbf{u}$, resulting in

$$
\frac{\partial f}{\partial t} + \mathbf{c} \cdot \nabla f = -\frac{\partial}{\partial \mathbf{c}} \cdot (\mathbf{a}_1 f) + \frac{1}{2} \frac{\partial^2}{\partial \mathbf{c}^2} (a_2 f), \quad (5.4)
$$

where we define the moments

$$
\mathbf{a}_1(\mathbf{c}) = \lim_{\Delta t \to 0} \frac{1}{\Delta t} \int K(\mathbf{c}, \mathbf{u}, \Delta t) \mathbf{u}\, \mathrm{d}^3 u, \quad (5.5)
$$

$$
a_2(\mathbf{c}) = \lim_{\Delta t \to 0} \frac{1}{3\Delta t} \int K(\mathbf{c}, \mathbf{u}, \Delta t) u^2\, \mathrm{d}^3 u, \quad (5.6)
$$

which are related, respectively, to the average jump size and the fluctuations around this average value.[2] The first moment can be writ-

[2]Note that indeed, in the limit, $a_2 = \lim_{\Delta t \to 0} \frac{\langle u^2 \rangle}{3\Delta t} = \lim_{\Delta t \to 0} \frac{[\langle u^2 \rangle - \langle u \rangle^2]}{3\Delta t}$.

ten as $\mathbf{a}_1 = \langle \mathbf{u}/\Delta t \rangle$, i.e. the average acceleration due to the action of the molecules, which is given by the Stokes friction. Then, $\mathbf{a}_1 = -\gamma \mathbf{c}/M$. Now, if we add at this stage an external force $\mathbf{F}_{\text{ext}}$, we obtain $\mathbf{a}_1 = (\mathbf{F}_{\text{ext}} - \gamma \mathbf{c})/M$.

The term associated with the second moment of K takes account of the fluctuations around the average velocity change. At small velocities of the Brownian particles, the second moment can be expanded as $a_2(\mathbf{c}) = a_{20} + a_{22}c^2 + \ldots$, where only even powers in c are allowed by symmetry.[3] We can convince ourselves that a_{20} is not vanishing because, even if the Brownian particle is at rest, there are fluctuating forces. Keeping only the first nonvanishing term, we finally obtain

$$\frac{\partial f}{\partial t} + \mathbf{c} \cdot \nabla f + \frac{\mathbf{F}_{\text{ext}}}{M} \cdot \frac{\partial f}{\partial \mathbf{c}} = \frac{\gamma}{M} \frac{\partial}{\partial \mathbf{c}} \cdot (\mathbf{c}f) + \frac{\Gamma}{2M^2} \frac{\partial^2 f}{\partial c^2}, \quad (5.7)$$

where we have adopted the usual notation of placing all terms associated with the fluid on the right-hand side. Finally, we have defined $\Gamma = M^2 a_{20}$, the noise intensity. This equation is one of a series of Fokker–Planck equations that we will find in this chapter. In this specific form, for the distribution function of position and velocity, it is called the Kramers equation.

If the term $\frac{\gamma}{M} \frac{\partial}{\partial \mathbf{c}} \cdot (\mathbf{c}f)$ is moved to the left, it gives the left-hand side of the first BBGKY equation for particles moving under the action of the average force, consistent with the interpretation of $\mathbf{a}_1$ as taking account of the deterministic part of the evolution. The right-hand side retains only the fluctuating terms. This form of the Fokker–Planck equation helps us to write, by inspection, the kinetic equations for other systems that present noise. All deterministic terms are written as a usual Liouville or BBGKY equation, and the fluctuating forces are added in the form of second derivatives on the right-hand side.[4]

5.3 Equilibrium solutions

5.3.1 Homogeneous equilibrium solution and the fluctuation–dissipation relation

Let us first consider the simple case of Brownian particles in the absence of any external force, placed homogeneously in a fluid. The distribution will remain homogeneous, and therefore the distribution function depends only on velocity and time. In Section 5.5, we will show that, under these conditions, the velocity distribution relaxes uniformly to a stationary stable distribution, which is the purpose of our analysis here. The stationary solution satisfies the equation

$$\frac{\Gamma}{2M^2} \frac{\partial^2 f}{\partial c^2} + \frac{\gamma}{M} \frac{\partial}{\partial \mathbf{c}} \cdot (\mathbf{c}f) = 0. \quad (5.8)$$

Imposing the condition that the distribution function should vanish for infinitely large velocities, one can directly find the solution,[5]

[3]The fluctuating force is created by the molecules in the fluid. Therefore, the relevant velocity scale against which one must compare is their thermal speed or, similarly, the sound speed in the liquid. We will see in the next section that the speed of the Brownian particles is always smaller than this scale.

[4]See, for example, Section 6.6.3.

[5]See Exercise 5.9.

$$f_0(\mathbf{c}) = A e^{-M\gamma c^2/\Gamma}, \tag{5.9}$$

where A is a normalisation constant.

We have obtained that, after a long time, the velocity of the Brownian particle reaches a Gaussian distribution with vanishing average and a finite time-independent variance. However, we know that a particle that is in contact for long times with a system of many degrees of freedom will reach thermal equilibrium at the same temperature as the bath; that is, the particle should reach a Maxwellian distribution with the same temperature as the liquid. This is precisely what we have obtained, but the width of the Gaussian distribution must be fixed to $\gamma/\Gamma = k_\mathrm{B}T/2$. We therefore obtain the relation

$$\Gamma = 2k_\mathrm{B}T\gamma. \tag{5.10}$$

We have found the fluctuation–dissipation relation, indicating that the intensity of the force fluctuations Γ is related to the intensity of the dissipative forces γ. In general, this establishes that, for finite temperatures, if there is any dissipative mechanism (the viscous friction in this case), there should also be a mechanism that injects energy via fluctuations.[6] If there were no fluctuations, the friction would dissipate all the energy and the asymptotic state would not correspond to an equilibrium where the thermal energy is finite. The intensity of the fluctuations is such that, once the system reaches equilibrium, the fluctuations are in accordance with its temperature.

Moreover, this relation between γ and Γ is not accidental or a pure result of imposing that the final state is in equilibrium. Indeed, both terms of the force (friction and fluctuation) are due to the same molecules; they have the same origin. In the process of modelling the Brownian motion, we have performed an arbitrary separation of the total force into two terms: one for which we know a simple expression and another, complicated term that we have modelled statistically. What the fluctuation–dissipation relation does is to reunify both terms.[7]

In summary, using the fluctuation–dissipation relation, it is found that the stationary solution is the Maxwell–Boltzmann distribution,

$$f_0(\mathbf{c}) = n_0 \left(\frac{m}{2\pi k_\mathrm{B}T}\right)^{3/2} e^{-mc^2/2k_\mathrm{B}T}. \tag{5.11}$$

Furthermore, the Fokker–Planck equation will be written from now on with Γ replaced by its fluctuation–dissipation value, i.e.

$$\frac{\partial f}{\partial t} + \mathbf{c} \cdot \nabla f + \frac{\mathbf{F}_\mathrm{ext}}{M} \cdot \frac{\partial f}{\partial \mathbf{c}} = \frac{\gamma}{M} \frac{\partial}{\partial \mathbf{c}} \cdot (\mathbf{c}f) + \frac{\gamma k_\mathrm{B}T}{M^2} \frac{\partial^2 f}{\partial c^2}. \tag{5.12}$$

5.3.2   Equilibrium solution under external potentials

Consider a particle that moves in three dimensions under the influence of an external force $\mathbf{F}_\mathrm{ext}$, that derives from a potential $\mathbf{F}_\mathrm{ext} = -\nabla U$,

[6]For an energetic analysis, see Exercise 5.6.

[7]Brownian motion is not the only case where a fluctuation–dissipation relation appears. Indeed, in any case where there is deterministic dissipative dynamics, there should be noise terms. This is the case for example with the viscous hydrodynamic equations, electric transport through resistive media, thermal plasmas, colloidal systems, molecular motors, etc. We also found this before in Section 3.8.4 for the Lorentz gas.

and that is immersed in a fluid at thermal equilibrium. The stationary distribution function $f(\mathbf{r}, \mathbf{c})$ obeys the Fokker–Planck equation, which in this case takes the form,

$$\mathbf{c} \cdot \nabla f - \frac{\nabla U}{M} \cdot \frac{\partial f}{\partial \mathbf{c}} = \frac{\gamma}{M} \frac{\partial}{\partial \mathbf{c}} \cdot (\mathbf{c} f) + \frac{\gamma k_B T}{M^2} \frac{\partial^2 f}{\partial \mathbf{c}^2}. \qquad (5.13)$$

[8]See Exercise 5.15.

We look for the stationary state predicted by this equation using the method of separation of variables. For notational simplicity, we consider the one-dimensional case, but the extension to three dimensions is direct.[8] We propose a solution of the form, $f(r, c) = R(r)C(c)$, which is substituted into the Fokker–Planck equation (5.13); after division by cRC, this results in

$$\frac{R'}{R} - \frac{U'}{M} \frac{C'}{cC} = \frac{\gamma}{McC} \left(C + cC' + \frac{k_B T}{M} C'' \right). \qquad (5.14)$$

The right-hand side depends only on c, while the left-hand side depends on both r and c. Compatibility is achieved if both sides are equal to a constant. This is possible if C'/cC is also a constant, which we denote by $-B$. The solution of this last equation is, up to a multiplicative constant, $C = e^{-Bc^2/2}$. Now we impose that the right-hand side must be independent of c, from which we obtain $B = Mk_B T$. The right-hand side vanishes identically, and the equation for R simplifies to $R'/R + U'/k_B T = 0$, with solution $R = Ae^{-U/k_B T}$. In summary, the equilibrium distribution is

$$f_{eq}(\mathbf{r}, \mathbf{c}) = A \exp \left[-\frac{Mc^2/2 + U(\mathbf{r})}{k_B T} \right], \qquad (5.15)$$

which corresponds to the stationary solution of a particle in contact with a thermal bath found in equilibrium statistical mechanics. Here, A is a normalisation constant. Note that the fluctuation–dissipation relation that gives the intensity of the force fluctuations, derived in the context of a homogeneous system, correctly takes account of the properties of the system in the presence of forces that derive from a potential.

Several observations can be made regarding the stationary solution. First, we note that it does not depend on the friction coefficient γ. Indeed, the equilibrium solution only depends on the conservative forces but not on the dissipative forces. We will see in the next chapter that, when nonequilibrium forces are considered (for example time-dependent forces or forces that do not derive from a potential), the stationary solution will then depend explicitly on γ. The friction coefficient is related to the rate at which the system reaches equilibrium, as measured by the time scale τ_B, but once in equilibrium, the effect of γ seems to disappear. This is only apparent though, because it is a result of Γ being proportional to γ; that is, varying the friction coefficient (for example by changing the liquid to glycerin) will also change the noise intensity, resulting in the same Maxwellian distribution.

For the Maxwellian velocity distribution, the variance of the velocity is $\langle c^2 \rangle = k_B T/M$. Therefore, the typical velocities of the Brownian particles are $c \sim \sqrt{k_{\mathrm{B}}T/M}$, being much smaller than the molecular velocities $c_{\mathrm{th}} \sim \sqrt{k_{\mathrm{B}}T/m}$ thanks to the mass contrast $M \gg m$. This justifies a posteriori the expansion for small velocities in the noise intensity a_2 made in Section 5.2.

5.4 Mobility under external fields

Consider the case of Brownian particles placed in a constant external field, a situation that can be achieved if, for example, the particles are charged and there is an external electric field, or if the particles are denser than the fluid and the whole system is placed under gravity. If there were boundaries in the direction of the force such that particles cannot move forever, the total force that results from the combination of the external field and the boundary condition can be written as derived from a potential, and the stationary distribution is given by the equilibrium expression found in the previous section. However, if there are no boundaries—or if on the time scale of the experiment the particles do not reach the boundaries—it is not possible to find an equilibrium solution. Indeed, imagine that the external force points in the positive x direction: $\mathbf{F}_{\mathrm{ext}} = F_{\mathrm{ext}}\hat{\mathbf{x}}$. The associated potential is $U = -F_{\mathrm{ext}}x$, resulting in an equilibrium distribution (5.15) that diverges at $x \to \infty$, being therefore impossible to normalise. The solution consists of imagining that there is a particle reservoir located at $x = -\infty$ that injects particles steadily, which are driven by the force in the positive direction. This injection occurs at a constant rate to produce a uniform distribution. We therefore look for solutions $f_0(\mathbf{c})$ of the Fokker–Planck equation (5.12) that are stationary and uniform,

$$\frac{\mathbf{F}_{\mathrm{ext}}}{M} \cdot \frac{\partial f_0}{\partial \mathbf{c}} = \frac{\gamma}{M}\frac{\partial}{\partial \mathbf{c}} \cdot (\mathbf{c}f_0) + \frac{\gamma k_{\mathrm{B}}T}{M^2}\frac{\partial^2 f_0}{\partial c^2}. \tag{5.16}$$

It is easy to verify that the solution to this equation is

$$f_0(\mathbf{c}) = n\left(\frac{M}{2\pi k_{\mathrm{B}}T}\right)^{3/2} \exp\left[-\frac{M(\mathbf{c} - \mathbf{F}_{\mathrm{ext}}/\gamma)^2}{k_{\mathrm{B}}T}\right]. \tag{5.17}$$

This is a Maxwellian distribution with density n, temperature T equal to that of the medium, and a nonvanishing average velocity, $\langle \mathbf{c} \rangle = \mathbf{F}/\gamma$. The Brownian particles, hence, move homogeneously in the direction of the force at an average velocity given by the deterministic force. The fluctuating part of the force produces a dispersion of the velocities with variance $\langle \Delta c^2 \rangle = k_{\mathrm{B}}T/M$.

Although this Maxwellian looks like an equilibrium solution, it corresponds to a nonequilibrium condition. Two features are characteristic of this regime. First, there is a uniform particle flux $\mathbf{J} = n\mathbf{F}_{\mathrm{ext}}/\gamma$, which can only be achieved if the system is in contact with two particle reservoirs: one that permanently injects particles, while the other absorbs them. The system is therefore not isolated but rather in contact with baths at different chemical potentials. The second feature is that the external field is permanently doing work in the system. The injected power per unit volume is

$$P = \int f_0(\mathbf{c})\mathbf{c} \cdot \mathbf{F} \, \mathrm{d}^3 c = F^2/\gamma > 0. \qquad (5.18)$$

The system is in a steady state thanks to the energy dissipation produced by the friction force. There is a permanent energy flux that follows a path from the external field to the particles and then to the molecules of the fluid. On average, no energy flux goes in the opposite direction.[9] This directed route violates the detailed balance condition, necessary to reach equilibrium, which states that, if one process takes place in a system, the inverse process must also be possible. The regime we have found is normally called a nonequilibrium steady state (NESS). Note, finally, that we cannot appeal to Galilean invariance to claim that the NESS solution (5.17) is equivalent to the equilibrium solution (5.11); here the suspending fluid fixes a preferential reference frame that makes the two solutions different.

[9]During short periods, it is possible, however, for energy to go from the molecules to the particle, transforming heat into work. The probability for this to happen decreases exponentially with the observation time, and it is the subject of study of the so-called fluctuation theorems. For long observation times, this process becomes impossible, in agreement with the second law of thermodynamics.

5.5 Long-time dynamics: diffusion

Up to now we have analysed the cases of equilibrium solutions and the nonequilibrium case of a homogeneous distribution of Brownian particles in an external field. We now aim to consider the general case of a distribution of particles that are nonhomogeneous in space and which evolve in time. Stated like this, the problem is extremely complex to study in general, but most importantly we will see that it is not relevant to find the most general solution. Indeed, we have discussed two time scales in the motion of Brownian particles: first the time between individual molecular collisions with the particles τ_{col}, and second the relaxation time of a particle in the fluid τ_{B}, with $\tau_{\mathrm{B}} \gg \tau_{\mathrm{col}}$. When particles are distributed inhomogeneously in space, a new time scale appears, much longer than the two mentioned above. In this section, we will find the temporal evolution of a system of Brownian particles on this time scale. We will find that, in this regime, the particle distribution obeys a diffusion equation, and we will be able to derive the diffusion coefficient D from the statistical properties of the fluid force on the particle. Specifically, we will derive the Einstein relation,

$$D = \frac{k_{\mathrm{B}}T}{\gamma}. \qquad (5.19)$$

Because of its importance and also because in different books diverse approaches are used for its derivation, in the next sections, four methods will be used to derive the diffusion equation. These are finally equivalent, but the spirit, rigour, and level of approximation vary.

5.5.1 Solution of the diffusion equation

Before performing the derivation of the diffusion equation from the Fokker–Planck model, we first derive some properties of the diffusion equation that will help with its derivation by recognising some of its features.

Consider a system of many particles that diffuse in a medium. The concentration of particles is dilute, hence they do not interact, and their density $n(\mathbf{r},t)$ is described by the diffusion equation,

$$\frac{\partial n}{\partial t} = D\nabla^2 n, \qquad (5.20)$$

where D is the diffusion coefficient. Initially, all particles are in a small region, which we model using a Dirac delta function as $n(\mathbf{r},0) = N_0\delta(\mathbf{r})$, where N_0 is the total number of particles.

To solve the equation, we use the Fourier transform method with

$$n(\mathbf{r},t) = \int e^{-i\mathbf{k}\cdot\mathbf{r}}\,\widetilde{n}(\mathbf{k},t)\,\mathrm{d}^3k, \qquad (5.21)$$

where $\widetilde{n}(\mathbf{k},t)$ are the amplitudes of the Fourier modes, with initial condition $\widetilde{n}(\mathbf{k},0) = N_0/(2\pi)^3$. Substitution into (5.20) results in

$$\dot{\widetilde{n}}(\mathbf{k}) = -Dk^2\,\widetilde{n}(\mathbf{k}). \qquad (5.22)$$

The eigenvalues of the diffusion operator are therefore Dk^2. This is a characteristic feature of a diffusion process, and indeed, whenever we identify relaxation eigenvalues which are quadratic in the wavevector, we will refer to the process as a diffusion process, as we did in Chapter 3. In the analysis of the Boltzmann equation in Chapter 4, viscosity was also associated with this kind of eigenvalue, corresponding to transverse diffusion of momentum.

The solution, using the initial condition for each mode, is $\widetilde{n}(\mathbf{k},t) = N_0 e^{-Dk^2 t}/(2\pi)^3$, which we reinsert into (5.21) to obtain an integral expression for the particle density that can be evaluated by a change of variable as

$$n(\mathbf{r},t) = \frac{N_0}{(2\pi)^3}\int e^{-Dt\left(\mathbf{k}+\frac{i\mathbf{r}}{2Dt}\right)^2 - \frac{r^2}{4Dt}}\,\mathrm{d}^3k = \frac{N_0}{(4\pi Dt)^{3/2}}\,e^{-r^2/4Dt}. \quad (5.23)$$

The initial, peaked distribution broadens in the form of a Gaussian distribution with a width that grows over time (Fig. 5.3). The mean square displacement can be obtained directly from the distribution as

$$\langle r^2(t)\rangle = 6Dt, \qquad (5.24)$$

which grows linearly with time.[10] This relation allowed Einstein to derive a formula that gives the diffusion coefficient in terms of the particle displacements as

$$D = \lim_{t\to\infty}\frac{\langle r^2(t)\rangle}{6t} \qquad (5.25)$$

or equivalently

$$D = \lim_{t\to\infty}\frac{1}{6}\frac{\mathrm{d}}{\mathrm{d}t}\langle r^2(t)\rangle, \qquad (5.26)$$

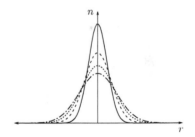

Fig. 5.3 Gaussian density distributions resulting from diffusion of an initially concentrated Dirac delta distribution—represented by an arrow. The width of the distribution grows as the square root of time, where the different curves in the figure are equidistant in time.

[10]In some systems, the mean square displacement grows with a different power of time, corresponding to sub- or superdiffusive processes that are not described by the diffusion equation.

which is usually used in experiments or computed simulations because it presents faster convergence. These expressions are indeed more general than they appear to be. They were derived by considering the initial condition of particles concentrated in a small region. However, we can extend the analysis to the case of a homogeneous system in equilibrium and tag a few of them, located in a small region. These tagged particles will mix with others by diffusion, but otherwise the system will remain homogeneous throughout the process. The density of the tagged particles will follow the Gaussian profiles, and the Einstein formulas can be equally applied. In summary, the Einstein formulas can be applied in equilibrium to a set of tagged particles or even to a single tagged particle.

5.5.2 Green–Kubo expression

Equation (5.26) in the case of a single tagged particle allows the derivation of another expression for the diffusion coefficient. Noting that the particle displacement is the integral of the velocity over time, we get

$$\langle r^2(t) \rangle = \int_0^t \int_0^t \langle \mathbf{c}(s_1) \cdot \mathbf{c}(s_2) \rangle \, \mathrm{d}s_1 \, \mathrm{d}s_2 = \int_0^t \int_0^t \mathrm{d}s_1 \, \mathrm{d}s_2 \, C(s_1 - s_2),$$

$$(5.27)$$

where the velocity correlation function, $C(s_1 - s_2) = \langle \mathbf{c}(s_1) \cdot \mathbf{c}(s_2) \rangle$, is computed as an average in equilibrium; it is an even function that depends only on the time lapse $s_1 - s_2$. To compute the diffusion coefficient, we must take the temporal derivative of this double integral. To do so we compute $\langle r^2(t + \Delta t) \rangle$, as shown in Fig. 5.4,

$$\langle r^2(t + \Delta t) \rangle = \langle r^2(t) \rangle + 2\Delta t \int_0^t C(t - s) \, \mathrm{d}s + \mathcal{O}(\Delta t^2).$$

$$(5.28)$$

Therefore, the diffusion coefficient is obtained from the following Green–Kubo expression:

$$D = \frac{1}{3} \int_0^\infty C(s) \, \mathrm{d}s,$$

$$(5.29)$$

which expresses the diffusion coefficient in terms of the equilibrium velocity correlation function. If the integral is finite, this means that velocities decorrelate after a finite time, losing their memory. After this, a random walk is obtained.

In the case of a Brownian particle, we can use this expression to estimate the value of the diffusion coefficient. Indeed, in equilibrium, we have that $C(0) = \langle c^2 \rangle = k_\mathrm{B}T/M$ and the particle velocities decorrelate on the time scale τ_B. We can, therefore, model the correlation function as a simple decaying function,

$$C(s) \sim \frac{k_\mathrm{B}T}{M} e^{-s/\tau_\mathrm{B}}.$$

$$(5.30)$$

Inserting this expression into the Green–Kubo formula results in

$$D \sim \frac{k_\mathrm{B}T\tau_\mathrm{B}}{M} \sim \frac{k_\mathrm{B}T}{\gamma}.$$

$$(5.31)$$

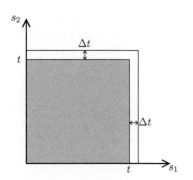

Fig. 5.4 Representation of the process to obtain the diffusion coefficient.

We will see in the next sections that this result is actually the exact expression for the diffusion coefficient.

5.5.3 Coarse-grained master equation

On time scales shorter than τ_B, a Brownian particle shows ballistic motion due to its inertia. This ballistic motion decorrelates on times of the order of τ_B, after which such particles adopt random directions—null on average if there is no external force—and with velocities following the Maxwellian distribution. On a longer time scale, the particles perform a random walk, originating from a succession of these uncorrelated ballistic trajectories. Consider a macroscopic time interval Δt, much longer than τ_B. If the particles are subject to an external force $\mathbf{F}_{ext}$, which remains constant during this time interval, they will displace on average a distance $\Delta\mathbf{r} = \mathbf{F}_{ext}\Delta t/\gamma$. However, the mentioned random walk implies that the displacements show a dispersion, and there is a kernel, $K(\mathbf{r},\Delta\mathbf{r},\Delta t)$, which gives the probability density that a particle experiences a displacement $\Delta\mathbf{r}$ in the time interval if it started at $\mathbf{r}$ (Fig. 5.5). It is then possible to write down the master equation for the particle density $n(\mathbf{r},t)$,

$$n(\mathbf{r},t+\Delta t) = n(\mathbf{r},t) + \int n(\mathbf{r}-\Delta\mathbf{r},t)K(\mathbf{r}-\Delta\mathbf{r},\Delta\mathbf{r},\Delta t)\,d^3\Delta r$$
$$- \int n(\mathbf{r},t)K(\mathbf{r},\Delta\mathbf{r},\Delta t)\,d^3\Delta r. \quad (5.32)$$

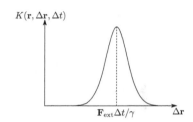

Fig. 5.5 Displacement kernel giving the probability that, in a time interval Δt, the particle moves $\Delta\mathbf{r}$. In the presence of an external force, the average displacement is $\mathbf{F}_{ext}\Delta t/\gamma$.

The kernel is smooth in its first argument, while it is sharp in its second. For small Δt, a Taylor expansion is performed in the first integral, resulting in

$$\frac{\partial n}{\partial t} = -\nabla\cdot(\mathbf{a}_1 n) + \frac{a_2}{2}\nabla^2 n. \quad (5.33)$$

The first moment is

$$\mathbf{a}_1 = \lim_{\Delta t\to 0}\frac{1}{\Delta t}\int K(\mathbf{r},\Delta\mathbf{r},\Delta t)\Delta\mathbf{r}\,d^3\Delta r = \lim_{\Delta t\to 0}\left\langle\frac{\Delta\mathbf{r}}{\Delta t}\right\rangle, \quad (5.34)$$

which is hence identified as the average velocity, $\mathbf{a}_1 = \mathbf{F}_{ext}/\gamma$. For the second moment, assuming isotropy, the Taylor expansion gives

$$a_2 = \lim_{\Delta t\to 0}\frac{1}{3\Delta t}\int K(\mathbf{r},\Delta\mathbf{r},\Delta t)\Delta\mathbf{r}^2\,d^3\Delta r = \lim_{\Delta t\to 0}\left\langle\frac{\Delta\mathbf{r}^2}{3\Delta t}\right\rangle, \quad (5.35)$$

which is precisely, up to a factor of 2, the Einstein definition of the diffusion coefficient. Hence, $a_2 = 2D$. In summary, on the coarse-grained scale, the density obeys the diffusion equation,

$$\frac{\partial n}{\partial t} = -\nabla\cdot(\mathbf{F}_{ext}n/\gamma) + D\nabla^2 n. \quad (5.36)$$

To obtain the value of the diffusion coefficient we adapt the idea originally presented by Einstein in 1905. Consider an ensemble of particles

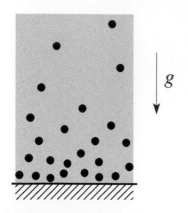

Fig. 5.6 Ensemble of suspended Brownian particles in a gravitational field, the setup used to derive the value of the diffusion coefficient.

suspended in a fluid limited from below by a solid surface, under the influence of the gravitational acceleration **g** (Fig. 5.6).

The diffusion equation (5.36) reads,

$$\frac{\partial n}{\partial t} = -\nabla \cdot \mathbf{J},\tag{5.37}$$

where the total flux is $\mathbf{J} = nM\mathbf{g}/\gamma - D\nabla n$. In the steady state, imposing that the flux at infinity must vanish results in a flux that must vanish everywhere. Hence, $nM\mathbf{g}/\gamma = D\nabla n$, with solution

$$n(\mathbf{r}) = n_0 e^{-Mgz/D\gamma}.\tag{5.38}$$

On the other hand, the external force is conservative, so we can use the equilibrium solution (5.15) of the Fokker–Planck equation, with $U = Mgz$. Integrating out the velocities, the equilibrium density profile is

$$n_{\text{eq}}(\mathbf{r}) = n_0 e^{-Mgz/k_{\text{B}}T},\tag{5.39}$$

the known barometric law, where n_0 is a normalisation constant. This result should be compared with eqn (5.38) to deduce that

$$D = \frac{k_{\text{B}}T}{\gamma}.\tag{5.40}$$

The Einstein relation $D = k_{\text{B}}T/\gamma$ is remarkable in many ways. It is the first expression in the history of physics where the Boltzmann constant k_{B} appears alone, not accompanied by another molecular quantity, in an expression for a macroscopic property—the diffusion coefficient.[11] Before this, for example in the Maxwellian distribution, it is the combination k_{B}/m that appears, which has macroscopic values. Note also that the Einstein expression mixes the three scales at play here: the macroscopic diffusion D, the microscopic friction of the Brownian particle γ, and the molecular agitation of the molecules $k_{\text{B}}T$.

5.5.4 Eigenvalue analysis

Let us analyse the Fokker–Planck equation (5.12) in the absence of external forces. By linearity, we consider spatial Fourier modes such that the distribution can be written as $f(\mathbf{r}, \mathbf{c}, t) = f_{\mathbf{k}}(\mathbf{c}, t)e^{i\mathbf{k}\cdot\mathbf{r}}$ and we have

$$\frac{\partial f_{\mathbf{k}}}{\partial t} = -i\mathbf{k}\cdot\mathbf{c}f_{\mathbf{k}} + \frac{\gamma}{M}\frac{\partial}{\partial \mathbf{c}}\cdot(\mathbf{c}f_{\mathbf{k}}) + \frac{\gamma k_{\text{B}}T}{M^2}\frac{\partial^2 f_{\mathbf{k}}}{\partial c^2},$$
$$= -L_{\mathbf{k}}f_{\mathbf{k}},\tag{5.41}$$

which defines the linear operator $L_{\mathbf{k}}$. The evolution of the distribution function is given by the eigenvalues of $L_{\mathbf{k}}$, which we will study in the limit of small wavevector (large wavelength).

We can proceed similarly to Section 3.8.3, where we dealt with the Lorentz model. Although they describe different physical situations, they share many properties. The most important is that, for vanishing

[11]This expression was derived by Einstein in 1905 in an article where he was looking for a method to measure the Avogadro number, republished in Einstein (1956).

wavevector, the linear operator is semidefinite positive and there is one nondegenerate vanishing eigenvalue, which is also associated with the conservation of the number of particles (mass conservation). There is also a finite gap to the first nonvanishing eigenvalue, which is on the order of $\tau_B^{-1} = M/\gamma$. Therefore, for small wavevectors, a perturbation analysis indicates that the eigenvalues will remain ordered and the smallest will correspond to the perturbation of the null eigenvalue. At long times, only the contribution of the smallest eigenvalue will remain, and therefore, if we are interested in the long-time dynamics, this is the only eigenvalue that is relevant. A perturbation analysis in **k** analogous to that applied in Section 3.8.3 shows that there is no contribution linear in the wavevector and the first contribution is quadratic,[12]

[12]See Exercise 5.4.

$$\lambda = \frac{k_B T}{\gamma} k^2. \tag{5.42}$$

This dependence on **k** is reminiscent of a diffusive process, with a diffusion coefficient $D = k_B T/\gamma$, and the particle density obeys the diffusion equation,

$$\frac{\partial n}{\partial t} = D\nabla^2 n. \tag{5.43}$$

We have found a new temporal scale $\tau_{\text{diff}} = 1/(Dk^2)$ that, for sufficiently small wavevector, is much longer than the other time scales described above. It depends on the length scale on which we observe and can be made as large as required.[13]

[13]For a particle with the characteristics of that of note [1], the diffusion coefficient is $D \sim 2.5 \times 10^{-13}\,\text{m}^2/\text{s} \sim 2.5 \times 10^{-1}\,\mu\text{m}^2/\text{s}$. The diffusion time depends on the considered scale; for one micron, $k = 2\pi/(1\,\mu\text{m})$ and $\tau_{\text{diff}} \sim 0.1\,\text{s}$, while for one millimetre $k = 2\pi/(1\,\text{mm})$ and $\tau_{\text{diff}} \sim 10^5\,\text{s}$. That is, pure diffusion would take more than one day to diffuse $1\,\text{mm}$. Under normal conditions, turbulence will enhance diffusion, making the process much faster.

5.5.5   Chapman–Enskog method

A more formal approach to deriving the diffusion equation is to perform a Chapman–Enskog treatment of the Fokker–Planck equation (5.12). The analysis here is very similar to that for the Lorentz model, worked out in detail in Section 3.9. As in the Lorentz case, the kinetic equation is linear and the mass is the only conserved quantity. Also, in the Fokker–Planck equation we have

$$\int \left(\frac{\gamma}{M} \frac{\partial}{\partial \mathbf{c}} \cdot (\mathbf{c}f) + \frac{\gamma k_B T}{M^2} \frac{\partial^2 f}{\partial c^2} \right) \mathrm{d}^3 c = 0, \tag{5.44}$$

which is the analogue to the collisional invariant relation.

We then proceed to formally expand the kinetic equation for spatial gradients and external forces that are small and consider multiple temporal scales. It is found that, at the dominant order, the distribution function is a Maxwellian with temperature fixed by the fluid and density n depending on space and time. The density field does not evolve on the fast temporal scales (on the order of τ_B), whereas on the slow scales it obeys the equation

$$\frac{\partial n}{\partial t} + \nabla \cdot (\mathbf{F}_{\text{ext}} n/\gamma) = D\nabla^2 n. \tag{5.45}$$

Furthermore, the Chapman–Enskog method gives the Einstein relation for the diffusion coefficient.[14]

[14]See Exercise 5.5.

5.5.6   Boundary conditions

There are two natural boundary conditions for the diffusion equation: reflecting and absorbing. A reflecting or hard boundary is impossible for the particles to cross; the flux over it must vanish. Then, if $\hat{\mathbf{n}}$ is the vector normal to the boundary, one must impose that

$$\mathbf{J} \cdot \hat{\mathbf{n}} = (n\mathbf{F}_{\text{ext}}/\gamma - D\nabla n) \cdot \hat{\mathbf{n}} = 0. \tag{5.46}$$

For the absorbing boundary condition, each particle that arrives is instantly absorbed, effectively disappearing from the system. This boundary condition models, for example, a reacting surface with high reactivity or a free surface from which particles can escape. As all particles disappear instantly, the associated condition is

$$n = 0. \tag{5.47}$$

5.6   Early relaxation

So far, we have analysed the long-time dynamics of Brownian particles, showing that the velocities follow a Maxwellian distribution and that the density obeys a diffusion equation. At short times, the dynamics is far more complex because the spatial and velocity distributions mix in a nontrivial way. To get a glimpse of this rapid dynamics, we will study an initially homogeneous distribution of particles which do not have a Maxwellian distribution and follow their relaxation to the equilibrium distribution. No external forces are considered either. The Fokker–Planck equation (5.12) for this case is

$$\frac{\partial f}{\partial t} = \frac{1}{\tau_{\text{B}}} \left[\frac{\partial}{\partial \mathbf{u}} \cdot (\mathbf{u}f) + \frac{1}{2}\frac{\partial^2 f}{\partial u^2} \right], \tag{5.48}$$

where the dimensionless velocity $\mathbf{u} = \mathbf{c}/\sqrt{2k_{\text{B}}T/M}$ has been defined. This equation trivially separates into each Cartesian component, corresponding to the eigenvalue equation of the Hermite polynomials—or equivalently to the Schrödinger equation for the harmonic oscillator. The full three-dimensional solution is

$$f(\mathbf{u}, t) = \sum_{n_x, n_y, n_z \geq 0} A_{n_x, n_y, n_z} e^{-u^2} H_{n_x}(u_x) H_{n_y}(u_y) H_{n_z}(u_z)$$
$$\times e^{-(n_x + n_y + n_z)t/\tau_B} \tag{5.49}$$

in terms of the Hermite polynomials H_n and the initial condition coefficients A. The distribution hence relaxes monotonically, and only the $\mathbf{n} = \mathbf{0}$ mode survives in the long term, corresponding to the Maxwellian distribution. All perturbations relax on the characteristic time scale τ_{B}; that is, an ensemble of Brownian particles adopts the Maxwellian in only a few microseconds.

5.7 Rotational diffusion

Colloidal particles immersed in a fluid are not only subject to translational Brownian motion. Also, the molecules of the fluid exert microscopic torques that generate rotation of the particles. This phenomenon is particularly relevant for objects that are not spherically symmetric (e.g. rods suspended in a fluid), particles that act on the system in a directional way (e.g. light-emitting quantum dots), or self-propelled swimmers that have a director, such as bacteria. For simplicity, we will consider spherically symmetric particles characterised by a single moment of inertia I. The equation of motion is then simply $I\dot{\boldsymbol{\omega}} = \boldsymbol{\tau}$, where $\boldsymbol{\tau}$ is the total torque and $\boldsymbol{\omega}$ is the angular velocity. The deterministic torque is given by the viscous friction $\boldsymbol{\tau} = -\mu\boldsymbol{\omega}$, which tends to slow down the particle. For spheres of radius R, $\mu = 8\pi\eta R^3$ (Landau and Lifshitz, 1987). Again, the stopping time is very short for small—micron sized—particles.[15] The fluctuating torque generates motion and, as we will see, rotational diffusion.

Considering the deterministic torque and the fluctuating torque, we are led to a Fokker–Planck equation for $f(\boldsymbol{\omega}, t)$,

$$\frac{\partial f}{\partial t} = \frac{\partial}{\partial \boldsymbol{\omega}} \cdot \left(\frac{\mu\boldsymbol{\omega}}{I} f\right) + \frac{\Gamma_r}{2I^2} \frac{\partial^2 f}{\partial \omega^2}, \qquad (5.50)$$

where Γ_r is the intensity of the rotational noise. The stationary distribution is easily found to be

$$f_0(\boldsymbol{\omega}) = A e^{-I\mu\omega^2/\Gamma_r}, \qquad (5.51)$$

which should be compared with the expected distribution from equilibrium statistical mechanics,

$$f_{\text{eq}}(\boldsymbol{\omega}) = A e^{-I\omega^2/2k_{\text{B}}T}, \qquad (5.52)$$

resulting in the fluctuation–dissipation relation,

$$\Gamma_r = 2k_{\text{B}}T\mu. \qquad (5.53)$$

We can now study the dynamics of the orientation of the particle. Its director $\hat{\mathbf{n}}$ obeys the equation $\dot{\hat{\mathbf{n}}} = \boldsymbol{\omega} \times \hat{\mathbf{n}}$, and we could write down a Kramers equation for the joint distribution $f(\hat{\mathbf{n}}, \boldsymbol{\omega}, t)$. Similarly to the case of translation, on long temporal scales, a diffusion equation is obtained for the orientation distribution function $f(\hat{\mathbf{n}}, t)$,

$$\frac{\partial f}{\partial t} = D_r \frac{\partial^2}{\partial \hat{\mathbf{n}}^2} f, \qquad (5.54)$$

where the rotational diffusion coefficient is given by the Einstein relation

$$D_r = \frac{k_{\text{B}}T}{\mu}. \qquad (5.55)$$

The Laplacian in the unitary vector $\hat{\mathbf{n}}$ requires a special explanation. In three dimensions, the director can be represented by two spherical

[15]For a particle with the characteristics of that of note [1], the rotational relaxation time is $\tau_\omega = I/(8\pi\eta R^3) \approx 2 \times 10^{-7}$ s, where $I = 2MR^2/5$ for solid spheres.

angles, and accordingly, the operator is the angular part of the Laplacian in spherical coordinates. In two dimensions, where a single angle is needed, $\mathbf{n} = (\cos\theta, \sin\theta)$, we have simply $\frac{\partial^2}{\partial\hat{\mathbf{n}}^2} f = \frac{\partial^2}{\partial\theta^2} f$.

[16]For a particle with the characteristics of that of note [1], the rotational coherence time is $\tau_r \sim 5.3\,\mathrm{s}$.

Dimensionally, the rotational diffusion coefficient has units of inverse time. We can interpret $\tau_r = D_r^{-1}$ as the time it takes the Brownian particle to adopt a direction completely uncorrelated with the original one. This gives the rotational coherence time.[16]

5.8   Application: light diffusion

In geometrical optics we learn that light propagates in straight lines at constant speed c. The direction of motion is specified at the source, i.e. by the initial conditions. In the presence of a slightly inhomogeneous medium, for example in air with a fraction of humidity or with temperature fluctuations, the refractive index is also inhomogeneous. As a consequence, light rays suffer weak scattering processes in random and uncorrelated directions. By weak we mean that the direction of motion is only slightly changed. In spite of being random and uncorrelated, these changes in direction accumulate, resulting in the rotational diffusion of the direction. This variation of the refractive index also modifies the propagation speed; in the case where a short light pulse is emitted, this effect will produce longitudinal diffusion of the pulse, elongating it. Here, we study the case of a persistent ray emitted by one source, where longitudinal diffusion becomes irrelevant.

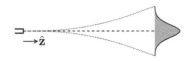

Fig. 5.7 A point source emits a ray along the $\hat{\mathbf{z}}$ direction. Rotational diffusion of the direction of motion provokes an effective widening of the ray.

The distribution of the light intensity at $\mathbf{r}$ pointing in the direction $\hat{\mathbf{n}}$, $f(\mathbf{r}, \hat{\mathbf{n}}, t)$—equal to the number of photons for monochromatic light—obeys the kinetic equation,

$$\frac{\partial f}{\partial t} + c\hat{\mathbf{n}} \cdot \nabla f = D_r \nabla_{\hat{\mathbf{n}}}^2 f, \qquad (5.56)$$

where D_r is the effective rotational diffusion coefficient that takes into account all the effects of inhomogeneity. This diffusion is not thermal, and hence it does not satisfy the Einstein relation.

Consider the situation of a ray of light that is steadily emitted in a specific direction, say $\hat{\mathbf{z}}$, from a point source. Rotational diffusion will slowly change the direction of motion of the ray, provoking its effective widening (Fig. 5.7). For times shorter than the rotational coherence time $\tau_r = D_r^{-1}$, the ray mainly persists in its original direction, with small deviations. Correspondingly, there is a correlation length $L = c/D_r$ such that, for $z \ll L$, the director has only suffered slight perturbations: $\hat{\mathbf{n}} = \hat{\mathbf{z}} + \delta\hat{\mathbf{n}}$, with $|\delta\hat{\mathbf{n}}| \ll 1$. Under these conditions, the unit sphere is locally flat (Fig. 5.8) and we can write $\hat{\mathbf{n}} = (n_x, n_y, 1)$, where the components $n_{x/y}$ simply perform Cartesian—not angular—diffusion. In summary, in the steady state, the kinetic equation reads,

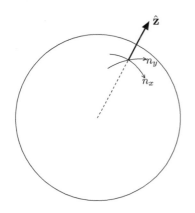

Fig. 5.8 The light director $\hat{\mathbf{n}}$ diffuses on the unit sphere. However, for short times, as the initial condition is concentrated along a specific direction, the diffusion takes place on a locally flat surface.

$$\frac{\partial f}{\partial z} = -n_x \frac{\partial f}{\partial x} - n_y \frac{\partial f}{\partial y} + \frac{1}{L}\left(\frac{\partial^2 f}{\partial n_x^2} + \frac{\partial^2 f}{\partial n_y^2}\right). \qquad (5.57)$$

This equation is similar to the Kramers equation, with the z coordinate playing the role of time, but without the friction term. The associated initial condition is $f(x, y, z = 0, n_x, n_y) = N_0 \delta(x)\delta(y)\delta(n_x)\delta(n_y)$. The dependences on x and y decouple, allowing us to concentrate on only one coordinate. There are several methods to solve eqn (5.57); for example, an explicit solution can be found by assuming that the distribution is Gaussian in x and n_x, or in Chapter 10, the associated Langevin equation is used in Excercise 10.7. Here, for illustrative purposes, we apply a moment method, which is described in detail in Chapter 10. We multiply the kinetic equation by n_x^2, $n_x x$, and x^2 and then integrate the result over x and n_x. This results in a series of differential equations in z for the moments: $\frac{\mathrm{d}}{\mathrm{d}z}\langle n_x^2 \rangle = 2/L$, $\frac{\mathrm{d}}{\mathrm{d}z}\langle n_x x \rangle = \langle n_x^2 \rangle$, and $\frac{\mathrm{d}}{\mathrm{d}z}\langle x^2 \rangle = 2\langle n_x x \rangle$, with solutions

$$\langle n_x^2 \rangle = 2z/L, \qquad \langle n_x x \rangle = z^2/L, \qquad \langle x^2 \rangle = 2z^3/3L. \qquad (5.58)$$

That is, the director n_x shows normal diffusion, but the width of the ray grows as $\Delta x \sim z^{3/2}$. Also, correlations between x and n_x build up, which is expected because rays that are above the central ray ($x > 0$) most probably also have a positive component of the normal vector. Finally, it is easily verified that the solution for the distribution function is

$$f(\mathbf{r}, \hat{\mathbf{n}}) = \frac{N_0 L \sqrt{3}}{2\pi z^2} e^{-\left[3Lr_\perp^2/z^3 - 3Lr_\perp \cdot \hat{\mathbf{n}}_\perp/z^2 + Ln_\perp^2/z\right]}, \qquad (5.59)$$

which can be guessed by assuming Gaussianity and the expressions for the moments. Here, $\mathbf{r}_\perp = (x, y)$ and $\hat{\mathbf{n}}_\perp = (n_x, n_y)$ are the perpendicular components.

5.9 Application: bacterial alignment

Some bacteria, such as *Escherichia coli*, are rod-like microorganisms that swim in water by the propulsive thrust generated by their flagella. At low concentrations, they can swim steadily in roughly straight lines, interrupted by random reorientation events. At higher concentrations, they interact with collision-like events, where their bodies come into contact. It has been observed that, because of their elongated shape, collisions tend to align the swimmers (Fig. 5.9). It has been proposed (Aranson and Tsimring, 2005) that this could be a mechanism responsible for the early development of swarm phases in dense bacterial suspensions. Consider a homogeneous bacterial suspension moving in two dimensions (for example over a surface), where each swimmer is labelled by its orientation θ. The relevant distribution function is $f(\theta)$, which gives the number of bacteria per unit area pointing towards θ. Two processes govern the dynamics of f. First, bacteria are microscopic objects and, hence, are subject to rotational diffusion, with a coefficient D_r. Second, we have the binary collision events. For simplicity, we consider the alignment to be total; that is, when two swimmers meet, both emerge with their average angle. The first process is described by a Fokker–Planck

Fig. 5.9 Alignment process of colliding bacteria.

term, and the second by a Boltzmann collision operator,

$$\frac{\partial f(\theta)}{\partial t} = D_r \frac{\partial^2 f(\theta)}{\partial \theta^2} + \nu \int_{-\pi}^{\pi} [f(\theta + \phi/2)f(\theta - \phi/2) - f(\theta)f(\theta + \phi)] \, d\phi,$$
(5.60)

where ν accounts for the cross section and the swim speed. We decompose into angular Fourier modes $f_m = \int d\theta f(\theta)e^{-im\theta}$ and $f(\theta) = (2\pi)^{-1} \sum_m f_m e^{im\theta}$, where $f_0 = n$ is the density. The kinetic equation admits an isotropic solution, $f(\theta) = f_0/2\pi$, and we are interested in studying the development of anisotropies (alignment or swarming). Projecting the kinetic equation onto the Fourier modes, we find

$$\dot{f}_m = -D_r m^2 f_m + \nu \left[\sum_p S((2m-p)\pi/2) f_m f_{p-m} - f_0 f_m \right], \quad (5.61)$$

where $S(x) = \sin(x)/x$. For $m = 0$, number conservation implies a vanishing right-hand side. For $m = \pm 1$, keeping only linear terms (weak alignment), we obtain

$$\dot{f}_{\pm 1} = [\nu n(4 - \pi)/\pi - D_r] f_{\pm 1}.$$
(5.62)

There is a critical concentration, $n_c = \pi D_r/[\nu(4 - \pi)]$ such that, for lower concentrations ($n < n_c$), the isotropic phase is stable, while in the opposite case the dipolar modes grow and, considering nonlinear terms, the final state can be obtained. Note that the diffusion and the loss term tend to reduce ordering, while the gain term creates order. Close to the threshold, higher multipoles ($|m| \geq 2$) are stable; therefore the distribution function can be expressed as $f(\theta) = n/2\pi + n_1 \cos(\theta - \phi)$, showing a preferred direction ϕ, which is chosen at random in a spontaneous symmetry break.

Further reading

The original papers of Einstein relating to Brownian motion are collected in Einstein (1956). The book by Risken (1996) is completely devoted to the study of the Fokker–Planck equation, with applications to different areas. In a broader approach, van Kampen (2007) and Gardiner (2009) present the theory of stochastic processes, with a detailed analysis of Brownian motion, using the Fokker–Planck and Langevin approaches. Finally, Brownian motion is presented in Kubo *et al.* (1998) in the context of nonequilibrium statistical mechanics. For the enthusiast of some specific issues covered in this chapter, Berg (1993) presents all sorts of random walks in biology, and reaction–diffusion systems are presented in Garcia-Ojalvo and Sancho (1999).

Exercises

(5.1) **Random walk.** The simplest model for diffusive motion is the random walk. Consider a particle that can move in 1D in discrete steps. At each time step, which are separated by Δt, the particle can jump to the right or to the left with equal probabilities, each time moving a distance Δx. After n time steps, show that the probability to have moved k steps to the right is given by a binomial distribution. Compute $\langle k \rangle$ and $\langle k^2 \rangle$ and show that the resulting motion is diffusive. Finally, changing variables to real time and space, compute the diffusion coefficient using eqn (5.25).

(5.2) **Fokker–Planck equation.** Derive eqn (5.4).

(5.3) **Mean square displacement.** Derive eqn (5.24).

(5.4) **Diffusive eigenvalue.** Perform the perturbation analysis and derive eqn (5.42).

(5.5) **Chapman–Enskog method.** Perform the Chapman–Enskog method and derive the diffusion equation (5.45).

(5.6) **Energetic analysis.** Consider a homogeneous distribution of Brownian particles in the absence of external forces. In unsteady conditions, they are described by the Fokker–Planck equation,

$$\frac{\partial f}{\partial t} = \frac{\gamma}{M}\frac{\partial}{\partial \mathbf{c}}\cdot(\mathbf{c}f) + \frac{\gamma k_{\mathrm{B}}T}{M^2}\frac{\partial^2 f}{\partial c^2}.$$

The evolution of the energy density e can be determined by integrating this equation over $\mathbf{c}$ after multiplying by $Mc^2/2$. Show that, if the fluctuation–dissipation condition is fulfilled, the energy becomes stationary when the particle thermalises with the bath.

(5.7) **Derivation of rotational diffusion.** Write down the Kramers equation for the joint distribution $f(\hat{\mathbf{n}}, \boldsymbol{\omega}, t)$ in the rotational diffusion process. Show that, for long times, the reduced distribution $f(\hat{\mathbf{n}}, t)$ obeys a diffusion equation and obtain the associated diffusion coefficient. Note: you can use either the eigenvalue method using an expansion in Fourier or spherical harmonic functions or the Chapman–Enskog method.

(5.8) **Relaxation in rotational diffusion.** Consider a spherical colloidal particle that has a labelled direction, which is immersed in a fluid. The director $\hat{\mathbf{n}}$ experiences rotational diffusion described by the equation

$$\frac{\partial f}{\partial t} = D_r \nabla_{\hat{\mathbf{n}}}^2 f.$$

Expanding the distribution function in Fourier modes in 2D or in spherical harmonics in 3D, show that the relaxation time of the slowest mode is proportional to $\tau_r = D_r^{-1}$. Compute the proportionality constant.

(5.9) **Solution for the homogeneous case.** Show that the equilibrium solution of the homogeneous Fokker–Planck equation (5.8) is effectively a Maxwellian distribution.

(5.10) **Forced rotor.** In Chapter 2 we studied the dynamics of a rotor forced by an external torque that additionally has internal motion that derives from a potential. Its equation of motion is $\dot{\theta} = \Omega + \omega(\theta)$, where Ω results from the external driving and $\omega = -\,\mathrm{d}U/\,\mathrm{d}\theta$. In the presence of noise, the angular distribution function satisfies the Fokker–Planck equation,

$$\frac{\partial F}{\partial t} = -\frac{\partial}{\partial \theta}\left[(\omega(\theta) + \Omega)F\right] + \frac{\Gamma}{2}\frac{\partial^2 F}{\partial \theta^2}.$$

Find the stationary solution and compute the average angular velocity $\langle \dot{\theta} \rangle$ for $\omega(\theta) = \sin\theta$. Compare with the case in the absence of noise studied in Exercise 2.11. The solution for F is quite involved, and the integral for the average may need to be done numerically.

(5.11) **Backlight due to light diffusion.** Consider a point source that emits light steadily and isotropically. The medium is slightly inhomogeneous, inducing diffusion. As a result, the light will not propagate radially; rather, far from the source, the rays will move in other directions, and eventually a fraction of them will direct back toward the source. The objective is to compute the fraction of light that goes back. The distribution is isotropic in space, and we need to model the distribution function $f(r, \hat{\mathbf{n}})$, which is the intensity of light at distance r, which moves along $\hat{\mathbf{n}}$, measured with respect to an axis that goes from the origin to the observation point (see Fig. 5.10).

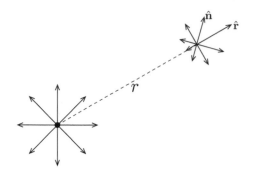

Fig. 5.10 Light is emitted isotropically at the source. At a distance r from the source, the light propagates in the direction pointed by $\hat{\mathbf{n}}$, which deviates from the radial direction $\hat{\mathbf{r}}$.

The distribution f has been integrated over the angular directions and, hence, does not show $1/r^2$ decay; rather, the total intensity at distance r remains constant. In the steady state, the distribution is described by the kinetic equation

$$c\hat{\mathbf{n}} \cdot \hat{\mathbf{r}} \frac{\partial f}{\partial r} = D_r \nabla_{\hat{\mathbf{n}}}^2 f,$$

with boundary condition $f(r = 0, \hat{\mathbf{n}}) = N_0 \delta(\hat{\mathbf{n}} - \hat{\mathbf{r}})$. This is a standard rotational diffusion equation with r playing the role of time. Solve it in either two or three dimensions and compute the fraction of light moving in the positive and negative directions as a function of distance from the source r.

(5.12) **Swimmer with rotational diffusion.** In Section 3.10 a model for microscopic swimmers that change direction at discrete events was introduced. Here, we consider a complementary process. A swimmer is characterised by its orientation $\hat{\mathbf{n}}$ and its position $\mathbf{r}$. The latter evolves as $\dot{\mathbf{r}} = V\hat{\mathbf{n}}$, where V is the swimming speed. A microscopic swimmer is also subject to rotational diffusion. Combining these two processes, the distribution function $f(\mathbf{r}, \hat{\mathbf{n}}, t)$ obeys the equation

$$\frac{\partial f}{\partial t} + V\hat{\mathbf{n}} \cdot \nabla f = D_r \nabla_{\hat{\mathbf{n}}}^2 f,$$

analogous to the light diffusion equation. Show that, for long times $t \gg D_r^{-1}$, the swimmers show diffusive motion in space and compute the corresponding diffusion coefficient. Before doing the full calculation, convince yourself that, combining the concepts of the coherence time and ballistic propagation during this time, the diffusion coefficient should scale as $D \sim V^2/D_r$.

(5.13) **Reaction–diffusion equations.** When chemical reactions take place between solute molecules in a liquid, those molecules can also be described by kinetic models that combine the Fokker–Planck and Boltzmann approaches. For example, in a simple annihilation process, $A + A \rightarrow \emptyset$, where two molecules disappear with probability λ when they meet, the appropriate kinetic equation for $f(\mathbf{r}, \mathbf{c}, t)$ is

$$\frac{\partial f}{\partial t} + \mathbf{c} \cdot \nabla f = J_{\text{FP}} + J_{\text{B}} + J_{\text{reaction}},$$

where the different terms are

$$J_{\text{FP}} = \frac{\gamma}{M} \frac{\partial}{\partial \mathbf{c}} \cdot (\mathbf{c}f) + \frac{\gamma k_B T}{M^2} \frac{\partial^2 f}{\partial c^2},$$

$$J_{\text{B}} = \int [f(\mathbf{c}')f(\mathbf{c}_1') - f(\mathbf{c})f(\mathbf{c}_1)] gb \, db \, d^3 c_1 \, d\psi,$$

$$J_{\text{reaction}} = - \int \lambda f(\mathbf{c})f(\mathbf{c}_1) gb \, db \, d^3 c_1 \, d\psi.$$

Show that, if $\lambda \ll 1$ or if the collision frequency is much smaller than γ/M, in the long-time limit the evolution reduces to a reaction–diffusion equation for the particle density,

$$\frac{\partial n}{\partial t} = D\nabla^2 n - \mu n^2.$$

Compute the reaction rate μ.

(5.14) **Reaction–diffusion in nuclear fission.** To produce nuclear fission a neutron must be absorbed by the isotope ^{235}U. However, after fission, three neutrons are emitted. Schematically, n + ^{235}U → X + 3n, where X are the products. If we track the neutrons, their concentration is described by the reaction–diffusion equation

$$\frac{\partial n}{\partial t} = D\nabla^2 n + 2\lambda n,$$

where λ is the reaction constant and the factor 2 accounts for the net production of two neutrons in the reaction. Consider a spherical pile of nuclear fuel of radius R from which neutrons can escape at the boundary. The associated boundary condition is $n(R, t) = 0$. Show that there is a critical radius above which the reaction becomes explosive and the number of neutrons increases exponentially (i.e. a chain reaction takes place).

(5.15) **Equilibrium solution under external potentials.** Find the solution of the stationary Fokker–Planck equation for the velocity distribution of particles in 3D under the action of an external potential (eqn 5.13). Show that the equilibrium solution (5.15) is recovered.

Plasmas and self-gravitating systems

6

6.1 Long-range interactions

Up to now, we have considered systems that evolve due to short-range interactions. For gases, these interactions took the form of binary collisions and were successfully modelled by the Boltzmann equation in Chapter 4. When a large particle was immersed in a fluid, the interactions were divided into an average force plus fluctuations. Because of the short duration of the interactions, the fluctuating force was considered to be random, resulting in the Fokker–Planck equation in Chapter 5. The Coulomb force between charged particles, which will be the subject of study in this chapter, lies in the opposite limit. It is a long-range force, resulting in scattering processes that take long times. In a system of charged particles, it is no longer possible to isolate individual collision processes, as each charge interacts with many others simultaneously.

When a system is dominated by Coulomb interactions, we will say that it is a plasma, which is normally generated by ionising neutral gases. In this process, positive and negative charges are produced while the plasma remains globally neutral. This feature gives rise to a new phenomenon: charge screening. Imagine that a large positive charge is placed in a plasma. Electrons will be attracted and ions repelled, and as a result the charge will be surrounded by a negative cloud, reducing its effective charge (to zero, as we will see). This screening occurs over a finite length, and our statistical approach will remain valid whenever the number of particles within this length is large.

Stars in a galaxy interact through the gravitational force, which shows the same distance dependence as the Coulomb force. Similarly, then, stars interact simultaneously with many others through persistent encounters. We will find that plasmas and self-gravitating systems are described by the same Vlasov kinetic equation. These two types of system share many properties, but although the kinetic equation is the same, there is an important difference: the gravitational force is always attractive, in contrast to the Coulomb case, where equal charges repel. This change of sign in the force implies that self-gravitating systems do not present screening. On the contrary, a large mass will attract others, rendering the complete system unstable, a process that finally gives rise to the formation of galaxies and other astrophysical structures.

Kinetic Theory and Transport Phenomena, First Edition, Rodrigo Soto.
© Rodrigo Soto 2016. Published in 2016 by Oxford University Press.

In both plasmas and self-gravitating systems, the Vlasov kinetic equation is reversible in time and there is no global tendency to equilibrium. To capture irreversible behaviour one needs to go beyond this description, and besides the collective interactions, we will need to model binary encounters as well. This granularity effect is small for large systems, and the result is that systems with long-range interactions present extremely slow relaxation processes. Although long to reach, for plasmas there is a well-defined thermal equilibrium state, but we will see that this is not the case for self-gravitating systems, for which the final state is still not well understood. This problem, which is of fundamental interest for the completeness of statistical mechanics, may seem irrelevant from a practical point of view because the relaxation times for galaxies can go far beyond the age of the universe. However, in the case of small stellar clusters, they can evolve on much shorter time scales.

6.2 Neutral plasmas

6.2.1 Introduction

Plasma is a state of matter in which atoms are partially or totally ionised, resulting in a mixture of ions, electrons, and neutral atoms. Consider first the hydrogen atom. The electron binding energy is $\varepsilon_0 = 13.6\,\text{eV}$. Atoms become ionised if the temperature is high enough, such that $T \sim \varepsilon_0/k_\text{B} \sim 1.2 \times 10^4\,\text{K}$.[1] Plasmas can also be obtained at low gas densities. Indeed, if the density of hydrogen atoms is low, it is entropically favourable to dissociate the atom into a proton and an electron, as the increase in configurational phase space compensates for the decrease of the momentum phase space. Carrying out a simple analysis using the tools of equilibrium statistical mechanics, it is found that the fraction of ionised protons n_{p+} among the total number of protons n_p is given by the Saha equation,[2]

$$\frac{n_{p+} n_e}{n_p} = \left(\frac{2\pi m_e k_\text{B} T}{h^2} \right)^{3/2} e^{-\varepsilon_0/k_B T}. \tag{6.1}$$

At high temperatures, the ionisation fraction increases, but it also does so with decreasing particle density. At very low densities, as in the interstellar medium, with a density that can be as low as 10^2 ions/m^3, the ionising temperature decreases. However, the effect is not dramatic due to the exponential factor. In fact, the gas in the interstellar or intergalactic medium is ionised by the large concentration of ultraviolet photons.[3]

The Saha equation (6.1) states that, at any temperature and proton density, the gas is composed of a mixture of neutral atoms, ions, and electrons. However, we will call the system a plasma when the ion and electron concentration is non-negligible. We will see that the presence of free charges in the system generates new phenomena and dramatically changes the physical properties compared with an uncharged gas. To

[1]Remember that $1\,\text{eV}=1.602 \times 10^{-19}$ J and that $k_\text{B} = 1.3806488 \times 10^{-23}$ J/K.

[2]See Exercise 6.1.

[3]In Chapter 7, we will study the radiation–matter interaction and show that if the radiation has the appropriate frequency, it can easily ionise matter.

clearly demonstrate the effects of having free charges, we will consider, for simplicity, the case of a fully ionised plasma (where all the atoms are ionised). In general, it is possible to describe plasmas made of any type of atom and not only hydrogen. Again, for simplicity, we will nevertheless consider that the plasma has only one type of ion, where each atom has been ionised Z times. Under these conditions, if n_{i0} is the average ion density, the average electron density is $n_{e0} = Zn_{i0}$, reflecting global neutrality.

In summary, in the simple neutral plasmas that we will be studying, the physical properties of ions and electrons are as follows:[4]

[4]The elementary charge is $e = 1.6 \times 10^{-19}$ C with the usual sign convention that the electron charge is $-e$.

	Mass	Charge	Average density
Electrons	m_e	$-e$	$n_{e0} = Zn_{i0}$
Ions	m_i	Ze	n_{i0}

The mass contrast between ions and electrons is large (in the case of the hydrogen atom, $m_i/m_e \approx 2000$, and larger values are obtained for other ions). The dynamical properties of the electrons and ions are therefore drastically different, with ions moving slowly while electrons can rapidly react to varying electric and magnetic fields. To present some of the concepts of plasma physics and to give results that otherwise would be unnecessarily involved, sometimes we will make the simplification that ions have infinite mass and are homogeneously distributed in space. They will not have any dynamical effect either, except for producing global charge neutrality. This immaterial homogeneous neutralising charge density is usually called jellium.

6.2.2 Debye screening

Consider a neutral plasma with ions modelled as having infinite mass with density n_{i0}. If an extra charge Q is placed in the plasma, electrons will move either toward the charge if it is positive or away from it in the opposite case. We aim to compute the resulting electron density profile, $n_e(\mathbf{r})$. The total charge density $\rho(\mathbf{r})$ in the system is given by the ion and electron densities plus the extra charge, which we model as a Dirac delta distribution,

$$\rho(\mathbf{r}) = Zen_{i0} - en_e(\mathbf{r}) + Q\delta(\mathbf{r}). \tag{6.2}$$

According to electrostatics, if there is a charge density, the electric potential is given by the Poisson equation,

$$\nabla^2\phi = -\frac{\rho}{\epsilon_0}, \tag{6.3}$$

where ϵ_0 is the electric constant.[5]

In thermal equilibrium, electrons are distributed with the Boltzmann weight,

$$n_e(\mathbf{r}) = n_{e0}e^{-V(\mathbf{r})/k_BT} = n_{e0}e^{e\phi(\mathbf{r})/k_BT}, \tag{6.4}$$

where we have used that the potential energy for electrons is $V = -e\phi$.

[5]In some contexts, it is more convenient to use other units, but as this is a general purpose book, we prefer to retain SI units and the factors ϵ_0 and $1/4\pi$. In SI units, the electric field and electrostatic potential produced by a charge are $\mathbf{E} = q\hat{\mathbf{r}}/4\pi\epsilon_0 r^2$ and $\phi = q/4\pi\epsilon_0 r$, respectively.

The coupled eqns (6.2), (6.3), and (6.4) constitute the so-called Poisson–Boltzmann system, allowing one, in principle, to obtain the electron distribution, although they are highly nonlinear.[6] To simplify them, we consider that the extra charge is small, and therefore the density perturbation is also small. We then write $n_e(\mathbf{r}) = n_{e0} + \delta n_e(\mathbf{r})$, with $\delta n_e(\mathbf{r}) \ll n_{e0}$. The resulting charge density is $\rho(\mathbf{r}) = -e\delta n_e(\mathbf{r}) + Q\delta(\mathbf{r})$, which is hence small. As a consequence, the electric potential is small as well, and eqn (6.4) can be expanded as $n_{e0} + \delta n_e(\mathbf{r}) = n_{e0}\left[1 + e\phi(\mathbf{r})/k_{\mathrm{B}}T\right]$. We thereby obtain the linearised Poisson–Boltzmann equations,

$$\nabla^2\phi = \frac{e\delta n_e}{\epsilon_0} - \frac{Q}{\epsilon_0}\delta(\mathbf{r}), \tag{6.5}$$

$$\delta n_e = n_{e0}e\phi(\mathbf{r})/k_{\mathrm{B}}T. \tag{6.6}$$

Substituting (6.6) into (6.5) yields a closed differential equation for ϕ that can be directly solved,[7] assuming that the potential must vanish at infinity. The result is the so-called Yukawa potential,

$$\phi = \frac{Q}{4\pi\epsilon_0}\frac{e^{-\kappa r}}{r}, \tag{6.7}$$

with

$$\kappa = \sqrt{\frac{n_{e0}e^2}{\epsilon_0 k_{\mathrm{B}}T}}. \tag{6.8}$$

We have found that, in a plasma, in contrast to what happens in vacuum or a dielectric material, the electric potential generated by an extra charge decays rapidly. The exponential decay implies that, if two charges were placed in the plasma, their mutual effective interaction would be short ranged instead of the usual long-range, $1/r$ Coulomb interaction. However, note that, at short distances ($\kappa r \ll 1$), the Coulomb potential is recovered. The origin of this dramatic change in the range of interaction is the mobility of the electrons.

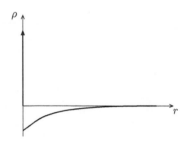

Fig. 6.1 Resulting charge density in a plasma when an extra Dirac delta charge is added at the origin. The plasma responds by building a screening cloud with a total charge that exactly compensates the added one.

The total charge density can now be computed by replacing the result obtained for the electron density, giving

$$\rho(\mathbf{r}) = Q\left[\delta(\mathbf{r}) - \frac{\kappa^2}{4\pi}\frac{e^{-\kappa r}}{r}\right], \tag{6.9}$$

[8]The screening of electric interactions does not take place only in plasmas. Indeed, the only requirement is that there are free-moving charges; for example, in water with dissolved salt (NaCl), the sodium and chloride ions act as free charges and the corresponding Debye length is $\lambda_D = \sqrt{\frac{\epsilon k_{\mathrm{B}}T}{(n_{\mathrm{Cl}}+n_{\mathrm{Na}})e^2}}$, where ϵ is the dielectric constant of water. Dissolving $1\,\mathrm{g}$ of salt in 1 litre of water gives $\lambda_D = 2.4\,\mathrm{nm}$. Electrons in metals, which now show quantum statistics, also screen electric interactions (see Section 8.9).

shown in Fig. 6.1. One can directly verify that the spatial integral of the electron density is exactly $-Q$; that is, electrons move in such a way as to screen the extra charge completely. In more detail, according to Gauss's law, the force on a test particle at a distance r from the added charge is obtained from the charge enclosed in a sphere of radius r,

$$Q_{\mathrm{enclosed}} = \int_0^r \rho(\mathbf{r}')\,\mathrm{d}^3r' = Q(1+\kappa r)e^{-\kappa r}. \tag{6.10}$$

At long distances ($\kappa r \gg 1$), the charge is no longer visible and the plasma behaves as if no extra charge had been added. The associated Debye screening length $\lambda_D = \kappa^{-1}$ is given by[8]

$$\lambda_D = \sqrt{\frac{\epsilon_0 k_\mathrm{B} T}{n_{e0} e^2}}. \qquad (6.11)$$

The screening length increases with temperature, because in this case electrons have more energy and the Debye cloud increases in size.

The screening process is thus as follows: when an additional charge is inserted, the free charges move, responding to the field. Their displacement tends to gradually reduce the field, and charges will continue to displace as long as a residual field is present and will stop when complete screening has been achieved. Only in the region inside the Debye length does the electric field remain finite.

The Debye length and the electron response were computed in the linearised regime. If we had kept the full, nonlinear Poisson–Boltzmann equation, in the near field the expressions would be different and the potential would not be of Yukawa type. However, screening will still take place. At sufficiently long distances (of the order of the Debye length), the effect of screening will be to reduce the potential, allowing use of the linear theory. The result is that the potential will decay at long distances as a Yukawa potential, $\phi = \frac{Q^*}{4\pi\epsilon_0} \frac{e^{-\kappa r}}{r}$, where the value of Q^* depends on the properties of the near-field solution.

For the statistical analysis of the screening to be valid, the electron density should be large enough. The relevant parameter is the number of electrons in the Debye volume, $N_D \equiv n_{e0}\lambda_D^3$, which must be large.[9] This assumption, valid under most plasma conditions, will be extensively used in the coming sections.

[9] Equivalently, the *plasma parameter*, defined as $g = 1/N_D$, should be small to have a good statistical description.

6.2.3 Vlasov equation

In the previous section we saw that charges (electrons and ions) interact via the Coulomb potential at distances shorter than the Debye length and with the effective Yukawa potential otherwise. If the number of particles N_D in a Debye volume is large (and using the fact that the Coulomb interactions are long ranged), it turns out that each charge interacts with many others simultaneously. Consider a point charge q that is surrounded by a homogeneous charge density ρ. The Coulomb force exerted by the charges located in an element of solid angle $\Delta\Omega$ between r and $r + \Delta r$ is $F_1 = q\rho\Delta\mathcal{V}/4\pi r^2$ (see Fig. 6.2). The volume is $\Delta\mathcal{V} = r^2 \Delta r \Delta\Omega$, and the force turns out to be independent of distance. Remarkably, charges at different distances exert the same force on the test particle. As a consequence, as long as $n_{e0}\lambda_D^3 \gg 1$, individual interactions are negligible compared with the total interaction within the Debye sphere. This is the key ingredient that we will use to build a kinetic theory for plasmas. We will deal accurately with the long-distance interactions in the Debye volume, whereas crude approximations will be made for the nearest charges. The error in so doing can be estimated based on the number of nearby charges compared with those that are correctly included, giving an error of the order of $1/N_D$.

Consider first a plasma composed of only electrons. We will add ions immediately after, but the derivation of the equation is clearer when

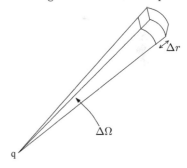

Fig. 6.2 The Coulomb force between a charge located at the origin and a charge density subtended by the solid angle $\Delta\Omega$ turns out to be independent of the distance r.

considering just one species. The Hamiltonian of the system is

$$H = \sum_i \frac{p_i^2}{2m_e} + \sum_{i<j} \varphi(\mathbf{r}_i - \mathbf{r}_j), \qquad (6.12)$$

where $\varphi(\mathbf{r}) = e^2/4\pi\epsilon_0|\mathbf{r}|$ is the Coulomb potential. It has the structure studied in Chapter 2. Therefore, the evolution of the electron distribution function is described by the BBGKY hierarchy. The first of these equations for $f(\mathbf{r}_1, \mathbf{p}_1, t)$ is

$$\frac{\partial f}{\partial t} + \frac{\mathbf{p}_1}{m_e} \cdot \frac{\partial f}{\partial \mathbf{r}_1} = -\int \left\{ \varphi(\mathbf{r}_1, \mathbf{r}_2), f^{(2)} \right\} \mathrm{d}^3 r_2 \, \mathrm{d}^3 p_2. \qquad (6.13)$$

The right-hand side of this equation takes into account the effect of adding all pair interactions into the motion of the particle located at $\mathbf{r}_1$. As previously discussed, the interactions are long ranged inside the Debye sphere, with each particle interacting with many others. If there were some correlations, they will only be present for neighbouring electrons. As we showed previously, the interaction with them is negligible compared with the rest. Distant electrons, on the other hand, dominate in the force and are expected to be uncorrelated. Therefore, we can assume that the electrons are completely uncorrelated in the integral term and the error will be only of order $1/N_D$; that is, we make the substitution,[10]

$$f^{(2)}(\mathbf{r}_1, \mathbf{p}_1, \mathbf{r}_2, \mathbf{p}_2, t) = f(\mathbf{r}_1, \mathbf{p}_1, t) f(\mathbf{r}_2, \mathbf{p}_2, t). \qquad (6.14)$$

Using this factorisation and the fact that the potential is momentum independent, the Poisson bracket in the integral term can be simplified to

$$\{\varphi(\mathbf{r}_1, \mathbf{r}_2), f(\mathbf{r}_1, \mathbf{p}_1) f(\mathbf{r}_2, \mathbf{p}_2)\} = -f(\mathbf{r}_2, \mathbf{p}_2) \frac{\partial \varphi}{\partial \mathbf{r}_1} \cdot \frac{\partial f(\mathbf{r}_1, \mathbf{p}_1)}{\partial \mathbf{p}_1}$$
$$- f(\mathbf{r}_1, \mathbf{p}_1) \frac{\partial \varphi}{\partial \mathbf{r}_2} \cdot \frac{\partial f(\mathbf{r}_2, \mathbf{p}_2)}{\partial \mathbf{p}_2}, \qquad (6.15)$$

where, to avoid overloading the notation, we have not written the temporal dependence. Once inserted into (6.13), the term with derivatives in $\mathbf{r}_2$ and $\mathbf{p}_2$ cancels by the divergence theorem because the distribution function vanishes for large momenta. We are therefore left with

$$\frac{\partial f}{\partial t} + \frac{\mathbf{p}_1}{m_e} \cdot \frac{\partial f}{\partial \mathbf{r}_1} = \frac{\partial f(\mathbf{r}_1, \mathbf{p}_1)}{\partial \mathbf{p}_1} \cdot \frac{\partial}{\partial \mathbf{r}_1} \int \varphi(\mathbf{r}_1, \mathbf{r}_2) f(\mathbf{r}_2, \mathbf{p}_2) \, \mathrm{d}^3 p_2 \, \mathrm{d}^3 r_2. \quad (6.16)$$

We note that $n(\mathbf{r}_2) = \int f(\mathbf{r}_2, \mathbf{p}_2) \, \mathrm{d}^3 p_2$ is the electron density, which generates the electrostatic potential,

$$\Phi_{\mathrm{mf}}(\mathbf{r}_1) \equiv -\frac{e}{4\pi\epsilon_0} \int \frac{n(\mathbf{r}_2)}{|\mathbf{r}_1 - \mathbf{r}_2|} \, \mathrm{d}^3 r_2, \qquad (6.17)$$

where the subscript 'mf' stands for mean field, as explained below. Then, the integral term equals the electrostatic energy of a charge located at

[10]If we want to make the same kind of approximation in a dense gas of charged spheres (for example, a charged colloidal suspension), the two-particle distribution should be approximated as in the Enskog model (see Section 4.8.1) for the term containing the short-range force, and as is done here for the term corresponding to the long-range force. That is, correlations are not completely discarded, but rather their effect on the integral term containing long-range interactions is discarded.

$\mathbf{r}_1$: $-e\Phi_{\mathrm{mf}}(\mathbf{r}_1)$. The kinetic equation then reads,

$$\frac{\partial f}{\partial t} + \frac{\mathbf{p}}{m_e} \cdot \frac{\partial f}{\partial \mathbf{r}} = \frac{\partial f}{\partial \mathbf{p}} \cdot (-e\nabla\Phi_{\mathrm{mf}}). \qquad (6.18)$$

We recall that the right-hand side term is a result of the pair interactions in the plasma. However, we can reinterpret this term by moving it to the left-hand side and defining the mean field electric field as $\mathbf{E}_{\mathrm{mf}} = -\nabla\Phi_{\mathrm{mf}}$, leading to

$$\frac{\partial f}{\partial t} + \frac{\mathbf{p}}{m_e} \cdot \frac{\partial f}{\partial \mathbf{r}} - e\mathbf{E}_{\mathrm{mf}} \cdot \frac{\partial f}{\partial \mathbf{p}} = 0. \qquad (6.19)$$

This equation can now be read as the Liouville equation for independent particles moving in an electric field (see Chapter 2). The electric field, however, is not externally imposed but rather is computed from the Poisson equation,

$$\nabla^2 \Phi_{\mathrm{mf}} = \frac{en}{\epsilon_0}; \quad n(\mathbf{r}) = \int f(\mathbf{r}, \mathbf{p}) \, \mathrm{d}^3 p. \qquad (6.20)$$

The kinetic equation (6.19) is the Vlasov equation and must be solved together with the Poisson equation (6.20). We note that the Vlasov–Poisson system[11] is nonlinear, similarly to what happens with the Boltzmann equation (see Chapter 4).

In the case of several species (including ions) with charges and masses q_j and m_j, respectively, and an externally imposed field $\mathbf{E}_{\mathrm{ext}}$, the Vlasov–Poisson system reads,

$$\frac{\partial f_j}{\partial t} + \mathbf{c} \cdot \frac{\partial f_j}{\partial \mathbf{r}} + \frac{q_j}{m_j}(\mathbf{E}_{\mathrm{mf}} + \mathbf{E}_{\mathrm{ext}}) \cdot \frac{\partial f_j}{\partial \mathbf{c}} = 0, \quad j = 1, 2, \dots, \qquad (6.21)$$

$$\nabla^2 \Phi_{\mathrm{mf}} = -\sum_j \frac{q_j n_j}{\epsilon_0}; \quad n_j(\mathbf{r}) = \int f_j(\mathbf{r}, \mathbf{c}) \, \mathrm{d}^3 c, \qquad (6.22)$$

where the velocity representation has been used. To simplify notation, if the plasma only consists of electrons, we will suppress the subindex 'e' of the distribution function and other terms; for example, the electron mass will be simply m.

We are now in a position to interpret the meaning of the mean field potential. The approximation made by Vlasov is that the free charges do not interact pairwise but rather via the potential that they collectively generate. This is the average potential observed by a test charge. Correlations are discarded, as well as the granularity related to the finite value of N_D. Binary encounters are neglected, and the mean field potential becomes a smooth function instead of a sum of millions of Coulomb potentials. Because of this, the Vlasov equation is sometimes referred to as the collisionless Boltzmann equation, although this is not strictly a correct interpretation. In fact, collisions are considered in the Vlasov equation, but the close binary collisions that would produce large deflections are not considered properly and are smoothed out (see Exercise 6.2

[11]Although we refer to the Vlasov–Poisson system, it should be clear that the distribution function is the only dynamical variable and the electric field is given at any instant in terms of the distribution function. When rapidly time-varying electromagnetic fields are involved, these will also become dynamical variables (see Section 6.4.1).

to estimate the rate of these encounters). In Section 6.6.1 we will model these encounters to estimate the dynamic friction on stars in a galaxy.

The Vlasov equation, in contrast to the other kinetic equations we have studied so far, is time reversible. Indeed, inverting all velocities will trace the system back to the initial condition.[12] In Section 6.6 we will discuss the irreversible phenomena that take place in plasmas.

6.2.4   Stationary solutions

Consider a plasma in the absence of external fields. Notably, if the plasma is neutral, any homogeneous distribution $f_j(\mathbf{c})$ is a solution of the Vlasov–Poisson system. Indeed, if the plasma is uniform, the electric field vanishes and eqn (6.19) corresponds to the Liouville equation for free particles. In Chapter 2 we showed that, for this case, any distribution function of the (kinetic) energy is a stationary solution. Furthermore, if the plasma is inhomogeneous but the charge density does not change, resulting in a constant electric field, the Vlasov equation (6.21) now corresponds to the Liouville equation for non-interacting particles moving in an external field. In this case, any function of the energy, kinetic plus potential, will be a stationary solution. This kind of solution will be analysed in detail in Section 6.5.2 when dealing with self-gravitating systems.

The apparent non-interacting character of the Vlasov equation is a result of the mean field approximation, because if we had kept the pairwise interactions, the only stationary solutions would be functions of the total Hamiltonian. Pair interactions would act as an energy interchange mechanism, allowing the particles to thermalise. In Section 6.6 we will discuss how the Vlasov description can be extended to take into account the relaxation of plasmas and other long-range systems toward equilibrium. These processes take place over long times and finally drive the plasma to thermal equilibrium.

Also, in real plasmas, a fraction of the atoms are not ionised. The interaction of the charged particles with them is described by the Boltzmann equation, which has the effect of thermalising the plasma, which will reach a Maxwellian distribution with the same temperature as the neutral gas.

Hence, under many conditions, it is appropriate to model the plasma using a homogeneous Maxwellian distribution, although other distributions will be used as well to reflect some transient states (stationary at the level of the Vlasov description, but not absolutely stationary when pairwise interactions are considered).

6.2.5   Dynamical response

In Section 6.2.2 we studied the static response of a plasma to a small external charge and found it to be screened by the moving charges. Now, we would like to describe this process in the context of the Vlasov equation. To do this, we will consider the dynamical response, which will also describe the transient formation of the screening charge or the stationary response to a moving charge.[13]

Consider a homogeneous plasma in thermal equilibrium. The charge densities of electrons and ions are equal, and therefore the plasma is neutral. For simplicity, we consider the heavy ion model such that only the electrons move. Besides the electrons and ions, we place a small external charge density, $\lambda \rho_{\text{ext}}(\mathbf{r}, t)$, where $\lambda \ll 1$ is a formal small parameter.[14] As a reaction to the external charge, the electron distribution function will be modified to

$$f(\mathbf{r}, \mathbf{p}, t) = f_0(\mathbf{c}) + \lambda g(\mathbf{r}, \mathbf{c}, t), \qquad (6.23)$$

where f_0 is the Maxwellian distribution and the perturbation g is of the same order as ρ_{ext}. The mean field potential is given in terms of the total charge density, and therefore it is also small. We write it as $\lambda \Phi_{\text{mf}}$, and

$$\nabla^2 \Phi_{\text{mf}} = -\frac{1}{\epsilon_0} \left[\rho_{\text{ext}}(\mathbf{r}, t) - e \int g(\mathbf{r}, \mathbf{c}, t)\, d^3 c \right]. \qquad (6.24)$$

Keeping up to linear order in λ in the Vlasov equation (6.21), we get

$$\frac{\partial g}{\partial t} + \mathbf{c} \cdot \frac{\partial g}{\partial \mathbf{r}} + \frac{e}{m} \nabla \Phi_{\text{mf}} \cdot \frac{\partial f_0}{\partial \mathbf{c}} = 0. \qquad (6.25)$$

The coupled eqns (6.24) and (6.25) can be solved by Fourier transforming them in space and time for functions of wavevector $\mathbf{k}$ and frequency ω. Using the conventions given in Appendix A, one has

$$-k^2 \widetilde{\Phi}_{\text{mf}} = -\frac{1}{\epsilon_0} \left[\widetilde{\rho}_{\text{ext}} - e \int \widetilde{g}\, d^3 c \right], \qquad (6.26)$$

$$-i\omega \widetilde{g} + i\mathbf{k} \cdot \mathbf{c}\widetilde{g} + \frac{ie}{m} \widetilde{\Phi}_{\text{mf}} \mathbf{k} \cdot \frac{\partial f_0}{\partial \mathbf{c}} = 0. \qquad (6.27)$$

From eqn (6.27), $\widetilde{g}$ is obtained in terms of $\widetilde{\Phi}_{\text{mf}}$. Substituting this result back into eqn (6.26) finally gives the potential as a function of the external charge. The complete solution is

$$\widetilde{\Phi}_{\text{mf}}(\mathbf{k}, \omega) = \frac{\widetilde{\rho}_{\text{ext}}(\mathbf{k}, \omega)}{k^2 \epsilon(\mathbf{k}, \omega)}, \qquad (6.28)$$

$$\widetilde{g}(\mathbf{c}, \mathbf{k}, \omega) = -e \left(\frac{\mathbf{k} \cdot \frac{\partial f_0}{\partial \mathbf{c}}}{\mathbf{k} \cdot \mathbf{c} - \omega} \right) \widetilde{\Phi}_{\text{mf}}(\mathbf{k}, \omega), \qquad (6.29)$$

where we have defined

$$\epsilon(\mathbf{k}, \omega) = \epsilon_0 \left[1 - \frac{e^2}{m\epsilon_0 k^2} \int \frac{\mathbf{k} \cdot \frac{\partial f_0}{\partial \mathbf{c}}}{\mathbf{k} \cdot \mathbf{c} - \omega}\, d^3 c \right]. \qquad (6.30)$$

To interpret the solution (6.28), we recall that, in dielectric media, the Poisson equation is $\nabla^2 \Phi = -\rho_{\text{ext}}/\epsilon$, where ϵ is the dielectric constant. In Fourier space this reads, $\widetilde{\Phi} = \widetilde{\rho}_{\text{ext}}/k^2 \widetilde{\epsilon}$. Therefore, we have found that the dynamic response of the plasma can be described by the dielectric

[13] This case corresponds, for example, to a heavy ion moving in a plasma.

[14] In this chapter, we will use λ as a small parameter, instead of ϵ used in other chapters, to avoid confusion with the electric constant.

function $\epsilon(\mathbf{k},\omega)$ given in eqn (6.30). The $\mathbf{k}$ and ω dependence of ϵ implies by the convolution theorem (see Appendix A) that, in real space, the response is nonlocal in space and time; that is, the electric potential at a given position and time will depend on the external charge density at other positions and times in the form of a convolution. Finally, recalling from the Poisson equation that the total charge density is given by $\tilde{\rho}_{\text{tot}} = k^2 \epsilon_0 \tilde{\Phi}_{\text{mf}}$, we obtain

$$\tilde{\rho}_{\text{tot}}(\mathbf{k},\omega) = \tilde{\rho}_{\text{ext}}(\mathbf{k},\omega)/\epsilon(\mathbf{k},\omega), \qquad (6.31)$$

which corresponds to dynamic screening, as we will see.

In the case of a thermal plasma, the plasma dielectric function simplifies to

$$\epsilon(\mathbf{k},\omega) = \epsilon_0 \left[1 + \frac{m\omega_p^2}{k_{\text{B}}Tk^2}\sqrt{\frac{m}{2\pi k_{\text{B}}T}} \int \frac{ck}{ck-\omega} e^{-mc^2/2k_{\text{B}}T}\, dc \right], \quad (6.32)$$

where

$$\omega_p = \sqrt{ne^2/(m\epsilon_0)} \qquad (6.33)$$

is the so-called *plasma frequency*, which is related to the Debye length by the thermal velocity, $\lambda_D \omega_p = \sqrt{k_{\text{B}}T/m}$.

Static case

To understand the properties of the dielectric function and the plasma response, we first consider the static case, that is, when $\rho_{\text{ext}}(\mathbf{r},t) = \rho_{\text{ext}}(\mathbf{r})$. Its Fourier transform is $\tilde{\rho}_{\text{ext}}(\mathbf{k},\omega) = \tilde{\rho}_{\text{ext}}(\mathbf{k})\delta(\omega)$. Therefore, the induced potential is $\tilde{\Phi}_{\text{mf}}(\mathbf{k},\omega) = \frac{\tilde{\rho}_{\text{ext}}(\mathbf{k})\delta(\omega)}{\epsilon(\mathbf{k},0)}$ and the dielectric function (6.32) is greatly simplified to $\tilde{\epsilon}(\mathbf{k},0) = \epsilon_0\left[1 + m\omega_p^2/(k_{\text{B}}Tk^2)\right]$. Hence, $\tilde{\Phi}_{\text{mf}}(\mathbf{k}) = \tilde{\rho}_{\text{ext}}(\mathbf{k})/(k^2 + \lambda_D^{-2})$. Going back to real space using the convolution theorem yields

$$\Phi_{\text{mf}}(\mathbf{r}) = \int \rho_{\text{ext}}(\mathbf{r}') \frac{e^{-\kappa|\mathbf{r}-\mathbf{r}'|}}{4\pi\epsilon_0|\mathbf{r}-\mathbf{r}'|}\, d^3r'. \qquad (6.34)$$

Note that we have recovered the Yukawa potential in the form of a convolution. To obtain this result we have neglected any time dependence. If the external charge moves slowly, this result will remain valid as long as $\omega_p\tau \ll 1$, where τ is the time scale that characterises the evolution of the charge density.[15]

Travelling charge

Let us now consider the case of an external charge that moves undeformed through the plasma at a constant velocity $\mathbf{v}$, as may be the case for a heavy ion. Then, $\rho_{\text{ext}}(\mathbf{r},t) = \rho_0(\mathbf{r}-\mathbf{v}t)$, with a Fourier transform, $\tilde{\rho}_{\text{ext}}(\mathbf{k},\omega) = \tilde{\rho}_0(\mathbf{k})\delta(\omega - \mathbf{k}\cdot\mathbf{v})$. This implies that, to evaluate the induced potential, we need to compute the dielectric function,

$$\epsilon(\mathbf{k},\omega = \mathbf{k}\cdot\mathbf{v}) = \epsilon_0 \left[1 + \frac{m\omega_p^2}{k_{\text{B}}Tk^2}\sqrt{\frac{m}{2\pi k_{\text{B}}T}} \int \frac{ck}{ck-\mathbf{k}\cdot\mathbf{v}} e^{-mc^2/2k_{\text{B}}T}\, dc \right]. \qquad (6.35)$$

The integral, however, is not defined because it presents a singularity at $c = \mathbf{k} \cdot \mathbf{v}/k$. The solution for this apparent inconsistency is given in the context of the propagation of waves in Section 6.3.2, and the application to the case under study here is worked out in Exercise 6.8.

6.3 Waves and instabilities in plasmas

6.3.1 Plasma waves

Plasmas can sustain a large variety of oscillatory phenomena. In contrast to gases, in which only sound waves are present, thanks to coupling with electromagnetic fields, different kinds of waves are present. These can be electrostatic or electromagnetic, with the electric field longitudinal or perpendicular to the wavevector, and the number of possibilities multiplies when external magnetic fields are applied. Our purpose here is not to study such wave phenomena in detail, but to give a glimpse of the new features that appear and the way kinetic theory can be used to analyse them. With this purpose in mind, we will consider a homogeneous plasma composed of thermal electrons and immobile ions in the absence of external electric and magnetic fields. To study the propagation of waves, we consider that the electrons are perturbed slightly such that their distribution function is $f(\mathbf{r}, \mathbf{c}, t) = f_0(\mathbf{c}) + \lambda g(\mathbf{r}, \mathbf{c}, t)$, where $\lambda \ll 1$ is used as a perturbation parameter and f_0 is the initial reference distribution function. Note that, in principle, the reference distribution function need not be that of equilibrium. Indeed, as the relaxation process to equilibrium is slow, nonequilibrium distribution functions can be considered as long as the time scale associated with the wave propagation is shorter than the relaxation rate.

To first order in λ, we recover the same equations as in Section 6.2.5, eliminating the external charge ρ_{ext}. Performing all the substitutions, the equation for the potential reduces to

$$k^2 \widetilde{\Phi}_{\text{mf}}(\mathbf{k}, \omega) \epsilon(\mathbf{k}, \omega) = 0, \qquad (6.36)$$

where $\epsilon(\mathbf{k}, \omega)$ is the dielectric function (6.30) found in the previous section and we considered perturbations with a well-defined wavevector.

Equation (6.36) indicates that the perturbation is not vanishing (that is, can evolve) if $\epsilon(\mathbf{k}, \omega) = 0$. As the initial perturbation was produced with a given set of wavevectors, this implies that the frequencies are not arbitrary but, rather, must be such that the dielectric function vanishes. This is an implicit equation for the frequencies that can give one or multiple solutions for each wavevector, labelled by the index n. The outcome is a dispersion relation for the wave, $\omega_n(\mathbf{k})$.

The structure of the equation $\epsilon(\mathbf{k}, \omega) = 0$ shows that it is not possible to have real solutions for ω, because if this were the case, the integral in eqn (6.30) would become undefined when the denominator vanishes. As noted by Landau in 1946, the frequencies must be complex, $\omega = \omega_R + i\omega_I$, with an interpretation that varies according to the sign of the imaginary part. Considering the sign convention used in the Fourier transform (see Appendix A), the temporal evolution is $e^{-i\omega t} = e^{-i\omega_R t + \omega_I t}$. Therefore, a negative imaginary part implies that the perturbations decay in time and we say that the system is stable or damped. On the contrary, if the imaginary part is positive, the perturbations grow exponentially and the system is unstable to small perturbations at the given wavevector.

Before going into the analysis by Landau to find the imaginary contribution, we first determine the real part in the case of small wavevectors, using an analysis originally carried out by Vlasov himself. The denominator in the integral for the Maxwellian case (6.32) is expanded in series for small k, giving

$$
\epsilon = \epsilon_0 \left[1 - \frac{m\omega_p^2}{k_B T k^2} \sqrt{\frac{m}{2\pi k_B T}} \int \left(\frac{ck}{\omega} + \left(\frac{ck}{\omega} \right)^2 + \dots \right) e^{-mc^2/2k_B T} \, dc \right]
$$

$$
= \epsilon_0 \left[1 - \frac{\omega_p^2}{\omega^2} - \frac{3k^2 k_B T \omega_p^2}{m\omega^4} - \dots \right], \tag{6.37}
$$

where only the even powers in c contribute to the integral and terms up to fourth order in (ck/ω) have to be retained. It is now possible to solve the equation $\epsilon(\mathbf{k}, \omega) = 0$, and one finds,

$$
\omega^2 = \omega_p^2 + \frac{3k_B T k^2}{m}. \tag{6.38}
$$

These electrostatic Langmuir waves are highly dispersive with frequencies that are bounded at vanishing wavevector by the plasma frequency. The interpretation is that ω_p corresponds to the natural frequency of oscillation when charges are displaced from their equilibrium position for very large wavelengths (Fig. 6.3). The wave we have studied is longitudinal; that is, the electric field is parallel to the wavevector (as one can see directly in Fourier space, where $\mathbf{E}_{mf} = i\mathbf{k}\Phi_{mf}$). Recall that, in vacuum, electromagnetic waves are transverse and furthermore an oscillating magnetic field also appears. Here, we can have pure electrostatic waves because there are free charges.

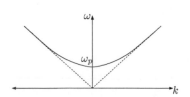

Fig. 6.3 Dispersion relation for electrostatic waves in plasmas. The slope of the asymptotes is $\sqrt{3k_B T/m}$.

6.3.2   Landau damping

To determine the imaginary part of the frequency, a careful analysis should be made using complex variables. The most direct approach is to consider that initially, at $t = 0$, a small perturbation is performed in a homogeneous plasma, and we will analyse its subsequent evolution. Noting that, for small perturbations, the Vlasov–Poisson system can be linearised, we will consider a single spatial Fourier mode. Hence, the distribution function is $f(\mathbf{r}, \mathbf{c}, t) = f_0(\mathbf{c}) + g(\mathbf{c}, t)e^{i\mathbf{k}\cdot\mathbf{r}}$, with initial condition, $g(\mathbf{c}, t = 0) = g_0(\mathbf{c})$. The electric potential is therefore also given by a single Fourier mode, $\Phi_{mf}(\mathbf{r}, t) = \Phi_{mf}(t)e^{i\mathbf{k}\cdot\mathbf{r}}$. To solve the time-dependent problem, we use the Laplace transform, which for any function of time $h(t)$ reads,

$$
\tilde{h}(s) = \int_0^\infty h(t)e^{-st} \, dt, \tag{6.39}
$$

being defined for any complex number s with positive real part. The inverse transform is

$$h(t) = \frac{1}{2\pi i} \int_{-i\infty+\sigma}^{i\infty+\sigma} \widetilde{h}(s)e^{st}\,\mathrm{d}s, \qquad (6.40)$$

where σ is such that all poles and singularities of $\widetilde{h}$ are located to the left of the contour of integration (see Fig. 6.4, left). We multiply the Vlasov equation,

$$\frac{\partial g}{\partial t} + i\mathbf{k}\cdot\mathbf{c}g + \frac{ie}{m}\Phi_{\mathrm{mf}}\mathbf{k}\cdot\frac{\partial f_0}{\partial \mathbf{c}} = 0, \qquad (6.41)$$

by e^{-st} and integrate from 0 to ∞. Integration by parts gives for the first term

$$\int_0^\infty \frac{\partial g}{\partial t}e^{-st}\,\mathrm{d}t = g e^{-st}\Big|_0^\infty - s\int_0^\infty g e^{-st}\,\mathrm{d}t = s\widetilde{g}(s) - g(0), \qquad (6.42)$$

while for the other terms the Laplace transform is directly obtained. Solving for the perturbation gives

$$\widetilde{g}(s) = \frac{g_0 - \frac{ie}{n}\Phi_{\mathrm{mf}}\mathbf{k}\cdot\frac{\partial f_0}{\partial \mathbf{c}}}{i\mathbf{k}\cdot\mathbf{c}+s}. \qquad (6.43)$$

Now, replacing this into the Poisson equation allows one to finally obtain the electric potential as

$$\widetilde{\Phi}(s) = -\frac{e}{k^2}\frac{\int \frac{g_0(\mathbf{c})\,\mathrm{d}^3 c}{i\mathbf{k}\cdot\mathbf{c}+s}}{\epsilon(\mathbf{k},\omega=is)}. \qquad (6.44)$$

Inserting this expression into the inverse transform (6.40) gives the definitive solution of the initial value problem. The solution is cumbersome, but some transformations can be made to render it more appropriate for analysis. Specifically, we consider long times, for which the slow or slowest modes will dominate the dynamics. For this purpose, we make use of the Cauchy integral theorem, which states that the contour of integration in the complex place can be deformed continuously, giving the same result for the integral if, during the deformation, no singularities of the integrand are found. Then, we aim to deform the contour in Fig. 6.4 (left) to that shown in the right part of the figure, where care has been taken not to cross any singularity. If this is achieved, at long times, the integral is dominated by all the enclosed singularities s_j of Φ,

$$\Phi(t) = \sum_j A_j e^{s_j t}, \qquad (6.45)$$

where the amplitudes A_j are the residues of the integrand; that is, the singularities of $\widetilde{\Phi}(s)$ are those that dictate the long-term dynamics. The singularities of $\widetilde{\Phi}$ appear when the denominator in (6.44) vanishes,[16]

$$\widetilde{\epsilon}(s) \equiv \epsilon_0\left[1 - \frac{ne^2}{m\epsilon_0 k^2}\int \varphi_0'(c_x)\frac{\mathrm{d}c_x}{c_x - is/k}\right] = 0, \qquad (6.46)$$

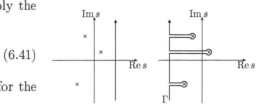

Fig. 6.4 Left: contour of integration used to evaluate the inverse Laplace transform. It runs in the complex s plane, parallel to the imaginary axis with a positive shift σ such that all singularities are to the left of the contour. Right: once the Laplace transform is analytically continued to negative real parts, the contour can be continuously deformed. The final contour Γ runs vertically with deviations to go around the poles.

[16]Singularities in $\widetilde{\Phi}$ can also appear in the numerator of (6.44), which depend on the initial perturbation. It has been shown that this is indeed the case for large k, but for small wavevectors the procedure we describe is essentially correct. This and other subtleties are the subject of research aiming to obtain complete understanding of the Landau damping phenomenon.

[17]For a Maxwellian distribution that is separable on the velocity components, φ_0 is also a Maxwellian.

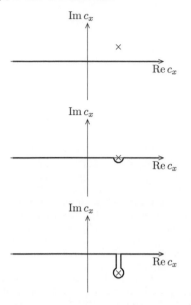

Fig. 6.5 The singularity for the integral (6.46) is located at $c_x = is/k$. Hence, when $\mathrm{Re}(s)$ changes sign, the integral encounters the singularity. To avoid the singularity and produce a continuous function in s, the contour of integration is deformed from the original one at the top, to that shown in the middle, and finally to the bottom one when $\mathrm{Re}(s) < 0$.

[18]Schematically, for an analytic function $h(z)$, the Plemelj formula is

$$\lim_{\delta \to 0^+} \int_{-\infty}^{\infty} \frac{h(z)\,dz}{z - z0 - i\delta}$$

$$= P \int \frac{h(z)\,dz}{z - z0} + i\pi h(z_0),$$

where P represents the principal value integral,

$$P \int \frac{h(z)\,dz}{z - z0} = \lim_{\delta \to 0^+} \left[\int_{-\infty}^{z_0 - \delta} \frac{h(z)\,dz}{z - z0} \right.$$

$$\left. + \int_{z_0 + \delta}^{\infty} \frac{h(z)\,dz}{z - z0} \right].$$

where φ_0' is the derivative of the reduced distribution that results from the integration of f_0 over c_y and c_z and dividing by the particle density.[17]

In the preceding development, we required that the contour of integration could be deformed to negative real values, without adding new singularities. However, in the definition of the Laplace transform, s was imposed to have a positive real part. We need to create a new function $\widetilde{\epsilon}(s)$ that for positive s takes the same value as (6.46) and changes continuously for negative arguments. The theory of complex functions tell us that, if one finds a protocol to do this, the generated function is unique. It is called the analytic continuation of the original function. If $\mathrm{Re}(s) > 0$, the integral in (6.46) is performed with c_x running from $-\infty$ to ∞. However, when $\mathrm{Re}(s)$ is decreased, the integral can encounter the pole located at $c_x = is/k$. To avoid this, we again take advantage of the Cauchy theorem to deform the contour, as shown in Fig. 6.5, without modifying the value of the integral. This protocol guarantees that the resulting function is continuous in s and no singularity is found during the integration. We finally obtain the result derived by Landau that the long-term dynamics is obtained from the solutions of the equation $\widetilde{\epsilon}(s) = 0$, where the contour of integration is that drawn in Fig. 6.5, that is, always below the pole.

We now proceed to find explicitly the solutions of the referred equation in the case of small wavevectors. We will assume that s will have a small real part, with a sign that is not yet settled. A posteriori we will confirm that it is indeed small. In this case, the singularity for c_x is close to the real axis and it is convenient to deform the contour as shown in Fig. 6.5 (middle). Specifically, c_x is integrated along the real axis to a finite distance δ from the pole, then a semicircle of radius δ is described below the singularity to later continue along the real axis to infinity. Everything is now properly defined, and the expansion for small wavevectors can be made. The integrals along the real axis are well defined and yield the principal value integral when the limit $\delta \to 0$ is taken. They can be expanded for small wavevectors, giving the contributions already computed in eqn (6.37), replacing ω by is. The integral over the semicircle C can be computed explicitly for small δ. It suffices to parameterise it as $c_x = c_{x0} + \delta e^{i\phi}$, where c_{x0} is the position of the pole on the real axis and the polar angle ϕ runs from $-\pi$ to 0,

$$\int_C \varphi_0'(c_x) \frac{dc_x}{c_x - is/k} = \varphi_0'(is/k) \int_C \frac{dc_x}{c_x - is/k} = i\pi\varphi_0'(is/k). \quad (6.47)$$

This development is an example of the Plemelj formula.[18]

Grouping all terms, the equation $\widetilde{\epsilon}(s) = 0$ (6.46) reads,

$$1 + \frac{\omega_p^2}{s^2} - \frac{3k^2 k_B T \omega_p^2}{m s^4} - \cdots - \frac{i\pi\omega_p^2}{k^2}\varphi_0'(is/k) = 0. \quad (6.48)$$

To solve for s, we write $s = -i\omega - \gamma$ (the signs are chosen for later convenience). At small wavevectors and assuming that γ is small, the equation can be solved iteratively, giving

$$\omega^2 = \omega_p^2 + \frac{3k_{\rm B}Tk^2}{m}, \tag{6.49}$$

$$\gamma = -\frac{\pi\omega_p^2}{2k^2}\varphi_0'(\omega_p/k) = \sqrt{\frac{\pi}{8}}\frac{\omega_p}{(k\Lambda_D)^3}e^{-\frac{1}{2(k\lambda_D)^2}}, \tag{6.50}$$

where in the second equality we have replaced φ_0 by a Maxwellian distribution. We now verify that the coefficient γ is indeed exponentially small for small wavevectors. Recall that the perturbations evolve in time as $e^{i(\mathbf{k}\cdot\mathbf{t}-\omega t)-\gamma t}$, and therefore they correspond to damped waves.

Note that the sign of γ is reversed and an instability appears instead of damping if $\varphi_0' > 0$. The classical picture to explain the damping and the instability makes the observation that the imaginary part of the dielectric function depends on φ_0' at $c_x = \omega/k$. Particles that move at the same speed as the wave (its phase speed) see it as stationary and resonate with it. This stationarity allows for efficient energy transfer mechanisms to take place. Particles moving at smaller velocities than the wave will be accelerated and gain energy, while those moving faster will lose energy (Fig. 6.6). If the number of slower particles is larger than the faster ones [i.e. $\varphi_0'(\omega/k) < 0$, as is the case for the Maxwellian], the net energy transfer will be from the field to the particles, giving rise to Landau damping. In the opposite case, particles inject energy into the field, which will increase its amplitude.

The emergence of this Landau damping phenomenon, which has been verified experimentally with great precision, is surprising because the Vlasov equation is time reversible, while damping is an irreversible process. In fact, there is no contradiction, because we have to remember that we started the system with a very particular initial condition, with energy allocated in the form of a coherent wave. The amplitude of the wave decays, but the energy is transferred to the particles. Actually, the distribution function shows growing oscillations in velocity space. If we were able to revert the distribution function, then the initial state would be recovered, consistent with the reversible character of the equations. Similarly to the Boltzmann explanation of the apparent irreversibility for Hamiltonian systems, the problem here relies on the practical impossibility of preparing such a detailed initial condition. Because of the reversibility of the Vlasov equation, even if the electrostatic wave has decayed, the plasma does not lose the information on the initial preparation (the entropy does not increase), and it can be accessed in spectacular plasma echo experiments (Ichimaru, 1973; Landau and Lifshitz, 1981).

When the wavevectors are not small or in other conditions of larger complexity (for example, in the presence of magnetic fields), it is not possible to find analytically the solutions of the equation $\epsilon(\mathbf{k},\omega) = 0$. Numerical methods can be applied where, for a fixed wavevector, the dielectric function (6.30) is computed in the complex plane, $\omega = \omega_r + i\omega_i$, where the integral should be made using the Landau contour (Fig. 6.5). The complex frequencies that solve the equation are those for which the real and imaginary parts of ϵ vanish simultaneously.

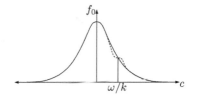

Fig. 6.6 The initial Maxwellian distribution interacts with the wave, showing a resonance close to $c = \omega/k$. Particles moving slightly slower than the wave are accelerated and those slightly faster are slowed down. As a result, the distribution function is distorted, as shown in the figure.

6.3.3   Instabilities

The analysis of Landau damping indicates that, if $\varphi_0'(\omega/k) > 0$, the propagating waves become unstable. Indeed, plasmas present innumerable unstable modes depending on the initial distribution function, the presence of external fields, and geometrical aspects. To illustrate this, we consider a simple one known as the beam–plasma instability. Consider an electron plasma of density n_0 and an electron beam with velocity $\mathbf{V}$ and density n_b directed toward the static plasma. We will assume that the thermal speeds of the plasma and the beam are small compared with V, so we can take both to be cold with $T = 0$. The initial distribution function is therefore

$$f_0(\mathbf{c}) = n_0\delta(\mathbf{c}) + n_b\delta(\mathbf{c} - \mathbf{V}), \tag{6.51}$$

where we have used that a Maxwellian distribution at vanishing temperature is a Dirac delta function. The temporal evolution of perturbations in this system is governed by the zeros of $\widetilde{\epsilon}(s)$. Considering the form of the distribution function, the integral can be performed directly without needing to compute principal values or the semicircular contour. Using $\omega = is$, the equation $\widetilde{\epsilon}(s) = 0$ reads,

$$1 - \frac{\omega_0^2}{\omega^2} - \frac{\omega_b^2}{(\omega - \mathbf{k}\cdot\mathbf{V})^2} = 0, \tag{6.52}$$

where $\omega_0 = \sqrt{n_0 e^2/(m\epsilon_0)}$ and $\omega_b = \sqrt{n_b e^2/(m\epsilon_0)}$. Defining $x = \omega/\omega_0$ and $q = \mathbf{k}\cdot\mathbf{V}/\omega_0$, this becomes

$$F(x) \equiv \frac{1}{x^2} + \frac{n_b/n_0}{(x-q)^2} = 1. \tag{6.53}$$

This is a quartic equation that can be solved graphically by intersecting $F(x)$ with 1, as shown in Fig. 6.7. For large values of q, the central part intersects 1 twice. Together with the other two intersections, for $x > q$ and negative x, we have the four real solutions of the quartic equation. However, for smaller q, the central part rises and two real solutions disappear and become complex. As the equation has real coefficients, the solutions should be complex conjugates, one with positive and one with negative imaginary part. The former gives rise to an instability. To reveal the threshold for this instability, we do not need to solve the equation analytically, but rather note that, at the threshold, F is tangential to 1; that is, $F(x^*) = 1$ and $F'(x^*) = 0$. These are two equations for the thresholds x^* and q^* that give

$$x^* = \left[1 + (n_b/n_0)^{1/3}\right]^{1/2}, \qquad q^* = \left[1 + (n_b/n_0)^{1/3}\right]^{3/2}. \tag{6.54}$$

This means that all modes with $\mathbf{k}\cdot\mathbf{V} < \omega_0\left[1 + (n_b/n_0)^{1/3}\right]^{3/2}$ will be unstable.

Note that it seems that the instability would also be present even if $n_b = 0$ (i.e. no beam). This is a singular limit where the equation is

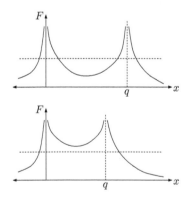

Fig. 6.7 The dimensionless frequencies x are obtained from the intersection of the function F with 1 (horizontal dashed line). For large q, there are four real solutions, while for small q only two real solutions appear.

of second order and all solutions are real. The singularity in this case can also be seen if we study the growth rate, equal to $\mathrm{Im}(\omega) = \omega_0 \mathrm{Im}(x)$. Close to the threshold, $q = q^* + \Delta q$ and $x = x^* + \Delta x$, gives

$$\Delta x = \frac{(n_b/n_0)^{1/6}}{\left[1 + (n_b/n_0)^{1/3}\right]^{1/4}} \sqrt{2\Delta q/3}. \tag{6.55}$$

For $\Delta q < 0$ (i.e. $q < q^*$), an imaginary part appears in the frequency, corresponding to the growth rate of the instability. Notably, it is proportional to $(n_b/n_0)^{1/6}$, clarifying the singular limit issue.

This and other instabilities have their origin in the (free) energy that is available to trigger dynamical processes. They are fundamentally out-of-equilibrium situations originating from the initial condition, boundary condition, or forcing. The last two cases have not been presented here but take place for example in the toroidal motion of plasmas confined by magnetic fields. In all cases, anyway, the unstable modes cannot grow exponentially forever. When the amplitudes become large, nonlinear effects take control and drive the subsequent dynamics.

6.4 Electromagnetic effects

6.4.1 Magnetic fields

The description of plasmas is not complete without resorting to magnetic fields. These can be externally imposed or generated dynamically by the plasma itself. Now we need to use the Maxwell equations for the fields, and the Vlasov–Maxwell system reads (written only for electrons to simplify notation),

$$\frac{\partial f}{\partial t} + \mathbf{c} \cdot \frac{\partial f}{\partial \mathbf{r}} - \frac{e}{m}\left(\mathbf{E} + \mathbf{c} \times \mathbf{B}\right) \cdot \frac{\partial f}{\partial \mathbf{c}} = 0, \tag{6.56}$$

$$\nabla \cdot \mathbf{E} = \frac{\rho}{\varepsilon_0}, \qquad \nabla \times \mathbf{B} - \frac{1}{c^2}\frac{\partial \mathbf{E}}{\partial t} = \mu_0 \mathbf{J}, \tag{6.57}$$

$$\nabla \cdot \mathbf{B} = 0, \qquad \nabla \times \mathbf{E} + \frac{\partial \mathbf{B}}{\partial t} = 0, \tag{6.58}$$

with

$$\rho = \rho_{\mathrm{ext}} - e\int f\,\mathrm{d}^3 c, \qquad \mathbf{J} = \mathbf{J}_{\mathrm{ext}} - e\int \mathbf{c} f\,\mathrm{d}^3 c. \tag{6.59}$$

As is quite evident from their structure, the equations also gain in complexity. We refer the interested reader to the books suggested at the end of the chapter, where detailed analysis of the behaviour under magnetic fields is presented.

6.4.2 Hydrodynamic equations

Similarly to the Boltzmann equation, it is possible to derive here hydrodynamic equations for plasmas. First the conservation equations for mass and momentum are obtained. The latter needs a closure for the pressure tensor, where several of the techniques we have discussed in previous chapters can be applied. The simplest is to assume that the distribution function remains close to a Maxwellian, hence the stress tensor is just given by the pressure of an ideal gas. Also, when rapid processes take place, the evolution is no longer isothermal but rather isentropic (as described by the Euler equations). In this case the pressure is rather given by $p = Cn^\gamma$, where C is a constant and $\gamma = c_p/c_v$ is the ratio of the specific heats. The hydrodynamic equations for the density n and velocity $\mathbf{v}$ of the electrons are[19]

[19]See Exercise 6.11.

$$\frac{\partial n}{\partial t} + \nabla \cdot (n\mathbf{v}) = 0, \tag{6.60}$$

$$mn\left(\frac{\partial \mathbf{v}}{\partial t} + (\mathbf{v} \cdot \nabla)\mathbf{v}\right) = -en(\mathbf{E} + \mathbf{v} \times \mathbf{B}) - \nabla p, \tag{6.61}$$

where the Lorentz term appears as the average of the Lorentz force in the Vlasov equation. These hydrodynamic equations must be analysed coupled with the Maxwell equations.

The hydrodynamic equations, which have a much simpler structure than the Vlasov kinetic equation, allow one to perform more detailed analysis of many situations that become complex at the kinetic level. See for example the study of waves in the presence of magnetic fields in Exercise 6.12. However, because of their simplified description, they miss some of the physics, and, for example, Landau damping is not captured at the hydrodynamic level, as shown in Exercise 6.13.

6.5 Self-gravitating systems

6.5.1 Kinetic equation

The stars in a galaxy interact through a potential that is very similar to the Coulomb potential, except that the masses (which replace the role played by the charges) are always positive and the interaction is attractive. These modifications simply imply a change of sign in the Poisson equation but, as we will see, have important effects on the dynamics. The most prominent is the absence of screening. Indeed, as all the interactions are attractive, the motion of some particles, attracted by a large intruder, will only reinforce the force instead of screening it. Also, a homogeneous collection of stars cannot be stable, and inhomogeneities spontaneously develop. As a result, galaxies form.

The Vlasov–Poisson equation for a self-gravitating system of particles of equal mass m is written as

$$\frac{\partial f}{\partial t} + \frac{\mathbf{p}}{m} \cdot \frac{\partial f}{\partial \mathbf{r}} + \mathbf{F}_{\mathrm{mf}} \cdot \frac{\partial f}{\partial \mathbf{p}} = 0, \tag{6.62}$$

$$\mathbf{F}_{\mathrm{mf}} = -m\nabla\Phi_{\mathrm{mf}}, \tag{6.63}$$

$$\nabla^2\Phi_{\mathrm{mf}} = 4\pi Gmn; \quad n(\mathbf{r}) = \int f(\mathbf{r}, \mathbf{p})\, \mathrm{d}^3p, \tag{6.64}$$

where $G = 6.674 \times 10^{-11} \, \mathrm{N \, m^2/kg^2}$ is the gravitational constant. Here we prefer to use the momentum instead of the velocity notation, as will be convenient in the next section. This system is sometimes referred to as the collisionless Boltzmann equation in the astrophysics community, implying that direct star–star close collisions have been discarded and only mean far-field interactions are included in the model.

6.5.2   Self-consistent equilibrium solutions

The Vlasov system can be written in a more compact form as a Poisson bracket,

$$\frac{\partial f}{\partial t} = -\{H_{\mathrm{mf}}, f\}, \qquad (6.65)$$

where H_{mf} is the mean-field one-particle Hamiltonian,

$$H_{\mathrm{mf}} = \frac{p^2}{2m} + m\Phi_{\mathrm{mf}}(\mathbf{r}). \qquad (6.66)$$

We know from classical mechanics (see Chapter 2) that any function of the Hamiltonian is a stationary solution of eqn (6.65). That is, in principle, we have found that the equilibrium distribution of stars in a galaxy can be written as

$$f_0(\mathbf{r}, \mathbf{p}) = g(H_{\mathrm{mf}}(\mathbf{r}, \mathbf{p})) \qquad (6.67)$$

for any function g. However, we also have to satisfy the Poisson equation, which is now written,

$$\nabla^2 \Phi_{\mathrm{mf}} = 4\pi Gm \int g\left(\frac{p^2}{2m} + m\Phi_{\mathrm{mf}}(\mathbf{r})\right) \mathrm{d}^3 p. \qquad (6.68)$$

This equation, for self-consistency of the mean-field potential, imposes restrictions on the possible functions g.

Let us first consider a simple model where we assume that the distribution function is that of thermal equilibrium at some temperature T. Then, $g(H_{\mathrm{mf}}) = n_0 (\frac{m}{2\pi k_{\mathrm{B}} T})^{3/2} e^{-H_{\mathrm{mf}}/k_{\mathrm{B}} T}$, where n_0 is a normalisation density. Note that there is no good reason to have a thermally equilibrated system, because we have neglected collisions, which are indeed extremely rare in astrophysical systems. In fact, we must interpret T as giving the dispersion in the speeds of the stars. The self-consistency equation (6.68) simplifies to

$$\nabla^2 \Phi_{\mathrm{mf}} = 4\pi Gm \int A e^{-\frac{p^2}{2mk_{\mathrm{B}} T}} e^{-\frac{m\Phi_{\mathrm{mf}}(\mathbf{r})}{k_{\mathrm{B}} T}} \mathrm{d}^3 p = 4\pi Gmn_0 e^{-\frac{m\Phi_{\mathrm{mf}}(\mathbf{r})}{k_{\mathrm{B}} T}}. $$
$$(6.69)$$

Assuming spherical symmetry, the differential equation is solved using the ansatz, $\Phi_{\mathrm{mf}} = a + b \ln r$, which results in $mb/k_{\mathrm{B}} T = 2$ and $b = 4\pi Gmn_0 e^{-ma/k_{\mathrm{B}} T}$. The expression for the mean-field potential is not very illustrative, but we note that the particle density is, except for

[20]Note that, if we were to introduce a radius cutoff in the distribution to solve this divergence, the resulting distribution function would not be a function of the Hamiltonian only, and therefore would not be a solution of the Vlasov–Poisson system.

[21]This is a fundamental unsolved problem in statistical mechanics because self-gravitating systems do not appear to be described by either the canonical or microcanonical distribution (Campa et al., 2014).

a factor of $4\pi Gm$, the right-hand side of eqn (6.69). Performing the substitutions, the density is given by the simple expression,

$$n = \frac{n_0 e^{-ma/k_BT}}{r^2};\qquad(6.70)$$

that is, the mass density in the isothermal model for a galaxy decays as $1/r^2$. This result is problematic because it would imply that the total mass of the galaxy (integrated to infinity) diverges. This is clearly not the case, as astronomical observations indicate.[20] The conclusion is thus that the isothermal model, although appealing because of its simplicity and because the self-consistency equation can be solved analytically, does not describe the stationary states of galaxies. This failure has several reasons, the principal one being that stars inside galaxies have not suffered enough close collisions to have equilibrated the system by means of momentum exchange processes; that is to say, they are not in thermal equilibrium.[21] However, galaxies appear to be roughly in steady state, and hence some kind of equilibration (although not thermal) has taken place.

As an improvement on the isothermal model, the Plummer model describes globular clusters with spherical shape rather well. It is defined by the function,

$$g(E) = \begin{cases} A|E|^{7/2} & E < 0 \\ 0 & E > 0 \end{cases},\qquad(6.71)$$

where A is a constant that will give the total mass of the cluster. To solve the self-consistency equation, it is more convenient to use $\Psi = -\Phi_{\mathrm{mf}}$, which will be positive. The mass density is then given by

$$\rho(\mathbf{r}) = A \int_{|\mathbf{p}|<p_m} \left(\frac{p_m^2 - p^2}{2m}\right)^{7/2} d^3p,\qquad(6.72)$$

where $p_m = \sqrt{2m^2\Psi}$ is the upper limit imposed in the Plummer model. The integral can be computed explicitly, giving $\rho(\mathbf{r}) = B\Psi(\mathbf{r})^5$, where B groups all the factors. The Poisson equation, $\nabla^2\Psi = -4\pi Gm\rho$, admits the simple solution,

$$\Psi(\mathbf{r}) = \frac{\Psi_0}{(r^2 + r_0^2)^{1/2}},\qquad \rho(\mathbf{r}) = \frac{\rho_0}{(r^2 + r_0^2)^{5/2}},\qquad(6.73)$$

with appropriate constants (Fig. 6.8). Notably, in the Plummer model the total mass is finite and there are no divergences at the origin either. The density decays sufficiently fast that, at far distances, the potential decays as $1/r$, as expected for a concentrated finite mass.

In galaxies and other stellar objects, besides energy, angular momentum is also conserved. Therefore, the Vlasov equation also admits arbitrary functions of the energy and angular momentum as stationary solutions, $f_0 = g(H_{\mathrm{mf}}, \mathbf{L})$. These models allow the description of non-spherical galaxies. For more detail on the different models for stellar objects, see some of the references at the end of the chapter.

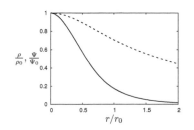

Fig. 6.8 Mass density ρ (solid line) and gravitational potential $\Psi = -\Phi$ (dashed line) of a self-gravitating cluster described by the Plummer model.

6.5.3 Jeans instability

As anticipated, a self-gravitating system cannot screen masses, but rather the gravitational attraction will generate a reinforcement of an initially added mass. To quantify this effect, we will use the hydrodynamic description for self-gravitating systems,[22]

[22]See Exercises 6.15 and 6.16 for a kinetic description of this phenomenon.

$$\frac{\partial \rho}{\partial t} + \nabla \cdot (\rho \mathbf{v}) = 0, \tag{6.74}$$

$$\left(\frac{\partial \mathbf{v}}{\partial t} + (\mathbf{v} \cdot \nabla)\mathbf{v}\right) = -\frac{\nabla P}{\rho} - \nabla \Phi, \tag{6.75}$$

$$\nabla^2 \Phi = 4\pi G \rho, \tag{6.76}$$

where $\rho = mn$ is the mass density.

Initially the system is at equilibrium with a homogeneous density ρ_0 and vanishing velocity. Consider that there are small density and velocity perturbations, ρ_1 and $\mathbf{v}_1$, that allow for linearisation of the equations and the use of spatial Fourier modes. The pressure, which depends on density, is expanded as $P = P_0 + c_s^2 \rho_1$, where c_s is the sound velocity. Keeping terms up to linear order, after eliminating the velocity, we obtain

$$\frac{\partial^2 \rho_1}{\partial t^2} = -c_s^2 k^2 \rho_1 + 4\pi G \rho_0 \rho_1. \tag{6.77}$$

The associated frequencies of the normal modes are $\omega^2 = c_s^2 k^2 - 4\pi G \rho_0$, which can become negative, and therefore an instability builds up when k is small enough. That is, for large wavelengths, the gravitational attraction wins over the restituting pressure, with a threshold given by the Jeans wavevector, $k_J = \sqrt{4\pi G \rho_0 / c_s^2}$. This result indicates that it is not possible to have a stable homogeneous self-gravitating system; rather, inhomogeneous configurations such as the Plummer model develop.

6.6 Beyond mean field

6.6.1 Velocity relaxation and dynamical friction

In 1943, Chandrasekhar studied the effect of binary encounters on the dynamics of self-gravitating systems in an attempt to describe the relaxation processes that take place in galaxies. The detailed calculations of these processes and their implications for the evolution of stellar systems can be found in Binney and Tremaine (2008). Here, we present a simple analysis that will help us to understand the difficulties of including the effects of granularity, but also indicates why the mean field description is so successful.

Consider a tagged star that moves in a galaxy of radius R, which is composed of N stars of the same mass m as the tagged one. The velocity of the tagged star is $\mathbf{c}$. After each encounter with another star it is deflected by $\delta\mathbf{c}$. The perpendicular component of the deflection is easily computed for the gravitational interaction (see Appendix C) and for small $\delta \mathbf{c}$ is given by

$$\delta c_\perp = \frac{2Gm}{bc}, \tag{6.78}$$

where b is the impact parameter. When the star crosses the galaxy, the number of encounters with impact parameter between b and $b + \mathrm{d}b$ is

$$\mathrm{d}n = \frac{N}{\pi R^2} 2\pi b \, \mathrm{d}b, \tag{6.79}$$

where we have assumed that the stars are homogeneously distributed in the transverse area, with an areal density of $N/(\pi R^2)$. On average, the encounters, which are randomly oriented, give a vanishing contribution to $\delta c_\perp$, but the mean square change is finite.[23] After crossing the galaxy, the accumulated value is

$$\Delta c^2 = \sum \delta c_\perp^2 = \frac{2N}{R^2} \int \left(\frac{2Gm}{bc} \right)^2 b \, \mathrm{d}b, \tag{6.80}$$

where we have assumed that subsequent interactions are uncorrelated. The upper integration limit is R, while the lower one is quite arbitrary, but we can take it as $b_{90} = 2Gm/c^2$, for which the deflection produced by one encounter is $90°$. This gives $\Delta c^2 = 8N(Gm/Rc)^2 \ln(R/b_{90})$. Now, using that for Keplerian orbits, $c^2 = GNm/R$, we obtain

$$\frac{\Delta c^2}{c^2} \approx \frac{8 \ln(R/b_{90})}{N}. \tag{6.81}$$

That is, the effect of granularity becomes extremely small when increasing the number of stars N, which explains the extraordinary success of the mean field descriptions.[24]

When the object under study is more massive, it is relevant to compute also the parallel component of the deflection, which does not vanish and indeed points opposite to the velocity. The result obtained by Chandrasekhar is a net dynamical friction which, in this case, does not depend on N and rather gives a finite value in the thermodynamic limit. The amplitude, however, depends on the Coulomb logarithm $\ln(R/b_{90})$, which reflects the long-range character of the force. This result is another indication of the difficulties in taking the thermodynamic limit in long-range systems. The dynamical friction can also be studied as a dynamic screening effect (Section 6.2.5) where, as an effect of motion, the screening cloud is left behind generating the drag force.

Combining the parallel and perpendicular contributions to the velocity deflection for the massive object, a Fokker–Planck equation can be written, using the fact that most of the encounters produce small deflections. This equation successfully describes the relaxation to equilibrium of objects falling into a galaxy.

6.6.2   Slow relaxation

The velocity relaxation in stellar systems evidences that non-mean-field effects are weak for large systems and the associated relaxation processes are slow. This problem has attracted great attention, and important progress has been made by considering simpler models that present

[23]In fact, for an inhomogeneous distribution of stars, the mean deflection does not vanish and rather is given by the mean-field Vlasov description. The mean square change computed here gives the deviation from mean field.

[24]In a typical galaxy, $N \sim 10^{11}$, resulting in enormous relaxation times, but for smaller objects, such as globular clusters with $N \sim 10^5$, the effects of granularity can be observed.

long-range interactions. The Hamiltonian mean field describes an abstract system of N particles with only angular coordinates θ_i with the Hamiltonian,

$$H = \sum_{i=1}^{N} \frac{p_i^2}{2} + \frac{\epsilon_J}{2N} \sum_{i,j} [1 - \cos(\theta_i - \theta_j)],\qquad(6.82)$$

where the sum runs over all pairs and ϵ_J is the coupling constant. Its mean feature is that, defining the magnetisation vector $\mathbf{m} = (m_x, m_y)$, with $m_x = \sum_i \cos\theta_i/N$ and $m_y = \sum_i \sin\theta_i/N$, the Hamiltonian can be written in the simple form,

$$H = \sum_{i=1}^{N} \frac{p_i^2}{2} + \frac{\epsilon_J N}{2}(1 - m^2),\qquad(6.83)$$

from where the label 'mean field' stems.

Extensive numerical simulations of this and similar models have shown that the evolution takes place in two stages. First, in what is called the violent relaxation phase, the system evolves, in a time scale that does not depend on the number of particles, from the initial state that generically will not be Vlasov stable. After the violent relaxation, the system stays for long times in quasistationary states that are Vlasov stable (see Fig. 6.9). Then, in the second stage, granularity effects apply and the much slower collisional relaxation takes place. During this stage, the system evolves through Vlasov-stable states until it reaches the equilibrium Boltzmann–Gibbs distribution. The duration of this phase depends on the number of particles and scales as N^δ, with an exponent δ that depends on the model, as in the case of the dynamical friction of Chandrasekhar, where we obtained $N/\ln N$.

In Section 4.3.3 we discussed the irreversibility problem and how this can be reconciled with the Poincaré recurrence in Hamiltonian systems. The conclusion was that irreversible behaviour is observed for large systems and observation times smaller than e^N. Here, we found that, in large systems, irreversibility only shows up for times longer than N^δ. Hence, the conclusion is that, for systems with long-range interactions, irreversible behaviour manifests in the time window $N^\delta \ll t \ll e^N$, which is very wide for large N.

6.6.3   Kinetic equations

Using the BBGKY hierarchy as a starting point, it is possible to build kinetic equations that go beyond the mean field Vlasov model. These equations include the fluctuations around mean field that result from correlations or, equivalently, from the ultimate granularity of the system. In plasmas, the most remarkable equations are those of Lenard–Balescu and Landau. These equations lie beyond the scope of the present book, but it is nevertheless instructive to mention that they are irreversible and describe the evolution of plasmas toward thermal equilibrium.

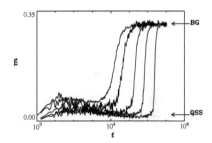

Fig. 6.9 Time evolution of the modulus of the magnetisation $m(t)$ in the Hamiltonian mean field for different particle numbers: $N = 10^3$, 2×10^3, 5×10^3, 10^4, and 2×10^4 from left to right ($\epsilon_J = 0.69$). In all cases, an average over several samples has been taken. The arrows indicate the magnetisation in the quasistationary state (QSS) and the Boltzmann–Gibbs equilibrium state (BG). Reprinted from Campa *et al.*, Copyright (2009), with permission from Elsevier.

For self-gravitating systems, there is the additional complexity that the reference states are nonhomogeneous (for example, galaxies). As the interactions are long range, it is not a matter of simply assuming local homogeneity. This issue makes it hard to build general kinetic equations including non-mean-field effects and, rather, ad hoc models are built as in the case of dynamical friction.

Finally, in a phenomenological approach, the Kramers–Chandrasekhar equation is a proposal to extend the Vlasov equation, considering the dynamical friction and transversal diffusion processes. It adds Fokker–Planck terms, giving

$$\frac{\partial f}{\partial t} + \mathbf{c} \cdot \frac{\partial f}{\partial \mathbf{r}} + \frac{\mathbf{F}_{\text{mf}}}{m} \cdot \frac{\partial f}{\partial \mathbf{c}} = \frac{\partial}{\partial \mathbf{c}} \left[D \left(\frac{\partial f}{\partial \mathbf{c}} + \beta m f \mathbf{c} \right) \right], \tag{6.84}$$

where D is the effective velocity diffusion coefficient and β is the friction coefficient such that the dynamical friction force is $\mathbf{F}_{\text{friction}} = -D\beta m\mathbf{v}$. Although this equation is appealing, it must be considered an approximate description as it is based on unjustified hypotheses, for example that the correlation times are small.

6.7 Application: point vortices in two dimensions

The concepts behind the Vlasov equation can be applied to many systems that present long-range interactions. The case of point vortices in two dimensions is an interesting example; in spite of being completely different of plasmas and self-gravitating systems, it is described by similar kinetic models.

In low viscous regimes, fluids are described by the Euler equation for the velocity field $\mathbf{v}$,

$$\frac{\partial \mathbf{v}}{\partial t} + (\mathbf{v} \cdot \nabla)\mathbf{v} = -\frac{\nabla p}{\rho}, \tag{6.85}$$

where ρ is the mass density and p is the pressure. For incompressible flows, the mass conservation equation reads $\nabla \cdot \mathbf{v} = 0$, which for two-dimensional flows implies that there exists a streamfunction $\psi(x, y, t)$ from where the velocity field reads,

$$\mathbf{v} = \frac{\partial \psi}{\partial y}\hat{\mathbf{x}} - \frac{\partial \psi}{\partial x}\hat{\mathbf{y}} = -\hat{\mathbf{z}} \times \nabla \psi. \tag{6.86}$$

The evolution of the vorticity field, defined by $\boldsymbol{\omega} = \omega\hat{\mathbf{z}} = \nabla \times \mathbf{v}$, is obtained directly from the Euler equation:

$$\frac{\partial \omega}{\partial t} + \mathbf{v} \cdot \nabla \omega = 0. \tag{6.87}$$

That is, vorticity is transported by the flow implying that the circulation $\Gamma = \oint \omega \, \mathrm{d}^2 r$ is conserved under transport (Kelvin theorem). It is direct to obtain that

$$\nabla^2 \psi = -\omega, \tag{6.88}$$

with the solutions written in integral form,

$$\psi(\mathbf{r}) = -\frac{1}{2\pi} \int \omega(\mathbf{r}') \ln|\mathbf{r} - \mathbf{r}'| \, \mathrm{d}^2 r', \quad \mathbf{v}(\mathbf{r}) = \frac{\hat{\mathbf{z}}}{2\pi} \times \int \omega(\mathbf{r}') \frac{\mathbf{r} - \mathbf{r}'}{|\mathbf{r} - \mathbf{r}'|^2} \, \mathrm{d}^2 r'.$$

$$(6.89)$$

The kinetic energy of the fluid can be written as

$$E = \frac{\rho}{2} \int v^2 \, \mathrm{d}^2 r = -\frac{\rho}{4\pi} \int \omega(\mathbf{r})\omega(\mathbf{r}') \ln|\mathbf{r} - \mathbf{r}'| \, \mathrm{d}^2 r \, \mathrm{d}^2 r'. \qquad (6.90)$$

Notably, it corresponds to a long-range interaction for the vorticity.

It is common in many experimental situations that the vorticity is highly localised. In three dimensions, vortex filaments form and in two dimensions, which is the case under study here, vorticity localises in points. When several of these are placed in the system, we write $\omega(\mathbf{r}) = \sum_i \gamma_i \delta(\mathbf{r} - \mathbf{r}_i)$, where γ_i is the circulation of each vortex. The point vortices generate a velocity field (eqn (6.89)) and therefore the velocity of the vortex i is $\mathbf{V}_i = -\sum_k \frac{\gamma_k}{2\pi} \hat{\mathbf{z}} \times \frac{\mathbf{r}_i - \mathbf{r}_k}{|\mathbf{r}_i - \mathbf{r}_k|^2}$. Kirchhoff noticed that this dynamics can be cast in a Hamiltonian form,

$$\gamma_i \dot{x}_i = \frac{\partial H}{\partial y_i}, \quad \gamma_i \dot{y}_i = -\frac{\partial H}{\partial x_i}, \quad H = -\frac{1}{2\pi} \sum_{i<k} \gamma_i \gamma_k \ln|\mathbf{r}_i - \mathbf{r}_k|. \quad (6.91)$$

This Hamiltonian structure allows the use of the techniques of statistical mechanics to study the equilibrium properties. Here x_i and y_i are conjugate variables and there is no kinetic energy term because of the point vortices' lack of inertia. The potential energy is long ranged, as anticipated before, indicating that the dynamics of the point vortex gas can be studied using a mean-field approach. Indeed, it is possible to write the BBGKY hierarchy for the reduced distribution functions and, assuming absence of correlations, a Vlasov-like equation is obtained. Alternatively, we consider the coarse-grained vorticity, $\overline{\omega}(\mathbf{r}, t)$ which, assuming lack of correlations between the point vortices satisfies the equation[25]

$$\frac{\partial \overline{\omega}}{\partial t} + \overline{\mathbf{v}} \cdot \nabla \overline{\omega} = 0, \quad \overline{\mathbf{v}} = -\hat{\mathbf{z}} \times \nabla \overline{\psi}, \quad \nabla^2 \overline{\psi} = -\overline{\omega}. \quad (6.92)$$

[25]The absence of correlation is used to perform the factorisation $\overline{\mathbf{v}\omega} = \overline{\mathbf{v}}\,\overline{\omega}$.

Similarly to the other cases studied in the chapter, this equation has been proved to be valid in the limit of the number of particles going to infinity. Otherwise, for finite systems, granularity effects appear and the system relaxes to equilibrium states via Fokker–Planck terms. At the level of the mean-field description, the point vortex gas shows also violent relaxation to Vlasov-stable states.

The Vlasov–Poisson system (6.92) admits the self-consistent stationary solutions, $\overline{\omega} = g(\overline{\psi})$, for any function g. It is easy to verify that the velocity field is perpendicular to the gradient of $\overline{\omega}$, hence the convective term vanishes. In this case, the streamfunction is obtained from the self-consistent equation,

$$\nabla^2 \overline{\psi} = -g(\overline{\psi}). \qquad (6.93)$$

If the Fokker–Planck terms were added, the equilibrium solution $g(\overline{\psi}) = Ae^{-a\overline{\psi}}$ is selected. It is possible to solve the resulting Boltzmann–Poisson equation for different geometries. For example, for an axisymmetric vortex distribution, the coarse-grained distribution is

$$\overline{\omega} = \frac{\omega_0}{\left(1 + \frac{\pi\omega_0 r^2}{\Gamma}\right)^2},$$

(6.94)

where Γ and ω_0 are free parameters. The distribution is peaked near the centre, corresponding to a 'supervortex'.

Further reading

The book by Chen (1984) is an excellent introduction to plasma physics. A more advanced text that focuses on the kinetic approach is Ichimaru (1973). The general books on kinetic theory by Landau and Lifshitz (1981) and Liboff (2003) also provide extensive analysis of plasmas. The dynamics of self-gravitating systems is nicely presented in Binney and Tremaine (2008).

Landau damping is presented in the texts mentioned above, but I also encourage reading the original article by Landau (1946) (in an English translation from the Russian) or in his collected papers (Ter Haar, 1965).

Finally, the relaxation processes of systems with long-range interactions and the point vortex gas model are presented in Chavanis (2002) and Campa *et al.* (2014).

Exercises

(6.1) **Saha equation.** Consider a gas of electrons, protons, and neutral hydrogen atoms in equilibrium. Model them as ideal gases, where hydrogen atoms have a negative energy $\epsilon_0 = -13.6\,\text{eV}$ besides their kinetic energy. Imposing the restrictions that the system is neutral $N_e = N_p$ and that the total number of particles is conserved $N_p + N_H = \text{cst.}$, obtain the equilibrium concentration of each species using the tools of thermodynamics or equilibrium statistical mechanics. Derive the Saha equation (6.1).

(6.2) **Importance of close encounters.** Use the results of the scattering theory in Appendix C to estimate the number of encounters per unit time that produce deflections larger than $\pi/2$ in a plasma of density n_0 and temperature T. Compare with the plasma frequency, which is the other relevant time scale.

(6.3) **Ratio of average energies.** Show that, for any tagged charge, the average potential energy divided by its kinetic energy is proportional to the plasma parameter, $g = 1/N_D$.

(6.4) **Yukawa potential.** Solve the equation for the screened potential in the linear regime [eqns (6.5) and (6.6)].

(6.5) **Reversibility of the Vlasov equation.** Show that the Vlasov–Poisson system is reversible when the velocities are inverted; that is, consider that $f(\mathbf{r}, \mathbf{c}, t)$ is a solution of the Vlasov–Poisson system which describes the evolution from an initial to a final state. Show that $\overline{f}(\mathbf{r}, \mathbf{c}, t) = f(\mathbf{r}, -\mathbf{c}, -t)$, which traces the system back, is also a solution of the kinetic equation.

(6.6) **Yukawa potential for nonequilibrium plasmas.** In Section 6.2.5 we derived the Yukawa potential as the static response of an equilibrium plasma to a small external charge. However, it is not necessary for the plasma to be at thermal equilibrium to develop a Yukawa-like response. Consider an external static charge density ρ_{ext}. Using the general expression for the dielectric function (6.30) and assuming that the reference distribution function f_0 is a monotonically decreasing function of the momentum, show that the electric potential is given by the expression (6.34), with κ defined in terms of the distribution function.

(6.7) **Dynamic response in quasistatic plasmas.** Show that, for small frequencies compared with the plasma frequency, the dielectric function can be approximated by the static one and the response potential is given by the Yukawa potential.

(6.8) **Dynamic response under a travelling charge.** Consider the dynamical response in plasmas in the case of a travelling charge as discussed in Section 6.2.5. A direct calculation results in an undefined integral. The origin of the problem lies in the consideration that the charge was travelling from the very beginning, and hence the unperturbed state was not properly defined. The solution consists in assuming that, for infinitely negative time, the plasma was in equilibrium and the charge was turned on gradually. This is achieved by writing $\rho_{\text{ext}}(\mathbf{r}, t) = \rho_0(\mathbf{r} - \mathbf{v}t)e^{\delta t}$, where δ is a small positive quantity that we will later consider to be vanishingly small. The divergence for large positive times is not of concern, because by causality it can only affect the future and not the observation time where the exponential factor remains finite (of order one in the limit $\delta \to 0^+$). Perform the Fourier transform to show that now $\omega = \mathbf{k} \cdot \mathbf{v} + i\delta$ and that the integral remains finite. Using the Plemelj formula evaluate the imaginary part of the dielectric function. Interpret its sign.

(6.9) **Landau damping in an electron–positron plasma.** Obtain the Landau damping rate for a plasma composed of electrons and positrons. They have the same mass but opposite charge.

(6.10) **Ion-acoustic wave.** Consider a plasma composed of electrons and ions, with both species at the same temperature. Because of the large mass contrast, it is possible to study frequencies in the range where the wave is slow for electrons and therefore they respond with static Debye screening, while it is fast for ions; that is, we require that $\sqrt{k_B T/m_i} \ll \omega/k \ll \sqrt{k_B T/m_e}$. Study the propagation of waves in this regime and find the

dispersion relation $\omega(k)$. Show that the real part is not dispersive, with a velocity that can be interpreted as the sound speed for a neutral gas. Find the damping rate.

(6.11) **Hydrodynamic equations.** Derive the hydrodynamic equations (6.60) and (6.61) from the Vlasov equation (6.56).

(6.12) **Hydrodynamic description of plasma waves.** Here we consider the propagation of electromagnetic plasma waves. Consider a uniform plasma with density n and small temperature that we take to be zero. Hence the pressure also vanishes. The modes to be considered are those where the electric and magnetic fields generated in the wave are perpendicular to the wavevector and there are no external charges or currents, except for the ions that are inert and neutralise the system. Show that, in this case, the total charge density vanishes and only currents develop. Write the hydrodynamic and Maxwell equations and find the dispersion relation $\omega(\mathbf{k})$ for the waves. Show that the phase velocity exceeds the speed of light but the group velocity is always smaller, in agreement with special relativity.

(6.13) **Landau damping and hydrodynamic description.** Analyse the electrostatic waves described in Section 6.3.1 using the hydrodynamic formalism. Show that the real part of the dispersion relation is correctly found but that the Landau damping is missing.

(6.14) **Plummer model with a black hole.** Consider a spherical galaxy that has a black hole at its centre. For the stellar dynamics the black hole is non-relativistic and only acts as an additional mass M_{bh} at the centre. The mass density is then

$$\rho(\mathbf{r}) = M_{\text{bh}}\delta(\mathbf{r}) + m \int f(\mathbf{r}, \mathbf{p})\, d^3c.$$

Derive the self-consistency equation and investigate its solutions. It may be necessary to apply numerical methods to find the solution.

(6.15) **Kinetic description of the Jeans instability. I.** To study the Jeans instability using the Vlasov equation we would need to compute the 'gravitational dielectric function' $\widetilde{\epsilon}_G$, which indeed has the same structure as the electric counterpart except for the sign change. The problem here is more complex because the poles change from real to imaginary when moving k. A simple example that can be solved analytically is obtained by assuming that the temperature is negligible and therefore the base distribution function is a Dirac delta. Show that, in this case, the self-gravitating gas is unconditionally unstable. Compare with the hydrodynamic prediction.

(6.16) **Kinetic description of the Jeans instability. II.** Although it is difficult to obtain the dispersion relation from the kinetic theory, the Jeans wavevector can be obtained directly. Indeed, at this wavevector, the frequency vanishes and this condition can be used to compute k_J. Therefore, the equation that must be solved is $\tilde{\epsilon}_G(k_J, 0) = 0$. Obtain k_J in terms of an integral over the reference distribution function. Evaluate it for a Maxwellian.

(6.17) **Vlasov description for the Hamiltonian mean field model.** The equations of motion for the Hamiltonian (6.83) are

$$\dot{p}_i = J(m_y \cos\theta_i - m_x \sin\theta), \qquad \dot{\theta}_i = p_i.$$

The magnetisation vector, which in principle is computed using the information of all particles, can be approximated as a mean-field magnetisation $\overline{m}$ that disregards granularity effects. Then, the equations of motion can be read as coming from the mean-field one-particle Hamiltonian,

$$H_{\mathrm{mf}} = \frac{p^2}{2} - \epsilon_J(\overline{m}_x \cos\theta + \overline{m}_y \sin\theta),$$

analogue to the gravitational case (6.66).

Derive the Vlasov equation for the distribution function $f(\theta, p, t)$ and express the mean-field magnetisation in terms of f. Show that the system admits self-consistent solutions, $f(\theta, p) = g(p^2/2 - \epsilon_J(\overline{m}_x \cos\theta + \overline{m}_y \sin\theta))$ for any function g. Finally, solve the self-consistent equation for the thermal solution $g(E) = g_0 e^{-E/k_B T}$.

(6.18) **Conservations in the point vortex gas model.** The Hamiltonian equations (6.91) conserve the energy H, the angular momentum $L = \sum_i \gamma_i r_i^2$ and impulse $\mathbf{P} = \sum_i \gamma_i \mathbf{r}_i$. Write the conserved quantities in terms of the coarse-grained vorticity $\overline{\omega}$ and show that the mean-field equations respect the conservations.

(6.19) **Rotating structures in the point vortex gas model.** Consider the pseudostationary solution in polar coordinates, $\overline{\omega}(r, \phi, t) = \overline{\omega}_r(r, \phi - \Omega t)$ that describes a structure rotating uniformly with angular velocity Ω. Show that it satisfies

$$\frac{\partial \overline{\psi}}{\partial \phi} \frac{\partial \overline{\omega}_r}{\partial r} - \left(\Omega r + \frac{\partial \overline{\psi}}{\partial r}\right) \frac{\partial \overline{\omega}_r}{\partial \phi} = 0$$

and that it admits the self-consistent solution, $\overline{\omega}_r = g(\psi_r)$, where $\overline{\psi}_r = \overline{\psi} + \Omega r^2/2$ is the relative streamfunction and g is an arbitrary function.

Quantum gases

7

7.1 Boson and fermion ideal gases at equilibrium

7.1.1 Description of the quantum state

In Chapter 4 we studied the dynamics of a classical gas in the Boltzmann–Grad limit. In this limit the interactions were small enough that the thermodynamic properties were those of an ideal gas, while the relaxation times (proportional to the mean free time) remained finite. In this regime we derived and analysed the Boltzmann kinetic equation, which provided an accurate description of the gas dynamics and allowed us to extract the hydrodynamic behaviour of gases.

Here, we aim to derive a similar equation for quantum gases, that is, systems of identical particles where quantum effects become important but particle–particle interactions remain small. For classical gases, the object under study was the distribution function of positions and velocities. In the quantum case we have to proceed differently because it is no longer possible to give simultaneously the positions and momenta of the particles due to the Heisenberg uncertainty principle. Instead, we should give the quantum state of the particles. To begin, we consider free particles in a rectangular box of dimensions $L_x \times L_y \times L_z$ and volume $\mathcal{V} = L_x L_y L_z$, which are described by one-particle wavefunctions of the form,

$$\psi_{\mathbf{k}}(\mathbf{r}) = \frac{1}{\sqrt{\mathcal{V}}} e^{i\mathbf{k}\cdot\mathbf{r}}, \qquad (7.1)$$

where for simplicity we use periodic boundary conditions.[1] For the wavevectors $\mathbf{k}$ to be compatible with these boundary conditions, we should use

$$\mathbf{k} = \left(\frac{2\pi n_x}{L_x}, \frac{2\pi n_y}{L_y}, \frac{2\pi n_z}{L_z} \right), \qquad (7.2)$$

with $n_i \in \mathbb{Z}$. Besides the wavevectors, we need to specify the z component of the spin s_z, so finally the quantum state for each particle is of the form, $|\mathbf{k}, s_z\rangle$.

For N particles we could, in principle, specify the state of the whole system by giving the state of each particle. However, when particles are identical, it is not possible to tell which particle is which, and therefore the total state function must retain symmetry properties under particle exchange. Here, Nature offers two possibilities: particles with semi-integer spin, such as electrons, are called fermions and their wavefunc-

[1] This election of boundary conditions is not strictly necessary, but it simplifies the calculations enormously because this will allow us to use exponential Fourier functions instead of sine and cosine functions. The final results are nevertheless the same.

Kinetic Theory and Transport Phenomena, First Edition, Rodrigo Soto.
© Rodrigo Soto 2016. Published in 2016 by Oxford University Press.

[2]In fundamental particle physics, particles of matter such as the electron, muon, tau, neutrinos, and quarks are also fermions. Bosons are the particles responsible for transmitting forces, e.g. the photon, the gluons, and the W and Z intermediate bosons, responsible for the weak force. Composite particles can be either bosons or fermions depending on the number of particles they are made of, with the total spin obtained from the rules of addition of angular momentum in quantum mechanics. As an example, protons and neutrons are fermions, while pions are bosons. Atoms also behave as bosons or fermions depending on their total number of electrons, protons, and neutrons. For example, for the isotopes of helium, ^{3}He is a fermion whereas ^{4}He is a boson.

[3]Sometimes there is confusion here, as in elementary atomic physics it is said that two electrons can be in each orbital. Indeed, these electrons have different spin, and consistently only one electron can be in each quantum state.

tion must be totally antisymmetric under particle exchange; whereas particles with integer spin, such as photons, are called bosons and their wavefunction must be totally symmetric under particle exchange.[2]

This symmetrisation is easily done by adopting the so-called second quantisation scheme, according to which, instead of providing the state of each particle, we indicate how many particles are in each one-particle quantum state. The symmetry properties of the wavefunction imply that, for fermions, up to one particle can be in each quantum state (this property is also referred to as the Pauli exclusion principle[3]); meanwhile, for bosons, the number of particles in each quantum state is unbounded. Associated with each quantum state $|i\rangle = |\mathbf{k}, s_z\rangle$ there is an occupation number n_i satisfying

$$n_i = \begin{cases} 0, 1 & \text{for fermions} \\ 0, 1, 2, \dots & \text{for bosons.} \end{cases} \tag{7.3}$$

In this language, the state of the whole system is specified by giving the occupation numbers of each one-particle quantum state,

$$|\Psi\rangle = |n_1 n_2 \dots n_i \dots\rangle. \tag{7.4}$$

When we wrote (7.1), we were considering a gas of free particles where the energy associated with each quantum state was $\varepsilon = \hbar^2 k^2/2m$ for nonrelativistic particles and $\varepsilon = \hbar c k$ for the ultrarelativistic case. The discussion and formalisms of the second quantisation are, nevertheless, not limited to free particles, and we can consider particles that interact with an external potential. For example, in Chapter 8, we will consider electrons moving in the periodic potential generated by ions, to model metals and insulators. Still, we do not allow the particles to interact with each other, to deal first with ideal gases. Weak interactions will be added in Section 7.3 and the strong electron–electron interactions in a solid will be worked out in the Hartree–Foch approximation (see Section D.2.4). Given an external potential, it is possible to solve the time-independent Schrödinger equation to obtain the one-particle quantum eigenstates $|i\rangle$ with energies ε_i. Then, for a gas described by the state (7.4), the total energy and number of particles in the gas are

$$E = \sum_i n_i \varepsilon_i, \tag{7.5}$$

$$N = \sum_i n_i, \tag{7.6}$$

where the independence of the particles has been used to write the total energy as a simple sum.

In the case of free particles, the sums run over s_z and $\mathbf{k}$. When large systems are considered, the wavevectors become quasicontinuous, and it is possible to transform the sums into integrals; for example,

$$N = \sum_{s_z, \mathbf{k}} n_{s_z}(\mathbf{k}) = \mathcal{V} \sum_{s_z} \int \frac{\mathrm{d}^3 k}{(2\pi)^3} n_{s_z}(\mathbf{k}), \tag{7.7}$$

where we have used that in the interval $\Delta k_x \Delta k_y \Delta k_z$ there is a total of $(L_x \Delta k_x/2\pi)(L_y \Delta k_y/2\pi)(L_z \Delta k_z/2\pi)$ states for each value of s_z.

7.1.2 Equilibrium distributions

For a quantum gas in thermal equilibrium, the occupation numbers fluctuate as particles move among quantum states. Their average values can be easily computed using the standard tools of equilibrium statistical mechanics. In the canonical ensemble, where the temperature, volume, and number of particles are fixed, summing over all states (7.4) to compute the partition function is complex, because we should impose that $\sum_i n_i = N$ to keep the number of particles fixed. Taking advantage of the equivalence of thermodynamic ensembles, we do better to use the grand canonical ensemble where, instead of the number of particles, the chemical potential μ is fixed. In this ensemble, the probability of any state is

$$P(|\Psi\rangle) = \Xi^{-1} e^{-(E-\mu N)/k_B T}, \tag{7.8}$$

where

$$\Xi = \sum_{n_1, n_2, \ldots,} e^{-(E-\mu N)/k_B T} \tag{7.9}$$

is the grand canonical partition function and E and N are given by eqns (7.5) and (7.6), respectively. Here, bosons and fermions differ in the range over which the occupation numbers are summed. Owing to the linearity of E and N, the exponential factorises into a product of terms for each quantum state. Consequently, the probability for the occupancy of each quantum state is simply

$$P(n_i) = \frac{e^{-(\varepsilon_i-\mu)n_i/k_B T}}{\sum_n e^{-(\varepsilon_i-\mu)n/k_B T}}. \tag{7.10}$$

The sum can be done directly for fermions and bosons, leading to

$$P(n_i) = \left[1 \mp e^{-(\varepsilon_i-\mu)/k_B T}\right]^{\pm 1} e^{-(\varepsilon_i-\mu)n_i/k_B T}, \tag{7.11}$$

where the top sign is for bosons and the lower one for fermions.

Finally, the average occupation of a quantum state, $\langle n_i \rangle = \sum_{n_i} n_i P(n_i)$, is

$$\langle n_i \rangle = f_{FD}(\varepsilon_i) = \frac{1}{e^{(\varepsilon_i-\mu)/k_B T} + 1} \tag{7.12}$$

for fermions and

$$\langle n_i \rangle = f_{BE}(\varepsilon_i) = \frac{1}{e^{(\varepsilon_i-\mu)/k_B T} - 1} \tag{7.13}$$

for bosons.

Equation (7.12) is the Fermi–Dirac distribution, shown in Fig. 7.1, which states that, for energies smaller than the chemical potential, the average occupation tends to one, while for energies larger than the chemical potential, the average occupation is small. In the limit of low temperatures, the Fermi–Dirac distribution approaches a step function and the chemical potential marks an abrupt separation between occupied and unoccupied states.

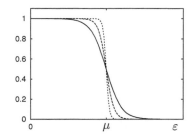

Fig. 7.1 Fermi–Dirac distribution, $f_{FD}(\varepsilon) = \frac{1}{e^{(\varepsilon-\mu)/k_B T}+1}$, presented for different temperatures, decreasing from the solid line to the dotted line.

The Bose–Einstein distribution in eqn (7.13) is not bounded from above, reflecting the fact that an arbitrary number of bosons can fit into any quantum state. The sum (7.10) for bosons converges only if $\varepsilon_i - \mu$ is positive, indicating that the chemical potential must always be smaller than the lowest energy level (ground state). When the chemical potential approaches the fundamental energy level, the occupation number of the fundamental level becomes large.[4]

The quantum effects are lost if the average occupation numbers are small, in which case there are no interference effects and the difference between bosons and fermions becomes irrelevant. In both cases, the occupation numbers are small if $\lambda = e^{\mu/k_{\mathrm{B}}T} \ll 1$. In this limit, the expressions (7.12) and (7.13) can be expanded in powers of λ to give in both cases

$$\langle n_i \rangle = f_{\mathrm{MB}}(\varepsilon_i) = e^{-(\varepsilon_i - \mu)/k_{\mathrm{B}}T}, \tag{7.14}$$

which corresponds to the usual Maxwell–Boltzmann classical distribution. In classical ideal gases, $\mu_{\mathrm{id}} = -k_{\mathrm{B}}T \log[\frac{\mathcal{V}}{N}(\frac{2\pi m k_{\mathrm{B}}T}{h^2})^{3/2}]$, and the condition for small occupancy can be written as $\Lambda \ll d$, where $\Lambda = h/\sqrt{2\pi m k_{\mathrm{B}}T}$ is the thermal de Broglie wavelength and $d = (\mathcal{V}/N)^{1/3}$ is the mean distance between particles. Hence, quantum effects on the statistics are obtained when wavepackets can overlap because either temperatures are low or densities are high.

For the three distributions (7.12), (7.13), and (7.14), the chemical potential is determined by imposing that the total number of particles is N. To perform this inversion, it is necessary to know the values of the energy levels, and therefore μ depends on the external potential. The case of free particles is worked out in Exercise 7.1.

7.2   Einstein coefficients

To build up kinetic equations we need to know how the quantum statistics (for example, the Pauli exclusion principle) are included in the dynamics. As an illustrative case, we will study the interaction of matter with light, aiming to build a kinetic model for photons, which, we recall, are bosons. This analysis was performed by Einstein in 1916 and, among other things, leads to the understanding of how lasers work.

Consider a gas of independent atoms that have two energy levels $|1\rangle$ and $|2\rangle$ with energies E_1 and E_2, respectively. At thermal equilibrium, the fraction of atoms in each of these states is given by the Boltzmann factor: $\langle N_1 \rangle/N_{\mathrm{total}} = Z^{-1}e^{-E_1/k_{\mathrm{B}}T}$ and $\langle N_2 \rangle/N_{\mathrm{total}} = Z^{-1}e^{-E_2/k_{\mathrm{B}}T}$, where $Z = e^{-E_1/k_{\mathrm{B}}T} + e^{-E_2/k_{\mathrm{B}}T}$ is the partition function. Therefore, the relative population is

$$\frac{\langle N_2 \rangle}{\langle N_1 \rangle} = e^{-(E_2 - E_1)/k_{\mathrm{B}}T}. \tag{7.15}$$

In the same chamber as the atoms, there are photons which are at thermal equilibrium with the atoms. Hence, their temperature is also T. Photons are bosons with vanishing chemical potential,[5] and hence

[4]In the case of a boson gas in three dimensions, the occupation number of the fundamental level can be a macroscopic fraction of the total number of particles, leading to a phenomenon called Bose–Einstein condensation.

[5]In a photon gas $\mu = 0$ because the number of photons is not conserved.

the average number of photons in each energy level ε_i is

$$\langle n_i \rangle = \frac{1}{e^{\varepsilon_i/k_\mathrm{B}T} - 1}, \tag{7.16}$$

known as the Planck distribution. The energy levels for photons are given by the Planck–Einstein relation, $\varepsilon_i = \hbar\omega_i$, where ω_i is the frequency of the photon.

We have said that the photons and atoms are in thermal equilibrium, but according to thermodynamics, to do so they should exchange energy. We can imagine that the relevant processes are photon absorption by an atom in state $|1\rangle$ or emission of a photon by an atom in state $|2\rangle$ (Fig. 7.2). Energy conservation implies that the frequency of the associated photon should be $\hbar\omega = E_2 - E_1$. Related to each of these processes, there are two microscopic coefficients (A and B) that measure the affinity of the quantum states of the atom with electromagnetic radiation at the frequency ω, such that the rate of variation of the number of atoms in the state $|1\rangle$ can be written as

$$\frac{\mathrm{d}N_1}{\mathrm{d}t} = AN_2 - BN_1 n_\omega. \tag{7.17}$$

The first term accounts for spontaneous emission, which is proportional to the number of excited atoms in state $|2\rangle$, whereas the second term gives the absorption rate of photons, which is proportional to the number of photons and the number of atoms that can absorb the radiation.

At equilibrium, the right-hand side of eqn (7.17) should vanish. Substituting the equilibrium distributions (7.15) and (7.16) gives the relation, $B/A = 1 - e^{-\hbar\omega/k_\mathrm{B}T}$. This expression cannot be correct because it indicates that the microscopic coefficients—related to scattering cross sections—depend on temperature, which is a measure of the collective state of the system. Einstein realised that this inconsistency could be overcome if a third radiation–matter process is introduced, called stimulated emission (see Fig. 7.3). In this process, if an atom is in state $|2\rangle$, it can decay to state $|1\rangle$ with a rate proportional to the number of photons of frequency ω; that is, the presence of photons stimulates the decay of excited atoms. Considering this process, the kinetic equation (7.17) is corrected to become

$$\frac{\mathrm{d}N_1}{\mathrm{d}t} = AN_2 - BN_1 n_\omega + B'N_2 n_\omega, \tag{7.18}$$

where B' is the new microscopic coefficient. In the steady state, the vanishing of the right-hand side implies that $A = \langle n_\omega \rangle (B\langle N_1 \rangle/\langle N_2 \rangle - B')$. Substituting the stationary distributions at thermal equilibrium, one obtains

$$A = \frac{B}{e^{\hbar\omega/k_\mathrm{B}T} - 1}\left(e^{\hbar\omega/k_\mathrm{B}T} - \frac{B'}{B}\right). \tag{7.19}$$

Here, it is possible to satisfy this equation with temperature-independent coefficients if $B' = B = A$. The kinetic equation is then greatly simplified to

$$\frac{\mathrm{d}N_1}{\mathrm{d}t} = A\left(1 + n_\omega\right)N_2 - AN_1 n_\omega. \tag{7.20}$$

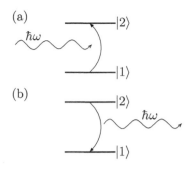

Fig. 7.2 Primary processes of matter–light interaction: (a) an atom in state $|1\rangle$ absorbs a photon of energy $\hbar\omega = E_2 - E_1$ and is excited to the state $|2\rangle$, and (b) an atom in the excited state $|2\rangle$ decays to the state $|1\rangle$ after emitting a photon of the same energy $\hbar\omega$.

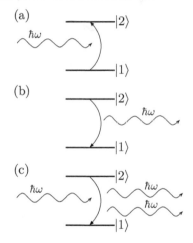

Fig. 7.3 Processes in the interaction of matter with light. Top: absorption of a photon provoking a jump from state $|1\rangle$ to $|2\rangle$. Middle: spontaneous emission of a photon with the corresponding decay of the atom from $|2\rangle$ to $|1\rangle$. Bottom: stimulated emission of a photon by the presence of another photon with the same frequency. The atom decays from $|2\rangle$ to $|1\rangle$ and the number of photons increases by one. Only by considering the three processes is it possible to have a consistent kinetic description of the equilibration of matter and light.

This equation is similar to (7.17), but where we should now consider that the reaction rate of the emission process is proportional to the number of photons that will be present in the system after the emission. That is, if there are photons already present, photon emission becomes more probable. This gregarious property of photons is shared by all bosons and is the opposite of what happens with fermions, where the existence of one particle in one state inhibits the reaction.

In Section 7.7 we will derive the spontaneous emission factor starting from the microscopic dynamics.

7.3   Scattering transition rates

We now aim to include interactions among identical particles. In the context of this book, we will only work with weak interactions, meaning that they can be worked out perturbatively. Typically, this will mean that the interaction potentials are short-ranged and that their intensity is weak. In a quantum gas, two possible approaches are possible to deal with interactions. We can use time-independent perturbation theory to compute the new eigenstates of the system and their corresponding energies. Alternatively, here we will use time-dependent perturbation theory (described in Appendix C), which uses the same quantum states as for the unperturbed system while the interactions generate transitions among them; that is, the interaction potential does not affect the quantum states or their energies, but rather enters through the scattering transition rates.

The interactions are described by a perturbation potential ΔU that is added to the free particle Hamiltonian. Time-dependent perturbation theory states that the effect of a perturbation ΔU is to produce transitions between eigenstates of the unperturbed Hamiltonian, with rates $\overline{W}_{\mathrm{i}\to\mathrm{f}}$ that depend on the initial and final state. Two cases are possible. If the perturbation potential is time independent, transitions only take place if the final and initial state have the same energy and

$$\overline{W}_{\mathrm{i}\to\mathrm{f}} = \frac{2\pi}{\hbar}\delta(\varepsilon_{\mathrm{f}} - \varepsilon_{\mathrm{i}})\left|\langle\mathrm{f}|\Delta U|\mathrm{i}\rangle\right|^2 . \qquad (7.21)$$

This expression is commonly known as Fermi's golden rule (see Appendix C for the derivation). When the perturbation depends on time, as is the case for external radiation, the perturbation is decomposed into Fourier modes, $\Delta U = \Delta U_\omega \sin(\omega t)$, and the transition rates are

$$\overline{W}_{\mathrm{i}\to\mathrm{f}} = \frac{\pi}{\hbar}\left[\delta(\varepsilon_{\mathrm{f}} - \varepsilon_{\mathrm{i}} - \hbar\omega) + \delta(\varepsilon_{\mathrm{f}} - \varepsilon_{\mathrm{i}} + \hbar\omega)\right]\left|\langle\mathrm{f}|\Delta U_\omega|\mathrm{i}\rangle\right|^2 ; \qquad (7.22)$$

that is, the final and initial states can differ by one quantum of energy $\hbar\omega$, which is provided to or taken away from the perturbation potential. This effect is normally represented as a vertex in a Feynman diagram (see Fig. 7.4).

Consider a box that contains particles that interact with weak potentials. We can analyse the scattering process where two particles arrive

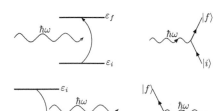

Fig. 7.4 Absorption (top) or emission (bottom) of one quantum on transition from the initial to final state induced by an external time-dependent potential. Energy level (left) and Feynman diagram (right) representations.

from asymptotic states $\mathbf{k}_1$ and $\mathbf{k}_2$ (for notational simplicity, we will not consider spin for the moment), exchange momentum and energy during a short time, and emerge with new asymptotic states $\mathbf{k}_1'$ and $\mathbf{k}_2'$ (Fig. 7.5). Using time-dependent perturbation theory, it is possible to compute the probability of this scattering process. It takes the form of a transition rate, $\overline{W}(\mathbf{k}_1, \mathbf{k}_2, \mathbf{k}_1', \mathbf{k}_1')$, such that $\overline{W}\,dt$ is the probability that the transition takes place in the time interval dt. The transition probabilities depend on the interaction potential and the incoming and outgoing states, and they are obtained using Fermi's golden rule (C.2.2).

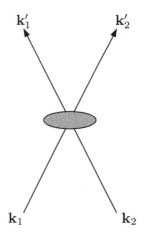

Fig. 7.5 Quantum scattering process for indistinguishable particles, where the labels '1' and '2' refer to states.

7.4 Master kinetic equation

We now write the kinetic equation for a quantum gas considering the statistical properties of the particles, with interactions described by transition rates $\overline{W}$. Here, we consider a gas of free particles with quantum states $|i\rangle = |\mathbf{k}, s_z\rangle$.

Let $f(i)$ be the coarse-grained number of particles in the state $|i\rangle$, whose equilibrium values are either the Bose–Einstein or Fermi–Dirac distributions. The coarse-graining process here consists in considering a small interval of wavevectors and averaging the occupation number in those intervals (Fig. 7.6). To do so, we will consider that the dimensions of the box are large, implying that the separation $2\pi/L$ between wavevectors (7.2) is small, allowing an average of neighbour states, which therefore have similar properties. The transition rates $\overline{W}$ are also averaged over those small intervals of wavevectors.

Fig. 7.6 Spatial coarse-graining procedure to construct the distribution function, schematised by an averaging window Δ, in this example encompassing five wavevectors. In practice, the averaging window should satisfy $2\pi/L \ll \Delta \ll 2\pi/a$, where a is a microscopic size; that is, the averaging takes place on a mesoscopic scale.

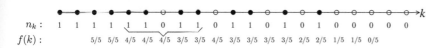

n_k : 1 1 1 1 1 1 0 1 1 0 1 1 0 1 0 1 0 0 0 0 0

$f(k)$: 5/5 5/5 4/5 4/5 4/5 3/5 3/5 4/5 3/5 3/5 3/5 3/5 2/5 2/5 1/5 1/5 0/5

In a large box, where the wavevectors are quasicontinuous, it is practical to integrate over them instead of performing a discrete sum. We will use the compact notation,

$$\int d1 \equiv \sum_{s_z} \int \frac{d^3 k_1}{(2\pi)^3} \qquad (7.23)$$

and analogous expressions for the other labels. In this context, the standard normalisation is that $\overline{W}(1, 2, 1', 2')\,d1\,d2\,d1'\,d2'\,dt$ is the number of scattering processes from states $1, 2$ to states $1', 2'$ in a time interval dt.

Without considering the quantum statistics of the particles, the average—coarse grained—number of scattering processes $|1\rangle|2\rangle \longrightarrow |1'\rangle|2'\rangle$ per unit time would be given by the transition rate times the occupation number of the initial states, i.e. $\overline{W}(1, 2, 1', 2')f(1)f(2)$, an expression that we used when we studied classical gases using the Boltzmann equation. Here, in the case of quantum gases, we must also include the occupation number of the final states. Indeed, consider the case of fermions. The final states $|1'\rangle$, $|2'\rangle$ can be either occupied or unoccupied, where the former case results in suppression of the transition according to the Pauli exclusion principle. When we coarse-grain in a wavevector interval, the fraction of successful transitions is $[1 - f(1')][1 - f(2')]$.[6] This factor must multiply the number of attempted scattering processes, resulting finally in $\overline{W}(1, 2, 1', 2')f(1)f(2)[1 - f(1')][1 - f(2')]$. In the case of bosons, we showed in Section 7.2 that the transition rates are amplified by a factor equal to the number of bosons in the final states, implying that the coarse-grained transition rate should be multiplied by $[1 + f(1')][1 + f(2')]$.

Analogously to the derivation of the classical Boltzmann equation in Section 4.2 or the master equation in Section 2.10, two processes are relevant when writing the kinetic equation. The occupation number $f(1)$ decreases by the direct scattering process $|1\rangle|2\rangle \longrightarrow |1'\rangle|2'\rangle$, while it increases by the inverse process $|1'\rangle|2'\rangle \longrightarrow |1\rangle|2\rangle$. So, we are now ready to write the kinetic equation for a quantum gas:

$$\frac{\partial f(1)}{\partial t} = \int \mathrm{d}2\,\mathrm{d}1'\,\mathrm{d}2'\,\big\{\overline{W}(1', 2', 1, 2)f(1')f(2')[1 \pm f(1)][1 \pm f(2)]$$
$$-\overline{W}(1, 2, 1', 2')f(1)f(2)[1 \pm f(1')][1 \pm f(2')]\big\}, \quad (7.24)$$

where the top sign $(+)$ is for bosons and the lower one $(-)$ is for fermions.[7]

Microscopic reversibility implies that the transition rates are symmetric under exchange of initial and final states. Also, the indistinguishability of particles makes them also symmetric under exchange of the labels '1' and '2' in both the initial and final states, that is,

$$\overline{W}(1, 2, 1', 2') = \overline{W}(1', 2', 1, 2) = \overline{W}(2, 1, 1', 2') = \overline{W}(1, 2, 2', 1'). \quad (7.25)$$

Using this property, the kinetic equation can be simplified to

$$\frac{\partial f(1)}{\partial t} = \int \mathrm{d}2\,\mathrm{d}1'\,\mathrm{d}2'\,\overline{W}(1', 2', 1, 2)\,\{f(1')f(2')[1 \pm f(1)][1 \pm f(2)]$$
$$-f(1)f(2)[1 \pm f(1')][1 \pm f(2')]\}. \quad (7.26)$$

This kinetic equation, sometimes referred to as the quantum Boltzmann equation, was first derived by L. W. Nordheim in 1928 and, later in 1933, rederived by E. A. Uehling and G. E. Uhlenbeck.

The conservation laws are included in the transition rates in the form of vanishing values if the conservations are not satisfied; for example, when energy and momentum are conserved, they are included as Dirac delta functions,

$$\overline{W}(1', 2', 1, 2) = \widehat{W}(1', 2', 1, 2)\delta(\varepsilon_1 + \varepsilon_2 - \varepsilon_1' - \varepsilon_2')\delta(\hbar\mathbf{k}_1 + \hbar\mathbf{k}_2 - \hbar\mathbf{k}_1' - \hbar\mathbf{k}_2'),$$
$$(7.27)$$

[6]Here, we have used the hypothesis that there are no correlations between the occupation numbers of different states, consistent with the assumption of absence of correlations when the expression $\overline{W}(1, 2, 1', 2')f(1)f(2)$ was written in the first place.

[7]This notation will be maintained for the rest of the chapter: the top sign is for bosons and the lower one for fermions.

where now $\widehat{W}$ can be a smooth function. It is straightforward to show that the macroscopic conservation laws are a direct consequence of the inclusion of the microscopic conservations in $\overline{W}$.[8]

[8]See Exercise 7.2.

7.5 Equilibrium solutions

The kinetic equation (7.26) is highly nonlinear, making it extremely difficult to analyse and find solutions under general conditions. However, the stationary solutions are easily found by a method similar to that used for the classical Boltzmann equation. Associated with the kinetic equation there in an H-theorem, where the H function is defined here as[9]

[9]As in the Boltzmann case, the H function is related (with a minus sign) to the entropy; see Exercise 7.3.

$$H = \int \mathrm{d}1 \left\{ f(1) \log f(1) \mp [1 \pm f(1)] \log[1 \pm f(1)] \right\}. \tag{7.28}$$

The time derivative of H is given in terms of that of f, which in turn is obtained from the kinetic equation (7.26). Performing transformations similar to those used in the classical case and using the symmetries (7.25) of the transition rates, one obtains that

$$\frac{\mathrm{d}H}{\mathrm{d}t} \leq 0. \tag{7.29}$$

Again, H can decrease monotonically to minus infinity or can asymptotically reach a finite value. The former case is not possible, because this would result in a non-normalisable distribution function. In summary, the kinetic equation predicts a monotonic evolution of the system to a final steady state.

The stationary solution is reached when the integrand on the right-hand side of eqn (7.26) vanishes for all combinations of the wavevectors. The conservation laws (7.27) automatically give a vanishing value whenever the conservations are not satisfied. Therefore, it remains to impose that $f(1')f(2')[1 \pm f(1)][1 \pm f(2)] = f(1)f(2)[1 \pm f(1')][1 \pm f(2')]$ for all scattering processes satisfying the conservations; an expression that can be recast as

$$\log \left[\frac{f(1)}{1 \pm f(1)} \right] + \log \left[\frac{f(2)}{1 \pm f(2)} \right] = \log \left[\frac{f(1')}{1 \pm f(1')} \right] + \log \left[\frac{f(2')}{1 \pm f(2')} \right]. \tag{7.30}$$

We therefore obtain that, for all processes fulfilling the conservation laws, the term $\log[f/(1 \pm f)]$ must be a collisional invariant in the stationary state. For free particles, the only collisional invariants are the mass, momentum, and kinetic energy, or any linear combination of them, so we can write,

$$\log \left[\frac{f}{1 \pm f} \right] = \alpha + \boldsymbol{\gamma} \cdot \hbar \mathbf{k} + \beta m \hbar^2 k^2 / 2, \tag{7.31}$$

where α, β, and γ are constants. Solving for f and redefining the free parameters, we obtain that the equilibrium solutions are

$$f_0 = \frac{1}{e^{[\hbar(\mathbf{k}-\mathbf{k}_0)^2/2m - \mu]/k_\mathrm{B}T} \mp 1}, \tag{7.32}$$

which are the Bose–Einstein (minus sign) or Fermi–Dirac (plus sign) distributions with a Galilean velocity $\hbar\mathbf{k}_0/m$, temperature T, and chemical potential μ. As in the case of the classical Boltzmann equation, the symmetries of the equation and the conservation laws determine entirely the equilibrium distribution, independent of the particular interaction potential between particles. The latter enters into the nonequilibrium properties, namely the transport coefficients and the approach rates to equilibrium via the transition rates $\overline{W}$.

As the stationary solutions of the kinetic equation are the equilibrium distribution for ideal quantum gases, the resulting thermodynamic properties will be those of ideal gases as well. Hence, as advanced in Section 7.1, this kinetic equation describes the properties of a gas with weak interactions, which generate finite relaxation rates, but with the thermodynamic properties of an ideal gas.

In a mixture of fermions and bosons, the kinetic equations for both species can be written in an analogous way, where now transition rates should be introduced for the fermion–fermion, boson–boson, and fermion–boson interactions. It can be shown that, in the presence of cross interactions, both species will reach equilibrium states with Fermi–Dirac and Bose–Einstein distributions that have equal temperature and Galilean velocity, but the chemical potentials will naturally be different (see Exercise 7.5).

7.6   Where is the molecular chaos hypothesis?

In the classical Boltzmann equation, presented in Chapter 4, we clearly signalled the point where the approximation of absence of correlations was made, in the so-called molecular chaos hypothesis. In that case, we started from the BBGKY equation and we approximated that, for particles going to collide, the two-particle distribution function could be factorised into a product of single particle distributions. Here, in the quantum case, we did not start from an equivalent of the BBGKY equation, and hence the place where the approximation was made is less transparent. Nevertheless, the same kind of hypothesis was made, assuming lack of correlations between particles that arrive to a scattering event.

In an ideal gas, the quantum state that describes a system is $|\Psi\rangle = |n_1 n_2 \ldots n_i \ldots\rangle$ (7.4), where the quantum phases are not specified because they are irrelevant. However, when interactions are considered, interference effects could affect the dynamics, and it is insufficient to give just the occupation numbers; the phases should be given as well. Then, if the interactions are weak, meaning long flight times, the phases $e^{i\varepsilon t/\hbar}$ will present many oscillations, and particles that meet will arrive with random, uncorrelated phases. We arrive at the random phase approximation, which is the starting point to derive the quantum Boltzmann equation from a formal N-body kinetic equation. Under these conditions, the scattering rates depend only on the number of particles in the referred states, and not on their phases.

The quantum Boltzmann equation can be derived formally from the von Neumann equation for the density matrix, which is the quantum version of the Liouville equation, when working with the Wigner distribution functions. Also, the Green function formalism can be used to derive the Boltzmann equation. In both cases, besides the presented hypothesis, it is required that the gas be close to homogeneous. Otherwise, the wavefunctions explore the inhomogeneities, creating correlations.

In summary, the Boltzmann equation is valid under semiclassical conditions, that is, for particles interacting through weak, short-range potentials, under the condition of near homogeneity and long mean free paths to avoid quantum correlation effects.

7.7 Phonons

7.7.1 Ideal gas of phonons

Phonons are the quanta of lattice vibrations in crystalline solids. We will see that they behave as bosons and their number is not a conserved quantity. When lattice vibrations are studied in the harmonic approximation, phonons emerge as ideal, non-interacting particles. Including anharmonicities results in phonon–phonon interactions similar to those studied previously in this chapter, but with the peculiarity that the number of phonons is not conserved. Finally, if electrons are placed in the lattice, their interaction with the ions can be cast as a phonon–electron interaction.

For notational simplicity, a one-dimensional model will be analysed, with later indications regarding the modifications required when moving to three dimensions and more complex models of solids.[10] Consider N ions interacting with a pair potential $U_{\text{ion–ion}}$, which presents a single minimum at the equilibrium distance a. The ion positions are R_i, and periodic boundary conditions are used such that $R_{N+1} = R_1$. At low temperatures, the ions will displace slightly from their equilibrium positions $R_i^0 = ia$. We start by considering the harmonic approximation by keeping the quadratic term in $u_i = R_i - R_i^0$ in the Hamiltonian,

$$H_{\text{harm.ions}} = \sum_{i=1}^{N} \left[\frac{p_i^2}{2M} + \frac{K}{2}(u_{i+1} - u_i)^2 \right], \qquad (7.33)$$

where M is the ion mass and $K = U''(a)$ is the effective elastic constant. Two successive changes of variables are performed. First, we move to Fourier space,

$$p_j = \frac{1}{\sqrt{N}} \sum_k \widetilde{p}_k e^{ikR_j^0}, \qquad \widetilde{p}_k = \frac{1}{\sqrt{N}} \sum_j p_j e^{-ikR_j^0}, \qquad (7.34)$$

$$u_j = \frac{1}{\sqrt{N}} \sum_k \widetilde{u}_k e^{ikR_j^0}, \qquad \widetilde{u}_k = \frac{1}{\sqrt{N}} \sum_j u_j e^{-ikR_j^0}, \qquad (7.35)$$

[10] Appendix D presents some relevant concepts of solid-state physics.

where the wavevectors are $k = 2\pi l/(aN)$, $l = -N/2, N/2 - 1, \ldots, N/2$.[11] Second, we define creation $b_k^\dagger$ and annihilation b_k operators,[12]

$$b_k = \frac{1}{\sqrt{2}}\left(\frac{\widetilde{u}_k}{l_k} + i\frac{\widetilde{p}_k}{\hbar/l_k}\right), \qquad b_k^\dagger = \frac{1}{\sqrt{2}}\left(\frac{\widetilde{u}_{-k}}{l_k} - i\frac{\widetilde{p}_{-k}}{\hbar/l_k}\right), \qquad (7.36)$$

$$\widetilde{p}_k = \frac{i\hbar}{l_k\sqrt{2}}\left(b_{-k}^\dagger - b_k\right), \qquad \widetilde{u}_k = \frac{l_k}{\sqrt{2}}\left(b_{-k}^\dagger + b_k\right), \qquad (7.37)$$

[12]The properties of these operators are found in any quantum mechanics text, where the creation and annihilation operators are also called ladder operators.

with

$$l_k = \sqrt{\frac{\hbar}{M\omega_k}}, \qquad \omega_k = 2\sqrt{\frac{K}{M}}\,|\sin(ka/2)|. \qquad (7.38)$$

[13]See Exercise 7.7.

These operators satisfy the commutation relations,[13]

$$[b_k, b_q] = 0, \qquad [b_k^\dagger, b_q^\dagger] = 0, \qquad [b_k, b_q^\dagger] = \delta_{k,q}, \qquad (7.39)$$

and the Hamiltonian now reads,

$$H_{\text{harm.ions}} = \sum_k \hbar\omega_k(b_k^\dagger b_k + 1/2). \qquad (7.40)$$

This is the Hamiltonian of a set of N independent harmonic oscillators, labelled with the wavevector k and frequencies ω_k. We recall that wavevectors and hence frequencies are bounded by the first Brillouin zone (FBZ): $|k| \leq \pi/a$ and $|\omega| \leq 2\sqrt{K/M}$. Note that, for small wavevectors, known as the Debye approximation, $\omega_k = ck$, with $c = a\sqrt{K/M}$ the sound speed. The Debye approximation fixes a characteristic energy scale $\hbar c\pi/a$, the value at the first Brillouin zone border.[14] Associated with this, the Debye temperature $\Theta_D = \pi\hbar c/(k_B a)$ distinguishes between low and high temperature regimes.

[14]The election of the FBZ to label the states is arbitrary, and other primitive cells are possible. Indeed, the reciprocal space is periodic with period $2\pi/a$, as can be verified in the dispersion relation (7.38) or in the Fourier transforms (7.34). The Debye approximation should therefore be applied with care, because either it is not periodic in k or nonanalyticities are introduced at the first Brillouin zone border, $\pm\pi/a$. See Appendix D for more details about the FBZ.

Defining the number operators, $N_k = b_k^\dagger b_k$, the eigenstates of the harmonic oscillators are labelled by the quantum numbers, $n_k = 0, 1, 2, \ldots$, where $N_k|n_k\rangle = n_k|n_k\rangle$ and the energies are $E|n_k\rangle = \hbar\omega_k(n_k+1/2)|n_k\rangle$. Hence, for the complete system we find that, analogously to the ideal quantum gases, the eigenstates of the system can be written as

$$|\Psi\rangle = |n_1 n_2 \ldots n_k \ldots\rangle, \qquad (7.41)$$

such that the energy of the harmonic solid is

$$E = \sum_k \hbar\omega_k n_k + E_0, \qquad (7.42)$$

with $E_0 = \sum \hbar\omega_k/2$ the zero-point energy that must be added to the minimum of the potential $U_{\text{ion-ion}}$ and

$$N_k|n_1 n_2 \ldots n_k \ldots\rangle = n_k|n_1 n_2 \ldots n_k \ldots\rangle. \qquad (7.43)$$

We can thus interpret the system as an ideal gas of ideal particles with energy levels $\varepsilon_k = \hbar\omega_k$. The number of particles in each level n_k is unbounded, and we have boson statistics. Finally, the number of these particles, called phonons, is not fixed, implying that they have a vanishing chemical potential (recall that the quantum numbers n_k are not conserved). In equilibrium, the average number of phonons in each level is given by the Planck distribution,

$$n_k^{\text{eq}} = \frac{1}{e^{\hbar\omega_k/k_B T} - 1}. \quad (7.44)$$

The total number of phonons in equilibrium, $N^{\text{eq}} = \sum_k n_k^{\text{eq}}$, depends on the temperature, with the following assymptotic values, valid in the Debye approximation:[15]

[15]See Exercise 7.8.

$$N^{\text{eq}} \propto \begin{cases} T^d & T \ll \Theta_D \\ T & T \gg \Theta_D \end{cases}. \quad (7.45)$$

Here, $d = 1, 2, 3$ is the dimensionality of the space.

The action of the creation and annihilation operators on (7.41) is given by the relations,

$$b_k |n_1 n_2 \ldots n_k \ldots\rangle = \sqrt{n_k} |n_1 n_2 \ldots (n_k - 1) \ldots\rangle, \quad (7.46)$$

$$b_k^\dagger |n_1 n_2 \ldots n_k \ldots\rangle = \sqrt{n_k + 1} |n_1 n_2 \ldots (n_k + 1) \ldots\rangle. \quad (7.47)$$

That is, they destroy or create, respectively, a phonon at the specified level.[16]

Phonons carry energy and therefore contribute to the heat current,[17]

[16]These properties are shown in any quantum mechanics text for the harmonic oscillator.

[17]In Chapter 8 we will analyse the contribution of electrons to the heat current.

$$q = \frac{1}{\mathcal{V}} \sum_k V(k)\hbar\omega_k n_k, \quad (7.48)$$

where $V(k) = \frac{\partial\omega}{\partial k}$ is the phase velocity. In the harmonic approximation, the occupation numbers are constants of motion (N_k commute with the Hamiltonian), implying that the heat current is conserved as well. No relaxation process to equilibrium exists, and persistent currents are possible, meaning that the thermal conductivity diverges. Scattering processes are needed to give finite values to the phonon mean free paths and result in finite conductivities. Phonons will acquire their full character as particles when we consider nonharmonic lattices or the electron–lattice interaction, where we will be able to model these effects as phonon–phonon and phonon–electron interactions, respectively, modifying their mean free path.

7.7.2   Phonon–phonon interactions

When lattice vibrations are large, nonharmonic contributions must be considered in the Hamiltonian for the ions. Keeping terms up to third order,[18]

[18]We will see that, under some circumstances, it will be necessary to go up to fourth-order terms.

$$H_{\text{non-harm.ions}} = \sum_{i=1}^N \left[\frac{p_i^2}{2M} + \frac{K}{2}(u_{i+1} - u_i)^2 + \frac{\Lambda}{6}(u_{i+1} - u_i)^3 \right], \quad (7.49)$$

$$= H_{\text{harm.ions}} + U_3, \quad (7.50)$$

where $\Lambda = U'''_{\text{ion}-\text{ion}}(a)$, $H_{\text{harm.ions}}$ is the Hamiltonian for ideal phonons that we have studied (eqns (7.33) or (7.40)), and U_3 is the perturbation, which we will consider to be small. Writing U_3 in terms of the Fourier-transformed operators, we obtain

$$U_3 = \frac{i\Lambda}{3\sqrt{N}} \sum_{k_1, k_2, k_3, G} \widetilde{u}_{k_1} \widetilde{u}_{k_2} \widetilde{u}_{k_3} \left[\sin k_1 a + \sin k_2 a + \sin k_3 a\right] \delta_{k_1 + k_2 + k_3, G},$$

(7.51)

where we have extensively used the identity,

$$\sum_j e^{iqR_j^0} = N \sum_G \delta_{q,G},$$

(7.52)

where the reciprocal vectors are $G = 2\pi l/a$, with $l \in \mathbb{Z}$.

This perturbation will induce transitions between eigenstates (7.41) of the unperturbed Hamiltonian, with rates that are computed using Fermi's golden rule. We then need to compute the bra-kets $\langle f | \widetilde{u}_{k_1} \widetilde{u}_{k_2} \widetilde{u}_{k_3} | i \rangle = \langle n_1^f n_2^f \ldots n_k^f \ldots | \widetilde{u}_{k_1} \widetilde{u}_{k_2} \widetilde{u}_{k_3} | n_1^i n_2^i \ldots n_k^i \ldots \rangle$ between the initial and final states, for which it is convenient to write the Fourier operators in terms of the creation–annihilation operators as

$$\widetilde{u}_{k_1} \widetilde{u}_{k_2} \widetilde{u}_{k_3} = \frac{l_{k_1} l_{k_2} l_{k_3}}{\sqrt{8}} \Big[b^\dagger_{-k_1} b^\dagger_{-k_2} b^\dagger_{-k_3} + b_{k_1} b^\dagger_{-k_2} b^\dagger_{-k_3} + b^\dagger_{-k_1} b_{k_2} b^\dagger_{-k_3}$$
$$+ b_{k_1} b_{k_2} b^\dagger_{-k_3} + b^\dagger_{-k_1} b^\dagger_{-k_2} b_{k_3} + b_{k_1} b^\dagger_{-k_2} b_{k_3} + b^\dagger_{-k_1} b_{k_2} b_{k_3} + b_{k_1} b_{k_2} b_{k_3} \Big].$$

(7.53)

Consider first the term with three creation operators. The bra-ket will be nonvanishing if the bra has three more phonons than the ket. Those states differ in energy by $\Delta E = \hbar(\omega_{k_1} + \omega_{k_2} + \omega_{k_3})$, and hence this transition is forbidden by the energy conservation in Fermi's golden rule. An analogous analysis indicates that the term with three annihilation operators should also not be considered.

The mixed terms, with both creation and annihilation operators, are usually rewritten in *normal order*, with the creation operators on the left. This transformation is easily achieved using the commutation relations (7.39), leaving in the process new terms with just one creation or annihilation operator. The presence of the delta factor in eqn (7.51) means that these operators create or destroy phonons with no energy and are, hence, inert. We can therefore ignore these terms.

In summary, for the purpose of computing the bra-ket it is enough to consider

$$\widetilde{u}_{k_1} \widetilde{u}_{k_2} \widetilde{u}_{k_3} = \frac{l_{k_1} l_{k_2} l_{k_3}}{\sqrt{8}} \Big[b^\dagger_{-k_2} b^\dagger_{-k_3} b_{k_1} + b^\dagger_{-k_3} b^\dagger_{-k_1} b_{k_2} + b^\dagger_{-k_1} b^\dagger_{-k_2} b_{k_3}$$
$$+ b^\dagger_{-k_3} b_{k_1} b_{k_2} + b^\dagger_{-k_2} b_{k_3} b_{k_1} + b^\dagger_{-k_1} b_{k_2} b_{k_3} \Big].$$

(7.54)

Consider the first three terms. Their action is to destroy one phonon and create two at the respective wavevectors. Then, when bra-keted, the result will be nonvanishing only if the initial and final states differ precisely in those phonons. Furthermore, the Kronecker delta in eqn (7.51) imposes that k is conserved, modulo G, e.g. $k_1 = k_2 + k_3 + G$. This is reminiscent of the momentum conservation law, but it is not appropriate to identify $\hbar k$ as momentum: first, the expectation value of the momentum operator in the states (7.41) vanishes (the lattice is not displacing). Second, it is not a real conserved quantity.[19] It is preferred, rather, to call $\hbar k$ the crystal momentum, and its conservation must be understood to be valid modulo G. Also, Fermi's golden rule imposes energy conservation, e.g. $\hbar\omega_{k_3} = \hbar\omega_{k_1} + \hbar\omega_{k_2}$. In summary, we obtain nonvanishing transition rates for processes where a phonon decays into another two, respecting crystal momentum and energy conservation. For the last three terms, the situation corresponds to two phonons that merge into a single one, again conserving crystal momentum and energy (both cases are represented in Fig. 7.7).

Each of these six contributions to the transition rates can be computed separately. For illustrative purposes, let us work out in detail the first term,

$$\overline{W}_{i\to f} = \frac{2\pi\Lambda^2}{72N\hbar}\delta(\varepsilon_f - \varepsilon_i)\delta_{k_1+k_2+k_3,G}\left|\langle f|b^\dagger_{-k_2}b^\dagger_{-k_3}b_{k_1}|i\rangle\right|^2 \tag{7.55}$$

$$= \frac{\pi\Lambda^2}{36N\hbar}n^i_{k_1}(n^i_{-k_2}+1)(n^i_{-k_3}+1)\delta(\hbar\omega_{k_2}+\hbar\omega_{k_3}-\hbar\omega_{k_1})$$
$$\times \delta_{k_1+k_2+k_3,G}\delta_{n^f_{k_1},n^i_{k_1}-1}\delta_{n^f_{-k_2},n^i_{-k_2}+1}\delta_{n^f_{-k_3},n^i_{-k_3}+1}. \tag{7.56}$$

Notably, as a result of (7.46) and (7.47), the spontaneous and stimulated emission factors appear naturally in the transition rates, reflecting the bosonic character of phonons.

We have found that the anharmonic term in the description of ion vibrations is equivalent to interactions between phonons, now constituting a weakly interacting gas in the same spirit as we discussed before. That is, the thermodynamic equilibrium properties are those of an ideal gas, but the interactions are responsible for conducting the gas toward equilibrium, selecting the Planck distribution. The interactions exchange crystal momentum and energy, which are conserved quantities, but the number of phonons is not conserved, resulting in a vanishing chemical potential at equilibrium.

The extension to three dimensions is direct, although some issues must be considered. First, the crystal momentum conservation is vectorial, e.g. $\mathbf{k}_1 = \mathbf{k}_2 + \mathbf{k}_3 + \mathbf{G}$, with $\mathbf{G}$ a reciprocal vector. Second, the displacement vector can be either parallel or perpendicular to the wavevector, giving rise to three phonon polarisations, each with its associated dispersion relation. Also, in systems with several atoms per unit cell, each atom leads to three phonon branches. In summary, in a three-dimensional solid with p atoms per unit cell, there will be $3p$ phonon branches. Of these, three will have linear dispersion relations, $\omega_{1,2,3} = c_{1,2,3}|\mathbf{k}|$ for small $|\mathbf{k}|$, being called the acoustic phonons. The others, which go to finite values at vanishing wavevector, are called optical phonons. At low temperatures, the Planck distribution implies that only acoustic phonons are present.

[19]This lack of strict conservation is a result of the lattice imposing a preferential reference frame, breaking the Galilean invariance.

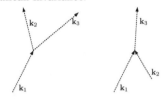

Fig. 7.7 Scattering processes involving three phonons induced by lattice anharmonicities. Left: one phonon decays into two; right: two phonons merge into one.

The transition rates are larger for smaller wavevectors, due to the term $(\sin k_1 a + \sin k_2 a + \sin k_3 a)^2 (l_{k_1} l_{k_2} l_{k_3})^2$. So, the dominant contribution to transport processes in principle should come from the case $\mathbf{G} = 0$ and small wavevectors. The triangular inequality implies that $|\mathbf{k}_1| \leq |\mathbf{k}_2| + \mathbf{k}_3|$, where the equals sign applies only for parallel vectors. However, energy conservation with $\omega_k = ck$ requires that $|\mathbf{k}_1| = |\mathbf{k}_2| + |\mathbf{k}_3|$. Hence, of all possible combinations of wavevectors, only parallel vectors are possible, reducing the efficiency of the scattering mechanism. Furthermore, when the full dispersion relation, $\omega_{\mathbf{k}} = 2\sqrt{\frac{K}{M}} |\sin(ka/2)|$ is used, energy conservation is impossible to achieve even for parallel vectors. In summary, three-phonon processes are suppressed for acoustic phonons except when $\mathbf{G} \neq 0$, called Umklapp processes, where at least one wavevector should be large (comparable to $2\pi/a$). Here, however, the factors l_k make the corresponding transition rates small. Therefore, three-phonon processes have small effective transition rates. Processes involving optical phonons are possible though.

The three-particle interactions appeared as a result of considering cubic anharmonicities in the ion interactions. It is possible to go then to quartic terms, which will produce four-particle scattering processes. These include one phonon decaying into three, the scattering of two phonons, and three phonons merging into one.

We can now write the Peierls or Boltzmann–Peierls equation for the phonon distribution function considering all of the three-phonon processes, and the reader is left to proceed similarly for quartic anharmonicities. To construct the collision term, consider that we study the evolution of $n_{\mathbf{k}_1}$. Then we should sum the transition rates for all combinations of $\mathbf{k}_2$ and $\mathbf{k}_3$. The Kronecker delta of crystal momentum conservation eliminates one sum, and the other can be cast as an integral, where a volume factor appears [see eqn (7.7)]. As the transition rates have a $1/N$ factor (7.56), the collision terms become intensive, allowing us to take the thermodynamic limit. The collision term is easily obtained if we use the following relations in the continuum limit: $\sum_{\mathbf{k}} = \mathcal{V} \int \mathrm{d}^3 k / (2\pi)^3$ and $\sum_{\mathbf{k}} \delta_{\mathbf{k},\mathbf{q}} = \int \mathrm{d}^3 k\, \delta(\mathbf{k} - \mathbf{q})$. Condensing all terms, the resulting kinetic equation is

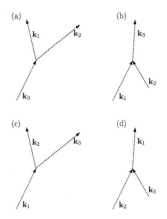

Fig. 7.8 Scattering processes involving three phonons that must be considered to construct the kinetic equation (7.57). $\mathbf{k}_1$ is the wavevector which labels the distribution function, while $\mathbf{k}_2$ and $\mathbf{k}_3$ are dummy wavevectors that are integrated out. (a), (b), (c), and (d) represent the first, second, third, and fourth terms, respectively, inside the curly braces of the right-hand side of eqn (7.57). In (c) and (d), $\mathbf{k}_2$ and $\mathbf{k}_3$ can be exchanged, producing the same result. To compensate this double counting during the integration, a factor of $1/2$ is introduced in the kinetic equation.

$$
\frac{\partial n_{\mathbf{k}_1}}{\partial t} + \mathbf{V}(\mathbf{k}_1) \cdot \nabla n_{\mathbf{k}_1} = \frac{\pi \Lambda^2}{36\hbar} \sum_{\mathbf{G}} \int \frac{\mathrm{d}^3 k_2}{(2\pi)^3} \frac{\mathrm{d}^3 k_3}{(2\pi)^3} \frac{(\sin k_1 a + \sin k_2 a + \sin k_3 a)^2}{\left(l_{k_1} l_{k_2} l_{k_3}\right)^2}
$$

$$
\times \Bigg\{ \delta(\mathbf{k}_1 + \mathbf{k}_2 - \mathbf{k}_3 - \mathbf{G}) \delta(\hbar\omega_1 + \hbar\omega_2 - \hbar\omega_3)
$$

$$
\times \left[(n_{\mathbf{k}_1} + 1)(n_{\mathbf{k}_2} + 1) n_{\mathbf{k}_3} - n_{\mathbf{k}_1} n_{\mathbf{k}_2} (n_{\mathbf{k}_3} + 1) \right]
$$

$$
+ \frac{1}{2} \delta(\mathbf{k}_2 + \mathbf{k}_3 - \mathbf{k}_1 - \mathbf{G}) \delta(\hbar\omega_1 - \hbar\omega_2 - \hbar\omega_3)
$$

$$
\times \left[n_{\mathbf{k}_1} (n_{\mathbf{k}_2} + 1)(n_{\mathbf{k}_3} + 1) - (n_{\mathbf{k}_1} + 1) n_{\mathbf{k}_2} n_{\mathbf{k}_3} \right] \Bigg\},
$$

$$
(7.57)
$$

where $\mathbf{V}(\mathbf{k}) = \frac{\partial \omega}{\partial \mathbf{k}}$ is the group velocity. Figure 7.8 shows the scattering processes that generate the different terms. The factor of $1/2$ in the third line overcomes the double counting produced by the integration over $\mathbf{k}_2$ and $\mathbf{k}_3$, which play identical roles in these terms. Note that, in the terms of the second line, $\mathbf{k}_1$ and $\mathbf{k}_2$ play identical roles in the scattering processes, but there is no double counting because $\mathbf{k}_1$ is the variable that is fixed, while $\mathbf{k}_2$ is integrated.

When the Planck distribution is placed into the kinetic equation, there is a one-to-one cancellation between the terms associated with the forward and backward processes [(a) with (b) and (c) with (d) in Fig. 7.8]. This one-to-one cancellation between a process and its time-reflected counterpart is another manifestation of the detailed balance principle for the equilibrium distribution.

Considering the complexity of the kinetic equation, as a simplified model, we take a relaxation time approximation where the relaxation time τ depends on the wavevector. Note that τ will also depend on temperature because the relaxation depends on the presence of phonons with which to scatter, having a population that depends on temperature. A detailed analysis on how to derive these relaxation times lies beyond the scope of this book. To illustrate some of the complexities involved, we present an analysis made by Peierls for the thermal conductivity. Consider the heat flux (7.48). At low temperatures, only the acoustic phonons are relevant and the Debye approximation can be made, leading to

$$\mathbf{q} = \frac{c^2 \hbar}{\mathcal{V}} \sum_{\mathbf{k}} \mathbf{k} n_{\mathbf{k}}, \tag{7.58}$$

which, we note, is proportional to the total crystal momentum.

Umklapp processes are rare at low temperatures because they need at least one high-energy phonon, and hence crystal momentum is exactly conserved. As a consequence, there is no relaxation mechanism for the heat current, which turns out to be conserved, corresponding to an infinite relaxation time for the current. Going beyond the Debye approximation, the problem persists. Indeed, imagine that an initial state is prepared with a finite heat current. There, the sum of wavevectors will generally differ from zero, and there will be no mechanism to approach the zero value that characterises the equilibrium state. For relaxation to take place, some processes must have $\mathbf{G} \neq 0$, together with others with $\mathbf{G} = 0$.[20] At low temperatures, the occurrence of these processes depends on the population of high-energy phonons, resulting in a global relaxation time that goes as $\tau \sim e^{-b\Theta_D/T}$, where b is a numerical coefficient of order one and Θ_D is the Debye temperature. Processes such as scattering with boundaries and impurities do not conserve crystal momentum either. In summary, two relaxation processes take place simultaneously: those three-phonon events that conserve crystal momentum, which lead to a relaxation toward a Planck distribution with a nonvanishing total crystal momentum, and the nonconserving processes that relax toward the equilibrium Planck distribution. Only by considering both is it possible to obtain correctly the thermal conductivity (Callaway, 1991). At very low temperatures, heat flux is limited by scattering with boundaries or impurities with $\kappa \sim T^3$.[21] At the other limit,

[20]Sometimes it is confusingly indicated that only Umklapp processes contribute to heat conductivity; in fact the distinction between Umklapp and normal processes depends on how the primitive cell is defined, with the FBZ being one of many possible choices. This choice cannot, however, affect a physical quantity (Maznev and Wright, 2014), implying that both processes should be considered simultaneously.

[21]See Exercise 7.12.

at high temperatures, where all wavevectors are populated by phonons, the relaxation depends on the global phonon density, giving $\tau \sim T^{-1}$.

7.7.3   Phonon–electron interactions

Electrons and ions interact through the potential $U_{\text{e-ion}}$, and we can write the Hamiltonian for the full system as

$$H = H_{\text{ions}} + H_{\text{e}} + \sum_i \sum_j U_{\text{e-ion}}(x_i - R_j), \qquad (7.59)$$

where H_{ion} has already been described, H_{e} considers the kinetic energy of electrons and their Coulomb interactions, and x_i are the coordinates of the electrons. Ions fluctuate around their equilibrium positions, allowing us to expand the interaction potential $U_{\text{e-ion}}$ to first order, giving

$$H = H_{\text{ions}} + H_{\text{e-lattice}} - \sum_i \sum_j u_j U'_{\text{e-ion}}(x_i - R_j^0), \qquad (7.60)$$

[22]See Appendix D for a discussion on how the electron–electron Coulomb interaction is considered in the framework of the Hartree–Fock approximation, where electrons are free particles in a mean field potential.

where $u_i = R_i - R_i^0$ and $H_{\text{e-lattice}}$ considers the interaction of the electrons with the ideal lattice, which will be the focus of attention in the next chapter. For the first two contributions of the Hamiltonian, we obtained the eigenstates, which describe an ideal gas of phonons and an ideal gas of electrons.[22] The third term is a perturbation that generates transitions between states, given by Fermi's golden rule. In principle, we need to compute the transition amplitudes between product states of the unperturbed Hamiltonian, $\Psi = |n_1 n_2 \ldots n_k \ldots\rangle_{\text{ph}} |k_1 k_2 \ldots k_n \ldots\rangle_{\text{e}}$. However, as electrons interact individually with the lattice deformations in (7.60), and at the interaction they can only change momentum, their number being conserved, we can work with the simpler product states of single electrons and a gas of phonons: $\Psi = |n_1 n_2 \ldots n_k \ldots\rangle_{\text{ph}} |k\rangle_{\text{e}}$. To proceed, we rewrite the perturbation in terms of the phonon creation–annihilation operators and in spatial Fourier components such that they operate adequately on the product states. The electron–ion potential is expanded as a Fourier series,[23]

[23]A function $U(x)$ that is periodic in the volume, can be expanded in a Fourier series. Here we chose the following normalisation:

$$U(x) = \frac{1}{\mathcal{V}} \sum_p \widehat{U}_p e^{ipx},$$

$$\widehat{U}_p = \int U(x) e^{-ipx}\, dx,$$

which gives Fourier coefficients that are independent of volume. Alternatively it is possible to place the volume factor in the second expression but the coefficients would be inversely proportional to $\mathcal{V}$.

$$U_{\text{e-ion}}(x) = \frac{1}{\mathcal{V}} \sum_p \widehat{U}_p e^{ipx}, \qquad (7.61)$$

where the Fourier coefficients $\widehat{U}_p$ are volume independent. The wavevectors can be decomposed as $p = q + G$, where q belongs to the FBZ and G is a reciprocal vector. Finally, for a single electron,[24]

$$\Delta U = -\frac{i\sqrt{N}}{\mathcal{V}} \sum_{q,G} \sqrt{\frac{\hbar}{2M\omega_{q+G}}} (q+G)\widehat{U}_{q+G} \left(b^\dagger_{-(q+G)} + b_{q+G} \right) e^{i(q+G)x}, \qquad (7.62)$$

[24]See Exercise 7.9.

where we have used that $\sum_j e^{i(k-q-G)R_j^0} = N\delta_{k,q+G}$. The transition rates between product states are therefore

$$\overline{W}_{i\to f} = \frac{\hbar N}{2M\mathcal{V}^2} \sum_{q,G} \frac{\left|p\widehat{U}_p\right|^2}{\omega_p} \left[(n_p+1)\delta_{k_f,k_i-p}\delta(\varepsilon_f - \varepsilon_i + \hbar\omega_p)\delta_{n_p^f,n_p^i+1}\right.$$

$$\left. +n_p\delta_{k_f,k_i+p}\delta(\varepsilon_f - \varepsilon_i - \hbar\omega_p)\delta_{n_p^f,n_p^i-1}\right], \quad (7.63)$$

where $\varepsilon_f = \varepsilon(k_f)$, $\varepsilon_i = \varepsilon(k_i)$, and for simplicity we have used plane waves for the electronic Bloch states (see Appendix D).

The nonvanishing transitions can be read as scattering processes where an electron changes from wavevector k_i to k_f by either emitting or absorbing one phonon (Fig. 7.9). In these processes, energy and crystal momentum are conserved, the latter up to an arbitrary reciprocal vector G. When emitting a phonon, the transition rates have the stimulated emission factor, while the rate for absorbing phonons is naturally proportional to the number of phonons. In summary, we have found that the crystal lattice and electrons can exchange energy and crystal momentum, and therefore thermalise, by inelastic scattering with phonons. These scattering processes can be understood and described in the context of kinetic theory, where electrons are fermions while phonons are bosons. When writing the corresponding collision term in the kinetic equation, all possible values of the phonon crystal momentum must be summed. The sum can be transformed into an integral, where a volume factor appears which, combined with the transition rates (7.63), gives an intensive term. In Chapter 8 we will write the kinetic equations associated with these transitions, and see that these processes modify the electrical conductivity of electrons at high temperatures, by reducing the electron mean free path.

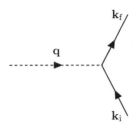

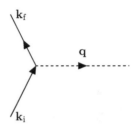

Fig. 7.9 Scattering processes between electrons and phonons. Top: a phonon is absorbed by an electron. Bottom: an electron emits a phonon. In both cases, energy and crystal momentum are conserved.

7.8 Application: lasers

In Section 7.2 we advanced that the stimulated emission process is the mechanism underlying lasers.[25] Consider the system with many two-level atoms (described in Section 7.2) in the presence of n_ω photons at the resonant frequency $\hbar\omega = E_2 - E_1$. The populations of the levels are N_1 and N_2. The occurrence of each of the three processes shown in Fig. 7.3 implies the net emission or absorption of one photon. Therefore, we can write an equation for the net rate of emitted photons analogous to eqn (7.18):

$$\dot{n}_\omega = A(1+n_\omega)N_2 - AN_1 n_\omega = A(N_2 - N_1)n_\omega + AN_2. \quad (7.64)$$

If we managed to generate a population inversion, where $N_2 > N_1$, then the number of photons would grow exponentially. In equilibrium, this is not possible because the relative populations are given by the Boltzmann factor (7.15), implying $N_2 < N_1$. It is then necessary to drive

[25]The term 'laser' is an acronym for light amplification by stimulated emission of radiation, introduced by G. Gould in 1959. The meaning of this acronym will become evident through this section. The first operating apparatus was a ruby laser in 1960 by T.H. Maiman.

the system out of equilibrium to attain population inversion; different mechanisms are possible, and here we discuss the three-level system, which is employed in ruby lasers.

To drive the system out of equilibrium, we have to inject energy into it. The naïve idea of illuminating the atoms directly is of no help, because for transitions to take place we would need light precisely at the resonant frequency, which is the one we want to emit. A second idea in complexity is to have three levels, ordered as $E_1 < E_2 < E_3$. Using light, we can now pump electrons from the first to the third level, then a nonradiative transition takes place, where the energy is released as heat or vibrations instead of light, driving the atoms to the second level and generating the desired population inversion. The difficulty here is to have a powerful light source with enough photons with frequency $(E_3 - E_1)/\hbar$. If the source has a broad spectrum, most of the energy will not be used, while a high-intensity monochromatic source raises the same problem that we had with two levels. However, if the two-level atoms are embedded in a crystal matrix, it is possible to engineer the electronic levels such that the first and second levels are discrete, while the third level is a broad band, as depicted in Fig. 7.10. This is actually achieved in crystals of ruby, which is an alumina crystal with some aluminium ions replaced by chromium ions. Now, an intense light, with a broad spectrum, can efficiently raise electrons from the first level to the continuous band. In the first ruby lasers, this light pumping was achieved by surrounding the ruby crystal with a flashtube.

The rate equations between level 1 and the broad band can be worked out as follows. The band is composed of K discrete levels, with $K \gg 1$. Each level has energy E_{3i} and population N_{3i}. The light source provides photons to excite the different transitions, with frequencies $\omega_i = (E_{3i} - E_1)/\hbar$, with n_{ω_i} being the population of each of these photonic modes. Using the appropriate Einstein coefficients, we have

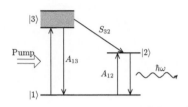

Fig. 7.10 Schematic representation of a three-level ruby laser. The illumination with a broad-spectrum source generates electronic transitions from the fundamental state $|1\rangle$ to the broad band $|3\rangle$, with characteristic amplitude A_{13}. The states in the band decay nonradiatively to the state $|2\rangle$ with a characteristic amplitude S_{32}. Finally, the transitions between states $|1\rangle$ and $|2\rangle$, which are relevant for laser emission, are characterised by the amplitude A_{12}.

$$\dot{N}_{3i} = A_{13i}N_1 n_{\omega_i} - A_{13i}N_{3i}(1 + n_{\omega_i}). \tag{7.65}$$

This equation can be summed up over the complete band, giving for large K,

$$\dot{N}_3 = A_{13}N_1 I - A_{13}N_3 I, \tag{7.66}$$

where A_{13} is the average Einstein coefficient, $N_3 = \sum_i N_{3i}$ is the total occupation of the band, and $I = \sum_i n_{\omega_i}$ is the total number of photons that can generate the transitions, which is proportional to the illumination intensity.

The second ingredient is that we need to store the radiation in a cavity for the exponential growth to take place, but we also want to use the emitted photons as a light source. For this purpose, the crystal is limited at its extremes by mirrors, which should be as parallel as possible to select a single wavevector. Finally, one of the mirrors is made partially transmissive, and the leakage through it will be the laser light that we will use (Fig. 7.11). If $t \ll 1$ is the transmission coefficient of the mirror and L is the length of the cavity, the probability per unit time for a

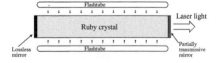

Fig. 7.11 Setup used in the ruby laser. The ruby crystal is limited by a lossless mirror on one side, while the mirror on the other side is partially transmissive, allowing for laser light to be emitted. The system is pumped by flashtubes that emit in a broad spectrum.

photon to be emitted is $\gamma = tc/2L$, where c is the speed of light and the factor of 2 accounts for the fact that the photon must go back and forth for any attempt to escape.

With these elements, it is possible to write down the rate equations, considering all transitions, for the populations of the three levels and photons:

$$
\begin{aligned}
\dot{n}_\omega &= A_{12}N_2(1 + n_\omega) - A_{12}N_1 n_\omega - \gamma n_\omega, \\
\dot{N}_2 &= A_{12}N_1 n_\omega - A_{12}N_2(1 + n_\omega) + S_{32}N_3, \\
\dot{N}_3 &= A_{13}N_1 I - A_{13}N_3 I - S_{32}N_3, \\
N_0 &= N_1 + N_2 + N_3 = \text{cst}.
\end{aligned} \tag{7.67}
$$

The last equation states that the electrons must be in any of the three levels.

For the purpose of determining whether it is possible to achieve stationary population inversion, we look first for steady states of the atomic populations. The three-level laser is normally operated under conditions in which the nonradiative transition is fast and S_{32} is large compared with the other rates. This implies that the broad band is emptied instantly and we can approximate $N_3 \approx 0$. Under these conditions in steady state,

$$
\frac{N_2 - N_1}{N_0} = \frac{A_{13}I - A_{12}}{A_{12}(1 + 2n_\omega) + A_{13}I}. \tag{7.68}
$$

Population inversion is possible as long as the illumination is intense enough, such that the effective transition rate to the state $|2\rangle$ via the broad band is faster than the decay rate from $|2\rangle$ to $|1\rangle$. Under population inversion conditions, the number of photons will grow exponentially, generating laser light. However, the denominator of (7.68) also grows, reducing the degree of inversion. To understand the final state, we now solve the stationary state equation for the number of photons, obtaining

$$
n_\omega = \frac{A_{12}N_2}{\gamma - A_{12}(N_2 - N_1)}, \tag{7.69}
$$

and the final state will be such that $A_{12}(N_2 - N_1)$ is just smaller than the loss rate γ. The solutions to eqns (7.68) and (7.69) have simple expressions in the relevant case of small loss rates:

$$
N_{2/1} = \frac{N_0}{2} \pm \left(\frac{A_{13}I - A_{12}}{A_{12}A_{13}I} \right) \gamma + \mathcal{O}(\gamma^2), \tag{7.70}
$$

$$
n_\omega = \frac{A_{13}N_0 I}{2\gamma} + \mathcal{O}(\gamma^0). \tag{7.71}
$$

Finally, the total emitted power is

$$
P = \hbar\omega\gamma n_\omega = \frac{\hbar\omega A_{13}N_0 I}{2} + \mathcal{O}(\gamma^1). \tag{7.72}
$$

We note the remarkable property of lasers that all this energy is emitted in a single monochromatic mode. Also, as the quantum states of light are defined by the wavevector, the emitted radiation has a unique wavevector and hence propagates coherently with an extremely small angular dispersion.

7.9   Application: quark–gluon plasma

The temperatures were so high in the early universe that it is believed that quarks and gluons formed a plasma instead of binding into protons and neutrons.[26] These conditions of extremely high energy densities are reproduced experimentally in hadron colliders, where protons and antiprotons collide at large energies. The scattering and particle production processes that take place can be described by kinetic equations like those studied in this chapter. In the first stages after the hadron collision, the plasma is mainly composed of gluons. Those are relativistic particles and are described by the four-momentum distribution function, $f(p^\mu, t)$, where $p^\mu = (p^0, p^i)$, $p^0 = E$ is the energy, and $m^2 c^4 = p_0^2 - \mathbf{p}^2 c^2$ is a Lorentz invariant. For notation simplicity we will use natural units with $c = \hbar = 1$. At high energies we can neglect the particle masses, and the integration in the four-momentum is therefore

$$\int \frac{\mathrm{d}^4 p^\mu}{(2\pi)^3} \delta(p_0^2 - \mathbf{p}^2) = \int \frac{\mathrm{d}^3 p}{2E(2\pi)^3}. \tag{7.73}$$

The kinetic equation for gluons is

$$\frac{\partial}{\partial t} f(\mathbf{p}_1) = J[f]$$

$$= \frac{1}{2} \int \frac{\mathrm{d}^3 p_2}{2E_2(2\pi)^3} \frac{\mathrm{d}^3 p_3}{2E_3(2\pi)^3} \frac{\mathrm{d}^3 p_4}{2E_4(2\pi)^3} \frac{1}{2E_1} W(1,2,3,4)$$

$$\times (2\pi)^4 \delta^{(4)}(p_1^\mu + p_2^\mu - p_3^\mu - p_4^\mu)$$

$$\times [f_3 f_4 (1 + f_1)(1 + f_2) - f_1 f_2 (1 + f_3)(1 + f_4)], \tag{7.74}$$

where the factor $1/2$ compensates double counting, similarly to the phonon case and the Dirac delta function imposes energy and momentum conservation. If we were studying the dynamics in the early universe one should also need to include an expansion term in the left-hand side, as in Section 4.10.

In quantum chromodynamics, collisions are dominated by small angle processes; that is, W is peaked when the transferred momentum is small: $\mathbf{p}_3 = \mathbf{p}_1 + \mathbf{q}$, with $\mathbf{q} \approx 0$. Momentum conservation implies $\mathbf{p}_4 = \mathbf{p}_2 - \mathbf{q}$.[27] For small momentum transfer the kinetic equation can be written as a Fokker–Planck equation. Indeed, each collision event produces a small jump of size $\mathbf{q}$. Summing all the collisions, there is a net flux in momentum space $\mathbf{Q}$ and the kinetic equation reads,

$$\frac{\partial}{\partial t} f(\mathbf{p}_1) = -\frac{\partial}{\partial \mathbf{p}_1} \cdot \mathbf{Q}(\mathbf{p}_1), \tag{7.75}$$

where the flux is given by the number of transitions times $\mathbf{q}$,

$$\mathbf{Q}(\mathbf{p}_1) = \int \frac{\mathrm{d}^3 p_2}{(2\pi)^3} \frac{\mathrm{d}^3 q}{(2\pi)^3} w(\mathbf{p}_1, \mathbf{p}_2, \mathbf{q}) \mathbf{q} f_1 f_2 (1 + f_3)(1 + f_4), \tag{7.76}$$

and momentum conservation has been used. The reduced transition rate, defined by

$$w(\mathbf{p}_1, \mathbf{p}_2, \mathbf{q}) = \frac{\pi}{16 E_1 E_2 E_3 E_4} \delta(E_1 + E_2 - E_3 - E_4) W(1,2,3,4), \tag{7.77}$$

[26] Quarks are fermions and bound states of three of them make up neutrons and protons. Gluons are bosons and are responsible for transmitting the strong nuclear force, which binds the quarks. Quantum chromodynamics is the theory that describes the interactions of quarks and gluons.

[27] As the particles are indistinguishable, W is also peaked for $\mathbf{p}_4 = \mathbf{p}_1 + \mathbf{q}$, with $\mathbf{q} \approx 0$. Therefore, a factor 2 should be added.

is peaked for $\mathbf{q} \approx 0$. Recalling that the transition rates should reflect the microscopic reversibility, the flux can be symmetrised using the inverse collision. Finally, for small $\mathbf{q}$,

$$
[f_1 f_2 (1+f_3)(1+f_4) - f_3 f_4 (1+f_1)(1+f_2)]_{\substack{\mathbf{p}_3 = \mathbf{p}_1 + \mathbf{q} \\ \mathbf{p}_4 = \mathbf{p}_2 - \mathbf{q}}}
$$
$$
\approx \mathbf{q} \cdot \left[h_1 \frac{\partial f_2}{\partial \mathbf{p}_2} - h_2 \frac{\partial f_1}{\partial \mathbf{p}_1} \right], \quad (7.78)
$$

with the shorthand notation $h = f(1+f)$. Putting all the terms together, the kinetic equation for gluons finally is

$$
\frac{\partial}{\partial t} f(\mathbf{p}_1) = -\frac{\partial}{\partial \mathbf{p}_1} \cdot \frac{1}{2} \int \frac{\mathrm{d}^3 p_2}{(2\pi)^3} \frac{\mathrm{d}^3 q}{(2\pi)^3} w(\mathbf{p}_1, \mathbf{p}_2, \mathbf{q}) \mathbf{q} \mathbf{q} \cdot \left[h_1 \frac{\partial f_2}{\partial \mathbf{p}_2} - h_2 \frac{\partial f_1}{\partial \mathbf{p}_1} \right],
$$
$$
(7.79)
$$

which has a structure similar to the Fokker–Planck equation.

In subsequent phases, quark–antiquark pairs are produced by the inelastic collision of two gluons: $gg \to q\bar{q}$. Thereafter, other $2 \leftrightarrow 2$ process take place; for example $qq \to qq$, $\bar{q}g \to \bar{q}g$, etc. It is possible to write the corresponding kinetic equation, with the Pauli factor for quarks. Small momentum transfers dominate again. Besides the corresponding Fokker–Planck term, the inelastic collisions give rise to additional source terms $S(\mathbf{p})$. The coupled equations, for the distribution functions of gluons f and quarks F read in a compact form,

$$
\frac{\partial}{\partial t} f(\mathbf{p}_1) = -\frac{\partial}{\partial \mathbf{p}_1} \cdot \mathbf{Q}_g(\mathbf{p}_1) + S_g(\mathbf{p}_1), \quad (7.80)
$$

$$
\frac{\partial}{\partial t} F(\mathbf{p}_1) = -\frac{\partial}{\partial \mathbf{p}_1} \cdot \mathbf{Q}_q(\mathbf{p}_1) + S_q(\mathbf{p}_1). \quad (7.81)
$$

These equations are highly nonlinear and numerical methods are used to analyse them.

An important question that emerges is whether gluons, which are bosons, present Bose–Einstein condensation. The condensation process of massive non-relativistic particles has been studied in the context of kinetic theory. Although the nucleation of the condensate is a phenomenon that is beyond the applicability of kinetic theory, kinetic equations can be used to describe the dynamics prior to and after the condensate has been formed. For the latter phase the distribution function is written as $f(\mathbf{p}, t) = \widetilde{f}(\varepsilon(\mathbf{p}), t) + (2\pi)^3 n_c(t) \delta(\mathbf{p})$, where $\widetilde{f}$ is the distribution function of the gas, assumed to be isotropic, and n_c is the density of the condensate. Numerical solution of the corresponding Boltzmann equation shows that for densities above the critical value, there is flux of particles toward zero momentum and the number of particles with small momentum increases in a self-similar way, as shown in Fig. 7.12.

An important element for the Bose–Einstein condensation to take place is the conservation of the number of particles, which fixes the chemical potential. In the quark–gluon plasma, the number of gluons is not conserved though. However, recent numerical simulations of the coupled quark and gluon kinetic equations, show that indeed during a transient time the condensate may form (Blaizot *et al.*, 2014).

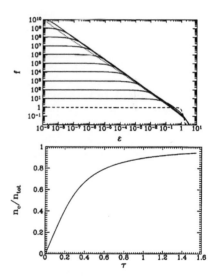

Fig. 7.12 Top: distribution function of the boson gas $\widetilde{f}(\epsilon)$ for different instants prior to condensate formation. Bottom: fraction of particles in the condensate after it has been formed. Reprinted figures with permission from Semikoz and Tkachev, Copyright (1997) by the American Physical Society.

Further reading

Quantum gases in equilibrium are analysed in any statistical mechanics book. In particular, they present the calculation of the chemical potential for ideal fermion and boson gases, including phonons, the latter in the context of the heat capacity of solids. Also, the Bose–Einstein transition is generally presented in detail in those books.

The kinetic theory of electrons and phonons is presented in great detail in some books devoted to kinetic theory (Liboff, 2003) or to solid-state physics (Ashcroft and Mermin, 1976; Ziman, 2001; Callaway, 1991; Jones and March, 1973; Bruus and Flensberg, 2004; Vasko and Raichev, 2006). The Landau theory of quantum liquids and the Green function formalism, theories that go beyond the scope of this book, and the more fundamental formulation of nonequilibrium quantum gases are found in Liboff (2003), Lifshitz and Pitaevskii (1980), and Kadanoff and Baym (1962).

The description of the Einstein coefficients and laser can be found in optics or atomic physics books (Fox, 2006). Lasers is also now a classic subject with many books covering different issues.

The application to cosmological problems, considering the production of elementary particles as well as compound objects like helium, can be found in Bernstein (2004). The kinetic description of quarks and gluons is found in Yagi *et al.* (2005). Finally, Bose–Einstein condensation for massive bosons is described in Semikoz and Tkachev (1997) and the application of kinetic theory to the description of nonlinear classical waves is presented in Zakharov *et al.* (1992).

Exercises

(7.1) **Chemical potential for free particles.** Consider a gas of free particles of spin s in a box of dimensions $L \times L \times L$. The energy levels are $\varepsilon = \hbar^2 k^2 / 2m$, where the wavevectors are given by $\mathbf{k} = 2\pi(n_x, n_y, n_z)/L$, with $n_{x,y,z} \in \mathbb{Z}$. For large systems, $\sum_{\mathbf{k}, s_z} = (2s+1)\mathcal{V} \int \mathrm{d}^3 k/(2\pi)^3$. Impose that the total number of particles is N to obtain an implicit equation for μ. Render it to a dimensionless equation for the fugacity, $z = e^{\mu/k_\mathrm{B}T}$.

(7.2) **Conservation laws.** Take the kinetic equation for a quantum gas (7.26) and derive the associated hydrodynamic equations; that is, multiply it by a function of $\mathbf{k}$ and integrate over the wavevector. Show that, for quantities that are microscopically conserved, expressed explicitly in the transition rates (7.27), the associated macroscopic equation also reflects the conservation.

(7.3) **Quantum entropy.** Show that, for a quantum gas described by the distribution function $f(\mathbf{k})$, the H function (7.28) is related to the entropy.

(7.4) **Entropy maximisation.** Consider the entropy for quantum gases,

$$S = -k_\mathrm{B} \int \mathrm{d}1\{f(1)\log f(1) \\ \mp [1 \pm f(1)]\log[1 \pm f(1)]\},$$

where the upper sign is for bosons and the lower sign for fermions. Show that maximising this entropy, fixing the total number of particles and energy with appropriate Lagrange multipliers, generates the Bose–Einstein and Fermi–Dirac distributions. Now, in the case of photons, where the number of particles is not fixed but only the total energy, show that the maximisation of the bosonic entropy,

$$S = -k_\mathrm{B} \int \mathrm{d}1\{f(1)\log f(1) \\ - [1 + f(1)]\log[1 + f(1)]\},$$

generates the Planck distribution.

(7.5) **Fermion–boson mixtures.** Consider a mixture of fermion and boson gases of masses m_F and

m_B, with transition rates W_{FF}, W_{BB}, and W_{FB} that respect energy and momentum conservation. Write down the kinetic equations for the distribution function of fermions $f(\mathbf{k})$ and bosons $n(\mathbf{k})$. Show that the collision terms vanish for each combination of wavevectors (that is, satisfy the detailed balance principle) if both gases are described by their equilibrium distributions, with equal temperature and Galilean velocity. Note that the chemical potentials are not fixed to be equal, as expected, because of the difference in the nature of the particles.

(7.6) **Electron–positron–photon equilibrium.** At high energies, photons can decay into an electron–positron pair, a reaction that can take place backwards through annihilation of an electron with a positron. Momentum conservation forbids these reactions in vacuum, but they can take place in matter, and consequently momentum is not conserved when considering photons, electrons, and positrons. The transition rate for the decay of a photon of frequency ω has a threshold at the energy necessary to produce the two particles, $2m_e c^2$. Hence, we can write $\overline{W}(\omega) = \theta(\hbar\omega - 2m_e c^2)\widehat{W}$. Find the equilibrium distribution.

(7.7) **Creation–annihilation operators.** Demonstrate the commutation relations (7.39) starting from the commutation relations for coordinates and momenta. Then, derive the Hamiltonian (7.40).

(7.8) **Number of phonons in thermal equilibrium.** Derive eqn (7.45) for the total number of phonons in equilibrium. For this, transform the sum over wavevectors into an integral over $k = |\mathbf{k}|$. Using the Debye approximation change variables to $x = \hbar c k/(k_B T)$ and, finally, approximate the integral in the limits of low and high temperatures.

(7.9) **Electron–phonon interactions.** Derive the expression (7.62) for the interaction potential written in terms of the creation and annihilation operators.

(7.10) **Equilibrium distribution for the three-phonon kinetic equation.** Show that the Planck distribution is indeed the equilibrium solution of the phonon kinetic equation (7.57).

(7.11) **Four-phonon processes.** Consider quartic anharmonicities for the lattice vibrations and show that these lead to four-phonon processes. Compute the transition rates and show that the stimulated emission and spontaneous absorption rates appear naturally from the calculation.

(7.12) **Mean free path model for phonon–lattice imperfection interactions.** If a solid has imperfec-

tions in the crystalline structure, for example surfaces, vacancies, or interstitial atoms, the phonons will scatter at them. An effective mean free path ℓ is established, depending on the concentration and type of imperfections. Instead of considering just six directions, as we did in Chapter 1, a continuous description is possible and, indeed, is also simple to implement. When a temperature gradient is imposed in the z direction, the energy flux through a surface in z is

$$J_e = \Phi(z - \ell) - \Phi(z + \ell) \approx -2\ell\frac{\partial\Phi}{\partial T}\frac{dT}{dz},$$

which defines the thermal conductivity $\kappa = 2\ell\frac{\partial\Phi}{\partial T}$, with Φ the energy flux crossing the surface in the specified direction,

$$\Phi = \int_{k_z > 0} \frac{d^3 k}{(2\pi)^3}(\hbar c k)(c k_z/k)n_{\mathbf{k}}.$$

Using the Planck distribution, $n_{\mathbf{k}} = 1/(e^{\hbar c k/k_B T} - 1)$, it is possible to compute the thermal conductivity. The result depends on how the temperature compares with the Debye temperature. Show that, at low temperature, where the integral in wavevectors can be extended up to infinity, $\kappa \sim T^3$.

(7.13) **Thermal conductivity by phonons: general expression.** Consider the relaxation time approximation for phonons. Show that the thermal conductivity can be written as

$$\kappa = \frac{N_{\mathrm{ph}}}{3}\int\frac{d^3 k}{(2\pi)^3}C(\mathbf{k})V(\mathbf{k})^2\tau(\mathbf{k}),$$

where $C(\mathbf{k}) = k_B(\hbar\omega_{\mathbf{k}}/k_B T)^2 n_{\mathbf{k}}^{\mathrm{eq}}(n_{\mathbf{k}}^{\mathrm{eq}} + 1)$ is the specific heat for each mode and N_{ph} is the number of acoustic phonon branches.

(7.14) **Thermal conductivity by phonons: special cases.** Different scattering processes give rise to various forms for the relaxation time. For scattering at point defects, one has $\tau_D^{-1} = Ak^4$, where A is independent of T. In a finite sample of size L, scattering at boundaries gives $\tau_B^{-1} = V/L$. Finally, three-phonon scattering gives $\tau_P^{-1} = BT^3\omega^2$, with B independent of T. Considering that these processes take place independently, the total relaxation time is (see, for example Richardson *et al.* (1992))

$$\tau^{-1} = \tau_D^{-1} + \tau_B^{-1} + \tau_P^{-1}.$$

Consider the low-temperature case, where the integral over wavevectors can be extended up to infinity and the Debye approximation can be used. Discuss the different limiting values for the conductivity when any of these mechanisms dominates.

(7.15) **Role of stimulated emission in laser operation.** The objective here is to show that stimulated emission plays a crucial role in the operation and efficiency of lasers. To do so, we can take the rate equations (7.67), changing the factor $(1 + n_\omega)$ to 1 to eliminate stimulated emission. Solve the stationary equations and show that, in the limit of low loss rate, the irradiated power is not an extensive quantity, that is, $\lim_{N_0 \to \infty} P/N_0 \to 0$. Also, explore the population inversion to show that level 2 is almost completely populated, but the decay rates are low because of the lack of stimulated emission, producing weak coherent radiation.

(7.16) **Four-level laser.** One problem with the three-level laser described in Section 7.8 is that the threshold for activation is relatively high. The flash illumination must be relatively intense to achieve lasing. To reduce the threshold, it would be desirable to depopulate level $|1\rangle$ to facilitate population inversion. This can be achieved by introducing a fourth level $|0\rangle$, with energy below $|1\rangle$, with $E_1 - E_0 \gg k_B T$, such that the first level is not populated in equilibrium (see Fig. 7.13). The system is built such that the transitions from $|3\rangle$ to $|2\rangle$ and from $|1\rangle$ to $|0\rangle$ are fast, while the relevant lasing transition from $|2\rangle$ to $|1\rangle$ is slow. Under these conditions, N_3 and N_1 are small. Compute the stationary populations and show that the threshold for inversion is lower than for the three-level laser.

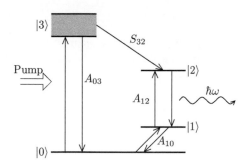

Fig. 7.13 Level scheme of the four-level laser. The transition coefficients S_{32} and A_{10} are large.

(7.17) **Fokker–Planck equation for gluons.** Consider the kinetic equation for gluons in the small angle scattering approximation (7.79). Show that it conserves the number of particles and the energy. Also, verify that the Bose–Einstein distribution is an equilibrium solution of the equation.

(7.18) **Classical nonlinear waves.** In classical waves, the squared amplitudes n_k of the Fourier modes are constant in time in the linear regime. However, when nonlinearities are included, interactions take place similarly to the phonon–phonon interactions. The form of these interactions depends on the nonlinearities and in the specific case of the nonlinear Schrödinger equation (NLS), three-wave scattering processes take place. Using the random phase approximation, the kinetic equation for waves in D dimensions reads,

$$\frac{\partial}{\partial t} n_{\mathbf{k}_1} = \int d^D k_2 \, d^D k_3 \, d^D k_4 W(\mathbf{k}_1, \mathbf{k}_2, \mathbf{k}_3, \mathbf{k}_4)$$
$$\times (n_{\mathbf{k}_3} n_{\mathbf{k}_4} n_{\mathbf{k}_1} + n_{\mathbf{k}_3} n_{\mathbf{k}_4} n_{\mathbf{k}_2} - n_{\mathbf{k}_1} n_{\mathbf{k}_2} n_{\mathbf{k}_3} - n_{\mathbf{k}_1} n_{\mathbf{k}_2} n_{\mathbf{k}_4}),$$

where momentum and energy conservation imply $W(\mathbf{k}_1, \mathbf{k}_2, \mathbf{k}_3, \mathbf{k}_4) = C\delta(\mathbf{k}_1 + \mathbf{k}_2 - \mathbf{k}_3 - \mathbf{k}_4)\delta(k_1^2 + k_2^2 - k_3^2 - k_4^2)$ and C measures the nonlinearity. Show that this equation has an H-theorem with $H = -\int d^D k \ln(n_{\mathbf{k}})$ and that in equilibrium the system is described by the Rayleigh–Jeans distribution,

$$n_{\mathbf{k}}^{\text{eq}} = \frac{T}{k^2 - \mu},$$

where T and μ are two constants fixed by the initial mass and energy.

Quantum electronic transport in solids

8

8.1 Electronic structure

In this chapter we will study the transport of electrons in crystalline solids, aiming to understand their electronic and thermal properties. We will show that the most prominent properties of solids emerge as a consequence of the Pauli exclusion principle combined with the periodic disposition of atoms. The latter generates a periodic potential for electrons, giving rise to a particular arrangement of electronic energy levels known as the band structure (Fig. 8.1). Appendix D presents a succinct description of band theory, with the objective of introducing the concepts necessary for the development of a kinetic theory of electronic transport in solids. For a comprehensive formulation of band theory and all its implications in solid-state physics, readers are directed to the books indicated at the end of the chapter. The main result of band theory for this chapter is that electrons moving in the periodic potential generated by the ions and other electrons, follow the Bloch theorem. They are described by wavefunctions $\Psi_{n,\mathbf{k}}(\mathbf{r})$ and energy levels $\varepsilon_{n,\mathbf{k}}$, with $n = 1, 2, \ldots$ labelling the bands and $\mathbf{k} = 2\pi\mathbf{n}/L$, a quasicontinuous wavevector called the crystal momentum. These vectors live in the first Brillouin zone of reciprocal space and in a volume $\mathcal{V}$, there are $\mathcal{V}g(\varepsilon)\,\mathrm{d}\varepsilon$ states in the energy interval $\mathrm{d}\varepsilon$, where g is the density of states.

8.2 Fermi–Dirac distribution, conductors, and insulators

Electrons are fermions, and accordingly, in equilibrium, they follow the Fermi–Dirac distribution (7.12), which gives the average number of electrons in each level. We remark that this distribution is independent of the structure of the energy levels and, in particular, the presence of energy bands and gaps. However, these two concepts when combined will allow us to understand qualitatively and quantitatively the main properties of the transport properties of solids. At vanishing temperature, the Fermi–Dirac distribution reduces to a step function; that is, all levels below the chemical potential μ are fully occupied, while the rest are empty. It is customary to call the chemical potential at vanishing temperature the Fermi level ε_F. If N_e is the total number of valence

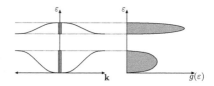

Fig. 8.1 Band structure in crystalline solids. Left: energy levels ε as a function of the wavevector $\mathbf{k}$. Two bands are shown with a gap in between. Right: corresponding density of states $g(\varepsilon)$, which vanishes in the bandgap.

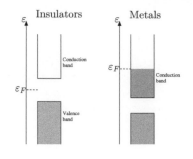

Fig. 8.2 Electron arrangement at zero temperature in the band structure of insulators (left) and metals (right). In grey are the occupied states.

[1]See Exercise 8.1.

[2]This happens for example in Ca, which is a metal, despite having two valence electrons. See Exercise 8.11 for an analysis of the magnetic properties of these metals.

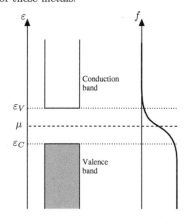

Fig. 8.3 At finite temperature but still with $k_B T \ll \Delta$, there is a small population of the lower conduction band levels while the top levels of the valence band are partially empty.

electrons in the system, using the relation (D.24) we get

$$N_e = \mathcal{V} \int_0^{\varepsilon_F} g(\varepsilon)\, \mathrm{d}\varepsilon, \qquad (8.1)$$

which gives the Fermi level implicitly in terms of the electron density $n_e = N_e/\mathcal{V}$ and the band structure, expressed in the density of states g.

One fact enormously simplifies the analysis of the position of the Fermi level in the band structure. On the one hand, in a solid with N unit cells, there are exactly N different wavevectors in the first Brillouin zone (see Appendix D), resulting in N energy levels for each band. Considering that the electrons are spin-1/2 fermions, we obtain that each band has exactly $2N$ quantum states to accommodate electrons. That is, the integral of $\mathcal{V}g(\varepsilon)$ over one band is $2N$. On the other hand, each cell contributes an integer number p of valence electrons, giving a total of $N_e = pN$ electrons in the crystal. Then, if the bands do not overlap at the Fermi level, two possibilities appear: either p is even, in which case $p/2$ bands are occupied, while the rest are empty. There is a gap between the last occupied level and the first unoccupied one, as presented in Fig. 8.2 (left). The chemical potential is located in this gap, and it can be shown that, in the low-temperature limit, it is located exactly in the middle of the gap.[1] The other case occurs if p is odd. Then, some bands are fully occupied and the last one is half filled. The Fermi level in this case lies inside this last band, as shown in Fig. 8.2 (right), with a precise location given by the inversion of eqn (8.1). In summary, two situations generically appear: the Fermi level lies either in between two bands or inside one band. We will see that this classification has a direct consequence for the properties of solids: the first case will correspond to insulators, and the second to metals. Finally, it is also possible that bands overlap at the Fermi level. In this case, for the purpose of electronic transport, we will see that such materials present metallic properties, because both bands are partially filled.[2]

When there is a gap at the Fermi level, the band below it (filled) is called the valence band and the band above it (empty) the conduction band. A quantity that will play an important role is the size of the gap, $\Delta = \varepsilon_C - \varepsilon_V$, where ε_V and ε_C are the limits of the valence and conduction band, respectively. At finite but small temperatures, the Fermi–Dirac distribution rounds off and presents the following properties: for $(\varepsilon - \mu) \gg k_B T$, the average number of electrons is exponentially small. Complementarily, for $(\mu - \varepsilon) \gg k_B T$, the average number of electrons is exponentially close to one. Only in the region of width $k_B T$ close to μ do the occupation numbers take values between 0 and 1, differing from the asymptotic distribution at zero temperature. Insulators have the Fermi level in the middle of the gap, therefore if $k_B T \ll \Delta$, this partial occupation close to μ is irrelevant, because no energy levels exist there and the solid will behave as an insulator at an effective vanishing temperature. However, if $k_B T \sim \Delta$, the top levels of the valence band will be partially unoccupied and the bottom levels of the conduction band partially occupied (see Fig. 8.3). This case of technological importance is called a semiconductor, where the transport properties depend strongly on the temperature and can be modified by doping. Table 8.1 gives the value of the gap for some materials and the temperature Δ/k_B at which the Fermi–Dirac distribution deviates notably from the zero-temperature asymptotic value at the band edges.

For metals, we define the Fermi temperature as $T_F = \varepsilon_F/k_B$. Normally, the Fermi temperature turns out to be much larger than room temperature (see Table 8.2). Therefore, it is reasonable to approximate metals as having low temperatures for the purpose of describing their electronic transport. At finite temperatures, the region of partially occupied states lies entirely in one band, which is called the conduction band. This is, in summary, the only relevant band for the study of the electronic properties of metals.

8.3 Boltzmann–Lorentz equation

8.3.1 Distribution function

Electrons in thermal equilibrium distribute in the energy levels according to the Fermi–Dirac distribution function. Electric fields, temperature gradients, or light illumination, for example, can alter this equilibrium state, and electrons will respond by adapting their distribution to the new conditions. We aim to characterise these nonequilibrium regimes and compute the macroscopic phenomena that emerge, electrical conduction being the most notable example.

In Chapter 7 we developed a theory for the coarse-grained distribution function, which gives the occupation numbers, averaged over small intervals of wavevector. Here, we will extend this concept in two ways. First, besides the wavevector, we include the band index n. Second, we are interested in describing inhomogeneous systems. The general theory of inhomogeneous transport in quantum systems is extremely complex, because electrons develop quantum correlations, and also, due to their wave nature, they explore the imposed fields nonlocally.[3] However, under many conditions, it is possible to adopt a simpler approach that is both qualitatively and quantitatively correct. For this, we will demand that the external perturbations do not vary rapidly in space.[4] In this case, we take advantage of the intensivity of the band structure of solids: namely, if a solid is divided, the resulting regions will have fewer valence electrons, but the number of energy levels decreases proportionally, leaving the band occupation structure unchanged. Hence, if the system is inhomogeneous on a large scale, we divide it into subsystems which are approximately homogeneous. In each of these, it is possible to define the distribution function, which will now depend on position $f_n(\mathbf{r}, \mathbf{k})$. Note that here we are not violating the uncertainty principle by giving simultaneously the positions and momenta of the electrons. Rather, we are giving the momenta of the electrons that are in a coarse-grained subvolume centred at $\mathbf{r}$. Certainly, for this to make sense, the division into subsystems should be such that they are large enough to

Table 8.1 Gap values measured at 300 K and the associated temperature for some insulators and semiconductors.

	Δ (eV)	Δ/k_B ($\times 10^4$ K)
Ge	0.66	0.77
Si	1.12	1.3
C	5.5	6.4
SiO$_2$	9	10

Table 8.2 Fermi level and temperature for some metals.

	ε_F (eV)	T_F ($\times 10^4$ K)
Cu	7.00	8.16
Au	5.53	6.42
Fe	11.1	13.0

[3] A more fundamental approach is to use the Wigner distributions.

[4] The electron mean free path in metals is of the order of tens of nanometres. Therefore, these variations should take place on scales not shorter than some tens of nanometres.

avoid inconsistencies with the uncertainty principle, again leading us to the condition of slightly inhomogeneous conditions. As we have learnt in other chapters, under nonequilibrium conditions, the system will relax to equilibrium, which is reflected by the fact that the distribution function will depend on time: $f_n(\mathbf{r}, \mathbf{k}, t)$. The purpose of this section is to find the equation that prescribes the evolution of f.

8.3.2   Scattering processes

We want to understand the mechanisms that make f change. For this purpose, we first consider the case of a homogeneous electron gas in the absence of external fields. The Bloch theorem shows that, if the total self-consistent potential is perfectly periodic, then the Bloch states are solutions of the Schrödinger equation (see Appendix D); that is, they are eigenstates of the system, and hence if the system is prepared with electrons occupying different levels, they will remain there. Consequently, any distribution function will remain unchanged if the potential is perfectly periodic.

Here, we must return to the discussion on the equilibrium distributions of classical systems presented in Section 2.4. A perfectly isolated system (in this case a gas of non-interacting electrons) cannot reach thermal equilibrium. As one learns from thermodynamics, energy exchange mechanisms must be present to attain equilibrium. In the present case, the Hartree–Fock approximation is not exact, and there is a residual interaction among electrons that, according to the description in Chapter 7, generates the equilibrium Fermi–Dirac distribution. Also, the ions are not static; rather, at finite temperature, they present thermal vibrations. In Section 7.7 we studied these interactions and showed that the vibrations, when quantised, behave as bosons—the phonons. Again applying the concepts of Chapter 7, the interactions of phonons with electrons thermalise the latter, providing an additional mechanism for the generation of the equilibrium Fermi–Dirac distribution.

So far, we have mentioned the following two interactions: (i) the interaction of electrons with the periodic potential produced by the ions and core electrons, and the self-consistent Coulomb interaction with other valence electrons. This periodic potential is responsible for generating the Bloch states; (ii) the residual electron–electron interaction and the phonon–electron interaction that thermalises electrons, selecting the Fermi–Dirac distribution for the Bloch states. On top of these, there are other interactions which are relevant for the transport processes. First, the crystalline lattice is never perfect. There are defects that appear either thermally or that were the result of the preparation. These defects can be of different natures, and for their detailed description and classification the reader is directed to any solid-state physics book. For the purpose of this book, we will consider the case of point defects, which normally consist of an ion that is missing, replaced by another element, or located at a wrong position, or an extra ion that is placed in the lattice. Also, as mentioned above, the ions present lattice vibrations which make the potential not perfectly periodic. All of these imperfections are modelled as a modification of the potential $U(\mathbf{r})$, which can be written as the periodic one plus a perturbation: $U = U_0 + \Delta U$. It is convenient to treat this perturbation in the context of time-dependent perturbation theory, which is presented in Appendix C; its use for quantum gases was described in Section 7.3.

The transition rates $\overline{W}_{i \to f}$ (eqns (7.21) and (C.24)) are the elements that account for the evolution of the distribution function. We adopt the master equation approach developed in Chapter 7. The loss term gives the rate at which electrons in state $|\mathbf{k}\rangle$ scatter to any state $|\mathbf{k}'\rangle$, where we have to sum over these states. Considering the Pauli exclusion principle for fermions, the loss term is

$$\left. \frac{\partial f(\mathbf{k})}{\partial t} \right|_{-} = \sum_{\mathbf{k}} \overline{W}_{\mathbf{k} \to \mathbf{k}'} f(\mathbf{k})(1 - f(\mathbf{k}'))$$

$$= \mathcal{V} \int \frac{\mathrm{d}^3 k'}{(2\pi)^3} \overline{W}_{\mathbf{k} \to \mathbf{k}'} f(\mathbf{k})(1 - f(\mathbf{k}')). \quad (8.2)$$

In the case of extended systems, the sum over final states is transformed into an integral with the use of (D.24) and it is convenient to define the rescaled transition rates, $W_{i \to f} = \mathcal{V} \overline{W}_{i \to f}$, which, as we will see, are independent of the volume $\mathcal{V}$ for point defects or lattice vibrations. Here there is no factor of 2, because the impurities are assumed to be nonmagnetic, and therefore no transitions between states with different spin are possible. If such transitions were possible, we should use transition rates $W_{\mathbf{k}s_z \to \mathbf{k}'s_z'}$ and sum over s_z' instead. The gain term has a similar structure, leading to

$$\frac{\partial f(\mathbf{k})}{\partial t} = \int \frac{\mathrm{d}^3 k'}{(2\pi)^3} \left[W_{\mathbf{k}' \to \mathbf{k}} f(\mathbf{k}')(1 - f(\mathbf{k})) - W_{\mathbf{k} \to \mathbf{k}'} f(\mathbf{k})(1 - f(\mathbf{k}')) \right].$$

$$(8.3)$$

Note that, to simplify notation, we have assumed that there is only one relevant band, but the extension to several bands is straightforward, and interband transitions will be considered when studying semiconductors in Chapter 9.

8.3.3 Semiclassical kinetic equation

In deriving the kinetic equation (8.3), we assumed that the electron gas was homogeneous and that there were no external forces besides those that generate the periodic potential and the transitions. If the gas is inhomogeneous, a convective term appears in the kinetic equation, similar to the classical cases. Here, we recall that electrons in Bloch states move with velocity $\mathbf{V}(\mathbf{k}) = \hbar^{-1} \frac{\partial \varepsilon}{\partial \mathbf{k}}$ [eqn (D.19)], leading to the streaming term,

$$\mathbf{V}(\mathbf{k}) \cdot \nabla f \quad (8.4)$$

on the left-hand side.

In the presence of weak external electric ($\mathbf{E}$) and magnetic fields ($\mathbf{B}$),[5] the crystal momentum obeys the equation of motion (D.20),

[5] By weak electric field we mean that the work done over one mean free path is small compared with the Fermi level, the bandgap, and the work function (see discussion at the end of Section 8.6.2). A weak magnetic field means that the associated Larmor radius is large, or equivalently that the Landau levels are quasicontinuous.

$\hbar\dot{\mathbf{k}} = \mathbf{F} = -e[\mathbf{E} + \mathbf{V}(\mathbf{k}) \times \mathbf{B}]$, where $q = -e$ is the charge of the carriers (electrons), with $e = 1.6 \times 10^{-19}$ C the elementary charge. The effect of the external fields is to induce migrations in $\mathbf{k}$-space, which are expressed as a conservation of the distribution function,

$$f(\mathbf{k}, t + \mathrm{d}t) = f(\mathbf{k} - \mathrm{d}t\mathbf{F}/\hbar, t). \tag{8.5}$$

Once expanded in $\mathrm{d}t$, this generates a streaming term $\frac{\mathbf{F}}{\hbar} \cdot \frac{\partial f}{\partial \mathbf{k}}$ on the left-hand side of the kinetic equation.

With all these ingredients, we can now write the transport equation for electrons in a crystalline solid as

$$\frac{\partial f(\mathbf{k})}{\partial t} + \mathbf{V}(\mathbf{k}) \cdot \nabla f(\mathbf{k}) - \frac{e}{\hbar}[\mathbf{E} + \mathbf{V}(\mathbf{k}) \times \mathbf{B}] \cdot \frac{\partial f(\mathbf{k})}{\partial \mathbf{k}} =$$
$$\int \frac{\mathrm{d}^3 k'}{(2\pi)^3} \left[W_{\mathbf{k}' \to \mathbf{k}} f(\mathbf{k}')(1 - f(\mathbf{k})) - W_{\mathbf{k} \to \mathbf{k}'} f(\mathbf{k})(1 - f(\mathbf{k}')) \right], \quad (8.6)$$

where, for simplicity, the dependence of f on position and time has not been written explicitly. This equation is a quantum version of the Lorentz model, studied in Chapter 3, and it is commonly called the Boltzmann or Boltzmann–Lorentz quantum transport equation. It is also referred to as the semiclassical transport equation.[6]

8.3.4   Linear collision operator

Whenever Fermi's golden rule (7.21) or (7.22) can be applied, the transition rates are symmetric: $W_{i \to f} = W_{f \to i}$, simplifying the kinetic equation, which now becomes linear:

$$\frac{\partial f(\mathbf{k})}{\partial t} + \mathbf{V}(\mathbf{k}) \cdot \nabla f(\mathbf{k}) - \frac{e}{\hbar}[\mathbf{E} + \mathbf{V}(\mathbf{k}) \times \mathbf{B}] \cdot \frac{\partial f(\mathbf{k})}{\partial \mathbf{k}} = -\mathbb{W}[f](\mathbf{k}). \quad (8.7)$$

The linear operator is defined as

$$\mathbb{W}[f](\mathbf{k}) = \int \frac{\mathrm{d}^3 k'}{(2\pi)^3} W_{\mathbf{k} \to \mathbf{k}'} \left[f(\mathbf{k}) - f(\mathbf{k}') \right], \tag{8.8}$$

where the sign has been chosen such that $\mathbb{W}$ is positive definite. We recall that the linearity here is an exact consequence of the symmetry of the transition rates; it is not an approximation obtained by being close to equilibrium.

For static impurities, transitions take place only between states of equal energy. Therefore, any distribution function that depends only on energy automatically cancels the right-hand side of (8.7) and becomes a solution of the kinetic equation. This is similar to what happened in the description of classical systems by the Liouville equation, where any function of the energy is a solution. Analogously to that case, the Fermi–Dirac distribution is selected among all the possibilities by scattering mechanisms other than those with static impurities. These can be electron–electron interactions or electron–phonon scattering processes, which provide mechanisms for energy exchange. We will assume that such mechanisms exist and, consequently, that the distribution function will initially be the Fermi–Dirac distribution.

[6]Note that the third term in the left-hand side can be interpreted as streaming in energy. Indeed, using that $\frac{1}{\hbar} \frac{\partial f}{\partial \mathbf{k}} = \mathbf{V} \frac{\partial f}{\partial \varepsilon}$, the referred term reads $P \frac{\partial f}{\partial \epsilon}$, where $P = -e(\mathbf{E} + \mathbf{V}(\mathbf{k}) \times \mathbf{B}) \cdot \mathbf{V}$ is the injected power associated to the Joule effect.

8.4   Time-independent point defects

8.4.1   Transition rates

The Boltzmann–Lorentz equation (8.6) must be complemented by a model for the transition rates W. The simplest case, which is already very instructive and represents many experimental conditions, is to consider static impurities. These defects can be interstitial atoms, vacancies, or substitutional atoms. If their concentration n_{imp} is small, they do not cooperate coherently in the scattering amplitude, which is then simply the sum of the effects of single impurities. Each impurity can then be modelled by a short-range potential centred around the defect,

$$\Delta U(\mathbf{r}) = \sum_i U_i(\mathbf{r} - \mathbf{R}_i), \qquad (8.9)$$

where $\mathbf{R}_i$ are translation vectors (D.1) pointing to the cells where the impurities are located. The non-rescaled transition rates are

$$\overline{W}_{\mathbf{k}\to\mathbf{k}'} = \frac{2\pi}{\hbar}\delta(\varepsilon-\varepsilon')\left|\langle\mathbf{k}'|\sum_i U_i|\mathbf{k}\rangle\right|^2 = \frac{2\pi}{\hbar}\delta(\varepsilon-\varepsilon')\sum_{i,j}\langle\mathbf{k}|U_i|\mathbf{k}'\rangle\langle\mathbf{k}'|U_j|\mathbf{k}\rangle. \qquad (8.10)$$

The double sum can be decomposed into a single sum with $i = j$ and a double sum with $i \neq j$. The double sum contains the terms,

$$\mathcal{V}^{-1}\sum_{i\neq j} e^{i(\mathbf{k}-\mathbf{k}')\cdot(\mathbf{R}_i-\mathbf{R}_j)}\int e^{i(\mathbf{k}-\mathbf{k}')\cdot(\mathbf{x}-\mathbf{x}')}U_i(\mathbf{x})U_j(\mathbf{x}')u_{\mathbf{k}}^*(\mathbf{x})$$

$$\times\, u_{\mathbf{k}'}^*(\mathbf{x}')u_{\mathbf{k}}(\mathbf{x}')u_{\mathbf{k}'}(\mathbf{x})\,\mathrm{d}^3x\,\mathrm{d}^3x', \quad (8.11)$$

where we have used that the Bloch wavefunctions (D.10) are $\Psi_{\mathbf{k}}(\mathbf{r}) = e^{i\mathbf{k}\cdot\mathbf{r}}u_{\mathbf{k}}(\mathbf{r})/\sqrt{\mathcal{V}}$, with $u_{\mathbf{k}}$ periodic and independent of the volume. If the impurities are distributed at uncorrelated positions, an average over their arrangement gives a vanishing contribution for the prefactor whenever $\mathbf{k} \neq \mathbf{k}'$.[7] We are then left with a term valid only when the wavevector does not change. This contribution cancels out exactly when considering the gain and loss terms of the kinetic equation. That is, an electron that continues with the same momentum after the scattering process does not affect transport. In summary, no contribution with $i \neq j$ is relevant in (8.10).

The bra-kets in the single sum are

$$\langle\mathbf{k}|U_i|\mathbf{k}'\rangle = \int \Psi_{\mathbf{k}}^*(\mathbf{r})U_i(\mathbf{r}-\mathbf{R}_i)\Psi_{\mathbf{k}'}(\mathbf{r})\,\mathrm{d}^3r, \qquad (8.12)$$

$$= \mathcal{V}^{-1}e^{-i(\mathbf{k}'-\mathbf{k})\cdot\mathbf{R}_i}\int e^{i(\mathbf{k}'-\mathbf{k})\cdot\mathbf{r}}u_{\mathbf{k}}^*(\mathbf{r})U_i(\mathbf{r})u_{\mathbf{k}'}(\mathbf{r})\,\mathrm{d}^3r, \quad (8.13)$$

$$= \mathcal{V}^{-1}e^{-i(\mathbf{k}'-\mathbf{k})\cdot\mathbf{R}_i}\widehat{U}_i(\mathbf{k},\mathbf{k}'), \qquad (8.14)$$

where we have defined the scattering amplitudes $\widehat{U}_i$. In this expression, we have used that the impurity potentials are short-ranged, making

[7] That is, when $\mathbf{k} \neq \mathbf{k}'$, $\langle\sin(\mathbf{k}-\mathbf{k}')\cdot(\mathbf{R}_i-\mathbf{R}_j)\rangle = \langle\cos(\mathbf{k}-\mathbf{k}')\cdot(\mathbf{R}_i-\mathbf{R}_j)\rangle = 0$.

$\widehat{U}_i$ volume independent and the full bra-ket inversely proportional to the volume. As a simple example, a short-range interaction can be approximated by a Dirac delta potential, $U(\mathbf{r}) = U_0 a_0^3 \delta(\mathbf{r})$, where U_0 is the potential intensity and a_0 its typical range. In this case, $|\widehat{U}|^2 = U_0^2 a_0^6$.

Now, if all impurities are equal, the rescaled transition rates are

$$W_{\mathbf{k} \to \mathbf{k}'} = \frac{2\pi}{\hbar} n_{\text{imp}} \delta(\varepsilon(\mathbf{k}') - \varepsilon(\mathbf{k})) \left| \widehat{U}(\mathbf{k}, \mathbf{k}') \right|^2, \qquad (8.15)$$

being, as advanced above, independent of volume.

The transition rates (8.15) fix that $\varepsilon(\mathbf{k}') = \varepsilon(\mathbf{k})$. The factors $f(1-f')$ and $f'(1-f)$ in (8.6) must then be evaluated at equal energies, but in equilibrium this factor is different from zero only in a narrow region of width $k_{\text{B}}T$ close to $\varepsilon = \mu$.[8] Therefore, the integrand can contribute to the evolution only if both $\mathbf{k}$ and $\mathbf{k}'$ are in the region close to the Fermi level; that is, in an electronic system, all the dynamical processes take place at the Fermi surface.[9] In metals, many levels exist at the Fermi surface, and hence the system can respond dynamically to external fields, for example in the conduction process. Insulators, in contrast, do not have states at the Fermi level, and consequently the collision term is vanishingly small. Insulators are thus inert in this context, being unable to respond to external perturbations.

8.4.2 Spherical models

In general, the Fermi surface has a complex geometry, but for some metals it is spherical.[10] As all the relevant physics takes place at the Fermi surface, we can use this spherical symmetry to simplify the analysis. In the relevant region, the dispersion relation has the symmetry $\varepsilon = \varepsilon(|\mathbf{k}|)$ and, consequently, the Bloch velocities are radial, $\mathbf{V}(\mathbf{k}) = V(k)\hat{\mathbf{k}}$. If, furthermore, the impurity potential has spherical symmetry, the transition rates will depend only on the angle between the incoming and outgoing wavevectors and their magnitudes,

$$\widehat{U}(\mathbf{k}, \mathbf{k}') = \widehat{U}(k, k', \hat{\mathbf{k}} \cdot \hat{\mathbf{k}}'), \qquad (8.16)$$

which for static impurities that conserve energy can be further simplified to

$$\widehat{U}(\mathbf{k}, \mathbf{k}') = \widehat{U}(k, \hat{\mathbf{k}} \cdot \hat{\mathbf{k}}'). \qquad (8.17)$$

In this last case, the Dirac delta function in (8.15) can be integrated out, rendering a simple form for the collision operator (8.8):

$$\mathbb{W}[f](\mathbf{k}) = \frac{n_{\text{imp}} k^2}{V(k)\hbar^2} \int \frac{\mathrm{d}^2 \hat{\mathbf{k}}'}{(2\pi)^2} \left| \widehat{U}(k, \hat{\mathbf{k}} \cdot \hat{\mathbf{k}}') \right|^2 \left[f(k\hat{\mathbf{k}}) - f(k\hat{\mathbf{k}}') \right]. \qquad (8.18)$$

In this model, there is a one-to-one correspondence between energies ε and wavenumbers $k = |\mathbf{k}|$. Depending on the context, we will use either of them as arguments of the various functions that appear in the calculations.

[8] Indeed, in equilibrium, the product $f_{\text{FD}}(\varepsilon)[1 - f_{\text{FD}}(\varepsilon)] = k_{\text{B}}T(-\frac{\partial f_{\text{FD}}}{\partial \varepsilon}) = \frac{1}{4\cosh^2[(\varepsilon-\mu)/k_{\text{B}}T]}$ is a function peaked around μ.

[9] The Fermi surface is defined by the wavevectors for which the energy is ε_F.

[10] This is the case for alkali metals such as Na or Cs.

8.5 Relaxation time approximation

To simplify estimates of transport processes, it is common to approximate the collision operator by a single relaxation time operator,

$$\mathbb{W}[f](\mathbf{k}) = \frac{f(\mathbf{k}) - f_{\mathrm{FD}}(\mathbf{k})}{\tau}, \qquad (8.19)$$

where f_{FD} is the equilibrium Fermi–Dirac distribution. The relaxation time τ helps in defining the mean free path for electrons as

$$\ell = \tau V_F, \qquad (8.20)$$

where V_F is the Bloch velocity at the Fermi level. Eventually, this approximation will be refined by allowing the relaxation time to depend on the wavevector $\tau(\mathbf{k})$.

Although this approximation is very useful, it should be used with caution, because it lacks particle number conservation (equivalently, charge is not conserved), in contrast to the collision operators (8.8) and (8.18). In particular, this model cannot be used to determine the diffusion coefficient of electrons.

8.6 Electrical conductivity

8.6.1 Qualitative description: metals and insulators

There are numerous interesting transport processes due to the motion of electrons that, for example, drive the microelectronics industry. Electrons react to the action of electric and magnetic fields, light, temperature gradients, and even deformations. In their motion, they interact with defects, boundaries, and surfaces, leading to a complex and rich phenomenology. This book does not intend to cover all of these processes, and the reader is directed to some of the books indicated at the end of the chapter for a comprehensive description. Certainly, the most important of all of these processes is the electrical conductivity.

We aim to describe the response of the electron gas to an applied electric field in order to understand the emergence of Ohm's law and the difference between metals and insulators, and to provide tools to compute the electrical conductivity. We start with the case with vanishing magnetic field and consider the electric field to be constant in time and homogeneous in space.[11] Under normal operating conditions, electric fields are weak, allowing us to use the semiclassical kinetic equation and perturbation theory. In this case, we write $\mathbf{E} = \epsilon \mathbf{E}_0$, where $\epsilon \ll 1$ is a formal parameter that will help us to order the calculation in a perturbation scheme. The distribution function is therefore expanded as $f = f_{\mathrm{FD}} + \epsilon f_1$, where f_{FD} is the Fermi–Dirac distribution with uniform temperature and chemical potential. Keeping the first order in ϵ in eqn (8.7), we obtain

$$-e\mathbf{E}_0 \cdot \mathbf{V}(\mathbf{k})\frac{\partial f_{\mathrm{FD}}}{\partial \varepsilon} = -\mathbb{W}[f_1](\mathbf{k}), \qquad (8.21)$$

[11] Consistent with the discussion on the definition of the coarse-grained distribution function, it is enough for the field to vary on scales longer than the mean free path.

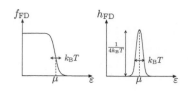

Fig. 8.4 Fermi–Dirac distribution and its negative derivative $h_{\mathrm{FD}}(\varepsilon) \equiv -\frac{\partial f_{\mathrm{FD}}}{\partial \varepsilon}$ at finite but small temperature.

where we have used the chain rule to write $\frac{1}{\hbar}\frac{\partial f}{\partial \mathbf{k}} = \mathbf{V}(\mathbf{k})\frac{\partial f}{\partial \varepsilon}$. The negative of the partial derivative of the Fermi–Dirac distribution will appear frequently in the rest of this chapter, and it is convenient, for further use, to define $h_{\mathrm{FD}}(\epsilon) \equiv -\frac{\partial f_{\mathrm{FD}}}{\partial \varepsilon}$, which is presented in Fig. 8.4. h_{FD} is a peaked distribution of width $k_{\mathrm{B}}T$, centred on μ, which converges to the Dirac delta function in the limit of vanishing temperature.

In the case of static impurities, the operator $\mathbb{W}$ acts on the energy shell. Then, if the left-hand side is a peaked function around μ, the perturbation f_1 must also be a peaked function. Schematically,

$$f_1(\mathbf{k}) \approx -\tau e \mathbf{E}_0 \cdot \mathbf{V}(\mathbf{k}) h_{\mathrm{FD}}(\varepsilon), \qquad (8.22)$$

where τ is a factor that has units of time and represents the effective action of the operator $\mathbb{W}^{-1}$; in the relaxation time approximation it is simply the associated time.

Here is where the difference between metals and insulators becomes evident. In an insulator, the density of states vanishes in a window of size Δ centred at μ. So, in insulators, the right-hand side of (8.22) vanishes for all wavevectors because there are no states within a window of size $k_{\mathrm{B}}T$ of the chemical potential. Consequently f_1 must vanish as well. For an insulator, as long as the scatters preserve the energy of the electrons and the thermal energy is much smaller than the gap, the electron distribution function is unchanged under the action of an electric field. In the case of semiconductors, the ratio of the thermal energy to the gap is not extremely small and the distribution function is slightly affected by the field. Metals, in contrast, have many levels in the vicinity of the Fermi level, making the perturbation of the distribution function finite.

Having obtained the stationary distribution function, the electric current can be computed as

$$\mathbf{J} = -\frac{2e}{\mathcal{V}} \sum_{\mathbf{k}} f(\mathbf{k})\mathbf{V}(\mathbf{k}). \qquad (8.23)$$

The factor of 2 accounts for the spin, because both components contribute equally to the current. By spherical symmetry, the contribution of f_{FD} to the electric current vanishes. In insulators, the distribution is unchanged, and consequently they show a vanishing electric current. In metals, the correction f_1 is finite and proportional to the electric field. Hence, the electric current is also proportional to $\mathbf{E}$, and we obtain Ohm's law.

In insulators, the poor conductivity is a result of the Pauli exclusion principle and the existence of the gap, which do not allow the electrons to respond to the electric field, being quenched to the equilibrium state. In metals, the Pauli exclusion principle does not limit the electronic response, as there are many unoccupied states available. In fact, in the presence of an electric field, the Fermi sphere[12] can migrate, with its centre taking higher and higher values; this migration is limited by the scattering processes, which tend to re-establish the equilibrium distribution. Hence, a poor conductor is so because there are too many scatters.

[12]The Fermi sphere corresponds to the set of wavevectors inside the Fermi surface.

We therefore conclude that the origin of the high resistivity in poor conductors is intrinsically different from that in an insulator.

8.6.2   Conductivity of metals

In the previous subsection, we showed qualitatively that metals respond to an imposed electric field by conducting electricity, while insulators are resistant to such perturbations. Here, we put this into more formal terms, as that analysis was mainly based on arguments. To solve (8.21), we start at vanishing temperature, for which h_{FD} becomes a Dirac delta function. The solution of this equation depends on the form of $\mathbb{W}$, which is a function of the particular scattering mechanisms that take place in the metal and are, therefore, specific to each solid and the impurities within it. To exemplify how to proceed, we consider first the relaxation time approximation (8.19) and later the spherical approximation (8.18). In the former case, the solution is simply given by[13]

$$f_1(\mathbf{k}) = -\tau(\mathbf{k})e\mathbf{E}_0 \cdot \mathbf{V}(\mathbf{k})\delta(\varepsilon(\mathbf{k}) - \varepsilon_F), \qquad (8.24)$$

with an electric current,

$$\mathbf{J} = \frac{2e^2}{\mathcal{V}} \sum_{\mathbf{k}} \tau(\mathbf{k})\mathbf{V}(\mathbf{k})\mathbf{V}(\mathbf{k})\delta(\varepsilon(\mathbf{k}) - \varepsilon_F) \cdot \mathbf{E}; \qquad (8.25)$$

that is, we obtain Ohm's law, from which we can extract the electrical conductivity. If the Fermi surface has spherical or cubic symmetry,[14] the conductivity tensor is isotropic and the scalar component has the simple expression,

$$\sigma = \frac{2e^2}{3\mathcal{V}} \sum_{\mathbf{k}} \tau(\mathbf{k})\mathbf{V}(\mathbf{k})^2\delta(\varepsilon(\mathbf{k}) - \varepsilon_F). \qquad (8.26)$$

Finally, for spherical symmetry, owing to (D.24), we can write,

$$\sigma = \frac{e^2}{3}\tau(k_F)V^2(k_F)g(\varepsilon_F) = \frac{e^2}{3}\ell(k_F)V(k_F)g(\varepsilon_F), \qquad (8.27)$$

where in the last expression we have used the definition (8.20) of the mean free path.

Several aspects of this important result deserve comment. First, it shows that the electrical conductivity remains finite at vanishing temperature, a result that is particularly relevant as we have discussed that, under normal conditions, metals are effectively at a very low temperature compared with the Fermi energy. Second, the conductivity depends only on the properties of the metal at the Fermi surface (or simply, the Fermi level, if this surface has spherical symmetry). As in the classical case, the conductivity is independent of the sign of the carrier, a sign that will appear explicitly only in the response to magnetic fields in Section 8.8.

The conductivity is proportional to the density of states at the Fermi level. Therefore, although the calculation was made considering metallic

[13]In Section 8.5, it was remarked that the relaxation time approximation does not guarantee charge conservation and that it should be used with care. In the conductivity problem, where f_1 (8.24) must be proportional to the vector $\mathbf{E}$, it results that $\int \mathrm{d}^3 k f_1(\mathbf{k}) = 0$. Hence, charge conservation is restored by symmetry, and we can safely work in this approximation.

[14]See Appendix B, where it is shown that, in systems with cubic symmetry, the second-order tensors are isotropic.

systems, we can extrapolate it to insulators by considering the limit of vanishing density of states. The result, advanced in the previous section, is a vanishing electrical conductivity for insulators.

Note also that the conductivity is proportional to $g(\varepsilon_F)$, which has units of particle density per unit energy. Hence, in contrast to the classical transport described in Chapter 3, the conductivity in metals is not proportional to the carrier density. This is a consequence of only the electrons at the Fermi level being responsive to external fields. In the next chapter, when studying semiconductors, we will see that the conductivity is then proportional to the carrier density, which is a concept that will be defined unambiguously in that context.

Now, using the more refined spherical approximation, f_1 is obtained from the equation,

$$\frac{n_{\text{imp}}k^2}{V(k)\hbar^2} \int \frac{\mathrm{d}^2\hat{\mathbf{k}}'}{(2\pi)^2} \left|\widehat{U}(k,\hat{\mathbf{k}}\cdot\hat{\mathbf{k}}')\right|^2 \left[f(k\hat{\mathbf{k}}) - f(k\hat{\mathbf{k}}')\right]$$
$$= -e\mathbf{E}_0 \cdot \hat{\mathbf{k}}V(k)\delta(\varepsilon(k) - \varepsilon_F). \quad (8.28)$$

Owing to the linearity and isotropy of $\mathbb{W}$, we can look for solutions, $f_1(\mathbf{k}) = \phi(k)\mathbf{E}_0 \cdot \hat{\mathbf{k}}$. Inserting this ansatz into the equation, the same solution (8.24) is found, now with

$$\tau(k) = \frac{1}{n_{\text{imp}}w_0(k)}, \quad (8.29)$$

expressed in terms of the scattering potential,

$$w_0(k) = \frac{k^2}{\hbar^2 V(k)} \int \mathrm{d}^2\hat{\mathbf{k}}' \left|\widehat{U}(k,\hat{\mathbf{k}}\cdot\hat{\mathbf{k}}')\right|^2 \left(1 - \hat{\mathbf{k}}\cdot\hat{\mathbf{k}}'\right). \quad (8.30)$$

Here, w_0 measures the persistence of the direction of motion after a scattering process: if the collisions do not change the direction ($\hat{\mathbf{k}}' \sim \hat{\mathbf{k}}$) appreciably, a small value of w_0 results. Note that the obtained result is formally equivalent to the relaxation time approximation, and the same result for the conductivity is obtained, but with the important advantage that it gives the effective relaxation time in terms of the microscopic scattering properties: $\sigma = e^2 V(k_F)^2 g(\varepsilon_F)/(3n_{\text{imp}}w_0(k_F))$.

Besides, it is based on more solid grounds, because in this model, the kinetic equation conserves charge, a property that is lacking from the relaxation time approximation. In the case of short-range potentials, discussed in Section 8.4.2, $\tau = \hbar^2 V^2/(4\pi k^2 U_0^2 a_0^6 n_{\text{imp}})$ (see Exercise 8.3). The case of a screened charged impurity is worked out in Exercise 8.4.

Recalling that, at vanishing temperature, $\mathbf{V}(\mathbf{k})\delta(\varepsilon(\mathbf{k}) - \mu) = -\hbar^{-1}\frac{\partial f_{\text{FD}}}{\partial \mathbf{k}}$, the distribution function in the presence of an electric field (8.24) can be written as

$$f(\mathbf{k}) = f_{\text{FD}}(\mathbf{k}) + \frac{e\tau}{\hbar}\mathbf{E} \cdot \frac{\partial f_{\text{FD}}}{\partial \mathbf{k}}, \quad (8.31)$$

which is nothing more than the Taylor expansion of a displaced Fermi–Dirac distribution,

$$f(\mathbf{k}) = f_{\text{FD}}(\mathbf{k} + e\tau\mathbf{E}/\hbar). \quad (8.32)$$

That is, the effect of the electric field is to displace the Fermi sphere without deforming it, as presented in Fig. 8.5.

We can now analyse the conditions for the perturbative scheme to be valid. The electric field must be weak such that the Fermi sphere is only slightly displaced, that is, $e\tau|\mathbf{E}|/\hbar \ll k_F$. Or, in terms of the mean free path, $e\ell|\mathbf{E}|/ \ll \hbar k_F V_F$. The right-hand side is of the order of ε_F implying, as we anticipated, that the work done over one mean free path should be small compared to the Fermi energy.[15]

8.6.3   Finite-temperature effects

The electrical conductivity of metals is only slightly modified at finite temperatures. However, insulators, which have vanishing conductivity at zero temperature, will now show finite conductivities, as can be anticipated by looking at eqn (8.21). Indeed, in the spherical approximation, the perturbation to the distribution function at finite temperature is

$$f_1(\mathbf{k}) = -e\tau(k)\mathbf{E}_0 \cdot \mathbf{V}(\mathbf{k})h_{\mathrm{FD}}(\varepsilon_{\mathbf{k}}), \qquad (8.33)$$

with τ given in (8.29), which gives rise to the following expression for the electrical conductivity:

$$\sigma = \frac{e^2}{3}\int g(\varepsilon)\tau(\varepsilon)V(\varepsilon)^2 h_{\mathrm{FD}}(\varepsilon)\,\mathrm{d}\varepsilon. \qquad (8.34)$$

In the case of metals, for which the density of states at the Fermi level is finite, a finite temperature will generate corrections to the conductivity, which can be calculated using the Sommerfeld method for degenerate Fermi gases.[16]

In contrast, the effect of finite temperatures on insulators is dramatic. The factor $h_{\mathrm{FD}}(\varepsilon)$ does not strictly vanish at the valence or conduction bands, resulting in a finite conductivity (Fig. 8.6). The distance from the Fermi level to the beginning of the conduction band is $\Delta_C = \Delta/2$, where Δ is the bandgap. If $k_B T \ll \Delta$, we can approximate $h_{\mathrm{FD}}(\varepsilon) \approx (k_B T)^{-1}e^{-(\varepsilon-\mu)/k_B T}$ for the purpose of performing the integration, and the conduction band contribution to the conductivity is

$$\sigma_C = \frac{e^2}{3k_B T}e^{-\Delta_C/k_B T}\int_0^\infty e^{-x/k_B T}\left[g(\varepsilon)\tau(\varepsilon)V(\varepsilon)^2\right]_{\varepsilon=\varepsilon_C+x}\,\mathrm{d}x. \quad (8.35)$$

Noting that, near the band edge, $g(\varepsilon) \sim \sqrt{\varepsilon - \varepsilon_C}$, it is possible to write $g(\varepsilon)\tau(\varepsilon)V(\varepsilon)^2 = \sqrt{\varepsilon - \varepsilon_C}u(\varepsilon)$, where u is a regular function that includes all the other factors. Approximating the integral for small temperatures, the resulting conductivity with contributions from both bands is

$$\sigma = \frac{e^2\pi^{1/2}(k_B T)^{1/2}}{6}\left[u(\varepsilon_C)e^{-\Delta_C/k_B T} + u(\varepsilon_V)e^{-\Delta_V/k_B T}\right]. \qquad (8.36)$$

The numerical value depends on properties at the band edges, but more importantly, the conductivity is dominated by the global exponential

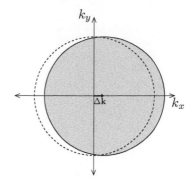

Fig. 8.5 When an electric field is applied to a metal at low temperatures, the electrons are distributed inside a Fermi sphere which is displaced with respect to the equilibrium distribution by $\Delta\mathbf{k} = -e\tau\mathbf{E}_0/\hbar$, but otherwise undeformed.

[15]When there is more than one relevant band and interband transitions can take place, the semiclassical description remains valid when $e\ell|\mathbf{E}|$ is much smaller than the associated energies. See Chapter 9 for the analysis of interband transitions.

[16]This method consists of noting that $h_{\mathrm{FD}}(\varepsilon)$ is approximately a Dirac delta function, allowing the expansion around μ of the other factors in the integral.

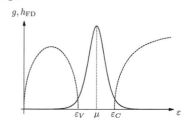

Fig. 8.6 At finite temperature, the factor $h_{\mathrm{FD}}(\varepsilon)$ (solid line) overlaps with the valence and conduction bands, which are represented via the density of states g (dashed line).

[17]Consider for example Si, where $\Delta = 1.12\,\text{eV}$, and hence $\Delta_C = \Delta_V = 0.56\,\text{eV}$. At $320\,\text{K}$ the exponential factors are roughly 10^{-9}.

prefactors that render the so-called intrinsic conductivity of semiconductors extremely small.[17] Semiconductors, which have smaller bandgaps, present larger intrinsic conductivities, but they are nevertheless quite small. In Chapter 9 we will show that, by adding a small concentration of specific impurities, it is possible to change the position of the chemical potential, reducing either Δ_C or Δ_V, and thereby increasing the conductivity of semiconductors.

8.6.4   Electron–phonon interactions

In Section 7.7.3 we showed that electrons interact with the lattice vibrations in a form that can be identified as scattering processes between electrons and phonons, where energy and crystal momentum are conserved but the number of phonons is not. Here, we consider three-dimensional lattices, but for simplicity, we will not make explicit reference to the different phonon branches and polarisations. Considering the transition rates (7.63), it is possible to write collision terms in the kinetic equations for the distribution functions of electrons $f(\mathbf{k})$ and phonons $n_\mathbf{k}$. In the case of electrons,

$$
\left.\frac{\partial f(\mathbf{k})}{\partial t}\right|_{\text{e–ph}} = \sum_\mathbf{G} \int \frac{d^3 q}{(2\pi)^3} \frac{d^3 k'}{(2\pi)^3} q_\mathbf{p}
$$
$$
\times \big\{ [f'(1-f)n_\mathbf{q} - f(1-f')(n_\mathbf{q}+1)]\,\delta(\mathbf{k}-\mathbf{k}'-\mathbf{p})\delta(\varepsilon-\varepsilon'-\hbar\omega_\mathbf{q})
$$
$$
+ [f'(1-f)(n_\mathbf{q}+1) - f(1-f')n_\mathbf{q}]\,\delta(\mathbf{k}-\mathbf{k}'+\mathbf{p})\delta(\varepsilon-\varepsilon'+\hbar\omega_\mathbf{q}) \big\},
$$
(8.37)

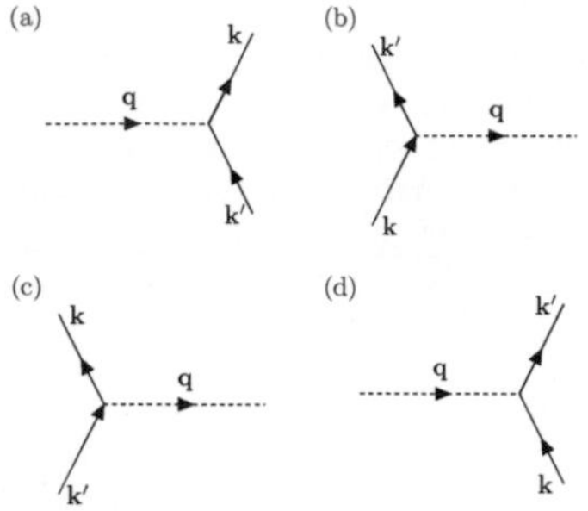

Fig. 8.7 Electron–phonon scattering processes associated with the different terms in the kinetic equation (8.37): (a) and (b) first and second terms of the second row, and (c) and (d) first and second terms of the third row.

where graphical representations of the different scattering processes are shown in Fig. 8.7. Here, $\mathbf{p} = \mathbf{q} + \mathbf{G}$, $g_\mathbf{p} = \hbar n |\mathbf{p}\hat{U}_p|^2/2M\omega_\mathbf{p}$ [see eqn (7.63)], and we adopt the shorthand notation, $f = f(\mathbf{k})$, $f' = f(\mathbf{k}')$, $\varepsilon = \varepsilon(\mathbf{k})$, and $\varepsilon' = \varepsilon(\mathbf{k}')$.

The phonon–phonon and phonon–impurity scattering are sufficiently efficient to keep an equilibrium distribution of phonons, allowing the use of the Planck distribution in (8.37). If this were not the case, we would need to work simultaneously with the coupled equations for phonons and electrons, where in the kinetic equation for phonons a term similar to (8.37) must be included.[18]

[18]See Exercise 8.12.

The electrical conductivity problem requires solving the equation, $e\mathbf{E} \cdot \mathbf{V}(\mathbf{k})h_{\text{FD}}(\varepsilon) = \left.\frac{\partial f(\mathbf{k})}{\partial t}\right|_{\text{e–ph}}$. Two difficulties enter into play. First, phonons only exist at finite temperature, and therefore even for metals we cannot take the zero temperature limit. Second, the scattering processes change the electron energies, and hence the modifications of the distribution function do not take place only at the Fermi surface. However, the energies of the phonons $\hbar\omega$ are limited from above by $k_B\Theta_D$, associated with the Debye temperature Θ_D, which is much smaller that the Fermi energy. Hence, any modification to the distribution function takes place close to the Fermi surface, implying that scattering with phonons generates the same phenomenology as before: solids with a nonvanishing density of states at the Fermi level will conduct, whereas electrons far from the Fermi surface are inert and cannot respond to electric fields, with the system presenting insulating behaviour.

In the linear response regime of $\mathbf{E} = \epsilon \mathbf{E}_0$ and $f(\mathbf{k}) = f_{\mathrm{FD}}(\mathbf{k}) + \epsilon f_1(\mathbf{k})$, with $\epsilon \ll 1$, one has

$$
e\mathbf{E} \cdot \mathbf{V}(\mathbf{k}) h_{\mathrm{FD}} = \sum_{\mathbf{G}} \int \frac{\mathrm{d}^3 q}{(2\pi)^3} \frac{\mathrm{d}^3 k'}{(2\pi)^3} q_{\mathbf{p}}
$$

$$
\times \left\{ [(n_{\mathbf{q}} + f_{\mathrm{FD}}) f_1' - (n_{\mathbf{q}} + 1 - f_{\mathrm{FD}}') f_1] \, \delta(\mathbf{k} - \mathbf{k}' - \mathbf{p}) \delta(\varepsilon - \varepsilon' - \hbar\omega_{\mathbf{q}}) \right.
$$
$$
\left. + [(n_{\mathbf{q}} + 1 - f_{\mathrm{FD}}) f_1' - (n_{\mathbf{q}} + f_{\mathrm{FD}}') f_1] \, \delta(\mathbf{k} - \mathbf{k}' + \mathbf{p}) \delta(\varepsilon - \varepsilon' + \hbar\omega_{\mathbf{q}}) \right\}.
$$
$$(8.38)$$

The collision integral does not have a simple form that can be transformed into a relaxation time approximation. It is not possible to find a close analytic solution, and we will perform a series expansion analogous to that used to compute the viscosity for a hard sphere gas in Section 4.7.2. We assume that the relevant band is spherical and that the phonon dispersion relation is isotropic. Furthermore, we will assume that Umklapp processes are not important, hence $\mathbf{G} = 0$. Under these conditions, by linearity and isotropy, $f_1(\mathbf{k}) = -e\mathbf{E} \cdot \mathbf{V}(\mathbf{k})\Psi(k)$. To motivate the expansion for Ψ, imagine that the collision term could be substituted by a relaxation time approximation, which would give $\Psi = \tau h_{\mathrm{FD}}(\epsilon)$. Hence, we propose,

$$
f_1(\mathbf{k}) = -e\mathbf{E} \cdot \mathbf{V}(\mathbf{k}) h_{\mathrm{FD}}(\epsilon) \sum_{n=0}^{\infty} \Psi_n k^{2n}, \tag{8.39}
$$

which, for analytic reasons, contains only even powers in k. To determine the coefficients of the expansion Ψ_n, eqn (8.38) is multiplied by $\mathbf{V}(\mathbf{k}) k^{2m}$ and integrated over $\mathbf{k}$. This results in a linear set of equations for Ψ_n, which is equivalent to the variational principle presented in some texts (see, for example, Ziman (2001)). The dominant order is captured by just one term in the expansion, Ψ_0. Solving for this and computing the electron current gives the conductivity as[19]

[19] See Exercise 8.13.

$$
\sigma = \frac{e^2 I_1^2}{3 I_2}, \tag{8.40}
$$

$$
I_1 = \int \frac{\mathrm{d}^3 k}{(2\pi)^3} V^2 H = V_F^2 g(\epsilon_F), \tag{8.41}
$$

$$
I_2 = \int \frac{\mathrm{d}^3 k}{(2\pi)^3} \frac{\mathrm{d}^3 k'}{(2\pi)^3} \frac{\mathrm{d}^3 q}{(2\pi)^3} g_{\mathbf{q}} \mathbf{V} \cdot \tag{8.42}
$$
$$
\left\{ [(n_{\mathbf{q}} + 1 - f_{\mathrm{FD}}') h_{\mathrm{FD}} \mathbf{V} - (n_{\mathbf{q}} + f_{\mathrm{FD}}') \mathbf{V}' h_{\mathrm{FD}}'] \, \delta(\mathbf{k} - \mathbf{k}' - \mathbf{q}) \delta(\varepsilon - \varepsilon' - \hbar\omega_{\mathbf{q}}) \right.
$$
$$
\left. [(n_{\mathbf{q}} + f_{\mathrm{FD}}') h_{\mathrm{FD}} \mathbf{V} - (n_{\mathbf{q}} + 1 - f_{\mathrm{FD}}') \mathbf{V}' h_{\mathrm{FD}}'] \, \delta(\mathbf{k} - \mathbf{k}' + \mathbf{q}) \delta(\varepsilon - \varepsilon' + \hbar\omega_{\mathbf{q}}) \right\}.
$$

To compute I_2 we make use of $k_{\mathrm{B}} \Theta_D \ll \varepsilon_F$ and hence $\varepsilon' \approx \varepsilon$, as well as $k' \approx k$. In all terms, we replace $k' = k$ and $\varepsilon' = \varepsilon$, except for f_{FD} and h_{FD}, which vary rapidly near the Fermi surface and are, therefore, sensitive to even small changes in energy. With this approximation, it is possible to compute I_2 explicitly,[20] resulting in the Bloch–Grüneisen expression for the phonon-limited electrical conductivity,

$$\sigma^{-1} = A \left(\frac{T}{\Theta_D}\right)^5 \int_0^{\Theta_D/T} \left|\widehat{U}(x)\right|^2 x^5 \frac{e^x}{(e^x - 1)^2} \, dx, \qquad (8.43)$$

where A contains temperature-independent factors. Room temperature is always much smaller than T_F, but it can be comparable to the Debye temperature (Table 8.3), making it appropriate to study the conductivity in the limits of low ($T \ll \Theta_D$) and high ($T \gg \Theta_D$) temperature. In the former case, the upper limit of the integral can be taken to infinity, resulting in $\sigma \sim T^{-5}$. In the opposite limit of high temperatures, the argument of the integral can be approximated for small values of x, giving $\sigma \sim T^{-1}$. This latter case can be understood by noting that, at high temperatures, the total number of phonons in the crystal scales as $N_{\rm ph} \sim T$ (see Section 7.7.1). Hence, the mean free path of the electrons and the corresponding relaxation time go as $\ell \sim \tau \sim T^{-1}$. Using eqn (8.27), we obtain the correct limiting behaviour for the conductivity. At low temperatures, the population of phonons scales as $N_{\rm ph} \sim T^3$. However, under these conditions, phonons have low energies ($U_{\rm ph} \sim T$), and therefore each phonon is rather inefficient at deflecting electrons. Many phonons are needed to limit the conductivity, and their proper counting gives the extra factor of T^2 needed to produce the result $\kappa \sim T^{-5}$ (see Ashcroft and Mermin (1976) for a detailed analysis of this effect).

Table 8.3 Debye temperature for various metals.

Element	Θ_D
Al	428 K
Au	170 K
Cs	38 K
Cu	344 K
Fe	470 K

8.6.5   Multiple scattering mechanisms and the Matthiessen rule

Under normal conditions, many scattering mechanisms coexist in a material; for example, phonons can be present simultaneously with different kinds of impurities. Under these conditions, when studying the electrical conductivity, we are led to solve the linear equation for the perturbation of the distribution function,

$$\sum_m \mathbb{W}^{[m]}[f_1] = -e\mathbf{E}_0 \cdot \mathbf{V}(\mathbf{k}) h_{\rm FD}(\varepsilon), \qquad (8.44)$$

where $\mathbb{W}^{[m]}$ are the linear operators associated with each of the scattering mechanisms. In the relaxation time approximation ($\mathbb{W}^{[m]} = 1/\tau^{[m]}$) the equation is readily solved,

$$f_1 = -\tau e\mathbf{E}_0 \cdot \mathbf{V}(\mathbf{k}) h_{\rm FD}(\varepsilon), \qquad (8.45)$$

where

$$\frac{1}{\tau} = \sum_m \frac{1}{\tau^{[m]}}. \qquad (8.46)$$

The resulting conductivity then satisfies Matthiessen's rule,

$$\frac{1}{\sigma} = \sum_m \frac{1}{\sigma^{[m]}}, \qquad (8.47)$$

where $\sigma^{[m]}$ corresponds to the conductivity for each mechanism alone. Noting that the resistivity is the reciprocal of the conductivity, this rule can be understood as placing all the resistances in series, because electrons must overcome all the possible scatters.

The Matthiessen rule, however, is not exact, being strongly based on the use of the relaxation time approximation. Indeed, to be concrete, consider the spherical model for impurity scattering. The solution of eqn (8.44) is again (8.45), but now τ depends on k,

$$\frac{1}{\tau(k)} = \sum_m \frac{1}{\tau^{[m]}(k)}. \qquad (8.48)$$

The conductivity is obtained as in (8.34) and can be written in a simplified form as

$$\sigma = \int \frac{G(\varepsilon)}{\sum_m \nu^{[m]}(\varepsilon)} \, d\varepsilon, \qquad (8.49)$$

where $\nu^{[m]} = 1/\tau^{[m]}$ are the relaxation rates of the different scattering mechanisms, and G includes all other factors. It can be proved that, for any positive functions G and $\nu^{[m]}$,

$$\left[\int \frac{G(\varepsilon)}{\sum_m \nu^{[m]}(\varepsilon)} \, d\varepsilon \right]^{-1} \geq \sum_m \left[\int \frac{G(\varepsilon)}{\nu^{[m]}(\varepsilon)} \, d\varepsilon \right]^{-1}, \qquad (8.50)$$

with the equality being achieved if either all $\nu^{[m]}$ do not depend on ε, i.e. the relaxation time approximation, or if all of them are proportional to each other, i.e. there is in fact a single scattering mechanism. The same analysis can be done for anisotropic and phonon scattering, leading us to the conclusion that, whenever the relaxation times depend on $\mathbf{k}$, the Matthiessen rule should be extended to the following inequality:

$$\frac{1}{\sigma} \geq \sum_m \frac{1}{\sigma^{[m]}}. \qquad (8.51)$$

8.7 Thermal conductivity and Onsager relations

It is a common experience that metals, in contrast to insulators, are also good thermal conductors.[21] This is certainly no coincidence, but rather is closely related to the electrical conductivity properties of each type of material. Thermal conductivity is associated with the process by which electrons in hotter regions move to colder ones, transferring their energy from one place to another. If electrons can move to transport charge, they will efficiently transfer energy as well.

Under inhomogeneous conditions, electrons migrate, transferring their energy but also their mass. In this case, according to the first law of thermodynamics, the transferred heat is given by $Q = \Delta E - \mu \Delta N$, where ΔN accounts for the number of electrons that migrate. Hence, in

[21]In Chapter 7 we showed that phonons can transfer heat, allowing electrical insulators to conduct heat. However, this mechanism is much weaker than the electronic transport of heat.

the formalism of band theory, the heat flux is

$$\mathbf{q} = \frac{2}{\mathcal{V}} \sum_{\mathbf{k}} \mathbf{V}(\mathbf{k})[\varepsilon(\mathbf{k}) - \mu]f(\mathbf{k}),\tag{8.52}$$

where the first term in the square brackets is the energy flux and the second term is the particle number flux.

Consider a material that is subject to a temperature gradient but is otherwise homogeneous. At first order, electrons are described by a Fermi–Dirac distribution at the local temperature, $f_0 = f_{\mathrm{FD}}(\mathbf{k}; T(\mathbf{r}))$. This distribution, however, is not the full solution of the Boltzmann equation (8.7), because the factor ∇f is not counterbalanced by another term. The deviation is on the order of the temperature gradient, and, for small gradients ($\nabla T \sim \varepsilon$) we can therefore propose a solution,

$$f(\mathbf{k}; \mathbf{r}) = f_{\mathrm{FD}}(\mathbf{k}; T(\mathbf{r})) + \varepsilon f_1(\mathbf{k}, \mathbf{r}).\tag{8.53}$$

To first order in ε, the Boltzmann equation reduces to

$$\nabla T \cdot \mathbf{V}(\mathbf{k})\frac{\partial f_{\mathrm{FD}}}{\partial T} = -\mathbb{W}[f_1],\tag{8.54}$$

where

$$\frac{\partial f_{\mathrm{FD}}}{\partial T} = \frac{\varepsilon - \mu}{T}h_{\mathrm{FD}}(\varepsilon).\tag{8.55}$$

The resulting equation is very similar to the conductivity case (8.21), and the same methods can be used to solve it. We will see that the vanishing temperature limit is uninteresting, so we proceed directly to finite temperatures, where we use the spherical approximation to obtain explicit results. The solution to (8.54) in this approximation is

$$f_1(\mathbf{k}) = -\tau(k)\nabla T \cdot \mathbf{V}(\mathbf{k})\left[\frac{\varepsilon - \mu}{T}h_{\mathrm{FD}}(\varepsilon)\right].\tag{8.56}$$

Notice that the resulting distribution function has polar symmetry (that is, it is proportional to a vector). Therefore, we obtain not only a heat flux but also a finite electric current. Indeed, it is a known phenomenon, related to the Seebeck and Peltier effects, that temperature gradients can induce electric phenomena. To isolate the thermal conduction effect, we need to work out the full problem of a solid simultaneously subject to an electric field and a temperature gradient. This being the case, the perturbation satisfies the equation,

$$\left(e\mathbf{E}_0 + \frac{\varepsilon - \mu}{T}\nabla T\right) \cdot \mathbf{V}(\mathbf{k})h_{\mathrm{FD}}(\varepsilon) = -\mathbb{W}[f_1].\tag{8.57}$$

As the equation is linear, the solution is also linear in both $\mathbf{E}_0$ and ∇T, which is simply the sum of (8.33) and (8.56). With the full solution, it is now possible to compute the charge and heat currents, which can be cast into the form,

$$\begin{aligned}\mathbf{J} &= L_{11}\mathbf{E} - L_{12}\frac{\nabla T}{T},\\[4pt]\mathbf{q} &= L_{21}\mathbf{E} - L_{22}\frac{\nabla T}{T},\end{aligned}\tag{8.58}$$

where the Onsager transport coefficients are given explicitly in the spherical approximation by

$$L_{11} = \frac{e^2}{3} \int g(\varepsilon)\tau(\varepsilon)V^2(\varepsilon)h_{\mathrm{FD}}(\varepsilon)\,\mathrm{d}\varepsilon, \qquad (8.59)$$

$$L_{12} = L_{21} = -\frac{e}{3} \int g(\varepsilon)\tau(\varepsilon)V^2(\varepsilon)(\varepsilon-\mu)h_{\mathrm{FD}}(\varepsilon)\,\mathrm{d}\varepsilon, \qquad (8.60)$$

$$L_{22} = \frac{1}{3} \int g(\varepsilon)\tau(\varepsilon)V^2(\varepsilon)(\varepsilon-\mu)^2 h_{\mathrm{FD}}(\varepsilon)\,\mathrm{d}\varepsilon. \qquad (8.61)$$

Here, we note that the cross transport coefficients L_{12} and L_{21} are equal. Notably, this property is not only a result of the spherical approximation. Indeed, Onsager showed that any pair of cross transport coefficients should be equal as a consequence of microscopic reversibility (de Groot and Mazur, 1984).

For metallic systems, the integrals can easily be performed in the low-temperature limit (see Exercise 8.5), resulting in

$$L_{11} = \frac{e^2 g(\varepsilon_F)\tau(\varepsilon_F)V^2(\varepsilon_F)}{3}, \qquad (8.62)$$

$$L_{12} = L_{21} = -\frac{e\pi^2(k_{\mathrm{B}}T)^2}{9}\left(g(\varepsilon)\tau(\varepsilon)V^2(\varepsilon)\right)'_{\varepsilon=\varepsilon_F}, \qquad (8.63)$$

$$L_{22} = \frac{\pi^2(k_{\mathrm{B}}T)^2 g(\varepsilon_F)\tau(\varepsilon_F)V^2(\varepsilon_F)}{9}. \qquad (8.64)$$

The thermal conductivity is measured under conditions in which there is no electric current. To obtain this, it is convenient to transform the relations (8.58) into the more convenient combinations,

$$\mathbf{E} = \rho\mathbf{J} + S\nabla T, \qquad (8.65)$$

$$\mathbf{q} = \Pi\mathbf{J} - \kappa\nabla T, \qquad (8.66)$$

with

$$\rho = L_{11}^{-1}, \qquad S = \frac{L_{12}}{TL_{11}}, \qquad (8.67)$$

$$\Pi = \frac{L_{21}}{L_{11}}, \qquad \kappa = \frac{1}{T}\left(L_{22} - \frac{L_{12}L_{21}}{L_{11}}\right). \qquad (8.68)$$

The coefficient ρ is the electrical resistivity, being the inverse of the conductivity (note that, in eqn (8.59), $L_{11} = \sigma$). The cross coefficients S and Π are called, respectively, the thermopower or Seebeck coefficient and the Peltier coefficient. These are relevant in the description of several phenomena, for example in the working principle of the thermocouple.[22] Finally, κ is the thermal conductivity, which relates the heat current to the temperature gradient in the absence of an electric current; that is, this corresponds to the experiment in which a conductor is placed in contact with two insulators at different temperatures. Note that, in this experiment, an electric field develops, being proportional to the Seebeck coefficient (see Fig. 8.8).

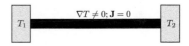

Fig. 8.8 When a conducting material is placed between two perfect insulators at different temperatures, a temperature gradient develops in the conductor, while there is a vanishing electric current. As a result, a heat current is established, being proportional to the temperature gradient. The proportionality constant is the thermal conductivity. Also, due to the Seebeck effect, an electric field is established.

[22]See Exercise 8.10.

In metals, at low temperatures, the cross coefficients give subdominant contributions to the thermal conductivity, which is therefore simply $\kappa = L_{22}$, vanishing at zero temperature. Similarly to the electrical conductivity, it is proportional to the density of states at the Fermi level. It is therefore expected that insulators will show vanishing thermal conductivity. To analyse the case of insulators in more detail, we should consider that there is a bandgap and, accordingly, approximate $h_{\mathrm{FD}}(\varepsilon)$ by an exponential close to the band edges. The result is that, similarly to the electrical conductivity, the thermal conductivity is exponentially small.[23]

[23]See Exercise 8.8.

8.7.1   Wiedemann–Franz law

The thermal and electrical conductivities (8.62) and (8.64) satisfy

$$\frac{\kappa}{\sigma T} = \frac{\pi^2 k_{\mathrm{B}}^2}{3e^2}, \tag{8.69}$$

[24]Named after G. Wiedemann and R. Franz, who in 1853 discovered that κ/σ has the same value for different metals at the same temperature. Later, in 1872, L. Lorenz reported the temperature dependence.

known as the Wiedemann–Franz law.[24] Although the conductivities were obtained in the spherical approximation, the Wiedemann–Franz relation is in fact more general. Actually, it remains valid at low temperatures whenever the scattering processes conserve energy.

To understand the origin of the Wiedemann–Franz law and its limitations, one must first realise that the charge and heat currents have very different characters. Charge is exactly conserved, and therefore the charge density obeys the conservation equation,

$$\frac{\partial \rho}{\partial t} = -\nabla \cdot \mathbf{J}. \tag{8.70}$$

The heat current, associated with the energy current, is not conserved, because there are energy exchange processes, for example with phonons. The energy density e, therefore, rather follows an equation of the form,

$$\frac{\partial e}{\partial t} = -\nabla \cdot \mathbf{q} + \frac{e_0 - e}{\tau}, \tag{8.71}$$

where for simplicity we have linearised about the steady energy density e_0, and τ is the characteristic time of the electron–phonon processes. The relaxation modes for the two densities are different, especially in the low wavevector limit. It is precisely this limit that is relevant, because we have obtained the conductivities for homogeneous or slightly inhomogeneous systems. Notably, however, if the scattering processes conserve energy, both densities follow similar dynamics and indeed their transport modes are intimately related. When scattering does not conserve energy, an additional relaxation mode appears in the heat transport, decoupled from the charge transport, invalidating the Wiedemann–Franz law.

At low temperatures, all the dynamics is limited to a single energy shell, where the relaxation times for both transport processes are the same, giving rise to the Wiedemann–Franz law. When the temperatures are finite, the conductivities result from averaging the relaxation processes of different energies. Each energy contributes differently to the conductivities, producing deviations from the Wiedemann–Franz law even if the scattering is elastic. At finite temperatures in metals, but still considering scattering only at impurities,

$$\frac{\kappa}{\sigma T} = \frac{\pi^2 k_B^2}{3e^2} \left[1 + \frac{16\pi^2 u''(\mu)}{15u(\mu)} (k_B T)^2 + \mathcal{O}(T^4) \right], \qquad (8.72)$$

where $u(\varepsilon) = g(\varepsilon)\tau(\varepsilon)V^2(\varepsilon)$.

Note also that no $\hbar$ is present in the Wiedemann–Franz law. This might suggest that it should also be valid if the carriers obey classical statistics instead of the Fermi–Dirac distribution. This is however not true because, for carriers obeying Boltzmann statistics, particles with many different energies are involved in the transport processes, invalidating the requirements for the law to be valid. For example, it can be shown that, if we consider the classical Lorentz model, described in Chapter 3, a completely different relation is obtained.[25] Also, in a semiconductor, where carriers (electrons in the conduction band and holes in the valence band) obey the Boltzmann distribution, the Wiedemann–Franz law is not satisfied either (see Exercise 9.7 in Chapter 9).

[25] See Exercise 8.9.

8.8 Transport under magnetic fields

The electric properties of materials are seriously affected by the presence of magnetic fields. Classically, they generate circular orbits perpendicular to the magnetic field, known as cyclotron orbits, with radii that are inversely proportional to the field. An important issue is that the magnetic force is perpendicular to the velocity, not doing work on the charges. When quantum effects are included, different regimes are possible depending on the intensity of the magnetic field. At high field intensities, the Landau quantised levels appear, a case that we will not consider here. For weaker intensities, we can use the semiclassical approximation,

$$\hbar \dot{\mathbf{k}} = \mathbf{F} = -e[\mathbf{E} + \mathbf{V}(\mathbf{k}) \times \mathbf{B}], \qquad (8.73)$$

and again, the magnetic field does not perform work on the system.

Different geometries are possible depending on whether the electric and magnetic fields are parallel or perpendicular, and on whether we observe the response in the direction of the field or perpendicular to it. Two cases are particularly relevant: the Hall effect and the magnetoresistance. In the Hall geometry, which was analysed using a mean free path approach in Exercise 1.9, a conductor is subject to perpendicular electric and magnetic fields, pointing in the $\hat{\mathbf{x}}$ and $\hat{\mathbf{z}}$ directions, respectively (Fig. 8.9). Here, we will show that, if we restrict the system to not develop currents in the $\hat{\mathbf{y}}$ direction, an electric field is developed in this direction, being sensitive to the carrier sign. For the transversal magnetoresistance effect, the same arrangement of fields is used, but here we measure the change in resistivity in the $\hat{\mathbf{x}}$ direction. It is seen experimentally that the resistivity is increased by the presence of the

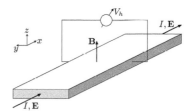

Fig. 8.9 Geometry used to measure the Hall effect and the transversal magnetoresistance for a conductor. An imposed electric field is placed along the $\hat{\mathbf{x}}$ direction, generating a current I in this direction. A magnetic field is placed oriented in the $\hat{\mathbf{z}}$ direction, perpendicular to the electric field. The magnetoresistance quantifies the modification of the current I, whereas the Hall effect accounts for the development of an electric field in the $\hat{\mathbf{y}}$ direction.

magnetic field. Notably, this effect vanishes for the spherical model, therefore being a test on the geometry of the Fermi surface.

8.8.1   Equilibrium solution

Let us first analyse the case of the magnetic field only, in the absence of the electric field. Recalling the definition of the Bloch velocity (D.19) and using the chain rule, we note that, for any distribution function that depends only on energy, $f(\varepsilon(\mathbf{k}))$, the left-hand side of the Boltzmann equation (8.6) vanishes. Indeed,

$$-\frac{e}{\hbar}[\mathbf{V}(\mathbf{k}) \times \mathbf{B}] \cdot \frac{\partial f}{\partial \mathbf{k}} = -e[\mathbf{V}(\mathbf{k}) \times \mathbf{B}] \cdot \mathbf{V}(\mathbf{k})\frac{\partial f}{\partial \varepsilon} = 0. \qquad (8.74)$$

This is a consequence of the fact that magnetic fields do not inject energy into the system. We can imagine classically that electrons move collectively, following the orbits described by (8.73). However, thanks to the indistinguishability of fermions, this is not necessary and we need only consider that the occupation numbers are not altered by the field.

Among all possible distributions, $f(\varepsilon(\mathbf{k}))$, the equilibrium Fermi–Dirac distribution is selected by the energy-exchanging scattering mechanism on the right-hand side of the Boltzmann equation, as we have already discussed in detail.

8.8.2   Linear response to electric fields

Similarly to what we did in the analysis of the conductivity, here we consider weak electric fields, which induce linear perturbations to the distribution function, $f = f_{\mathrm{FD}} + \varepsilon f_1$, and we look for stationary and homogeneous solutions. The magnetic field, however, is not assumed to be small. For notational simplicity, we consider the relaxation time approximation, with the possibility that the relaxation time depends on energy. The Boltzmann equation is arranged to

$$\frac{f_1}{\tau} - \frac{e}{m^*}(\mathbf{V} \times \mathbf{B}) \cdot \frac{\partial f_1}{\partial \mathbf{V}} = -eh_{\mathrm{FD}}(\varepsilon)\mathbf{V} \cdot \mathbf{E}, \qquad (8.75)$$

where we have used the parabolic approximation for the dispersion relation $\varepsilon = \hbar^2 \mathbf{k}^2/2m^*$, in which case $\mathbf{k} = m^*\mathbf{V}/\hbar$, and we can write everything in terms of the Bloch velocity. This linear and inhomogeneous equation implies that f_1 must be proportional to $\mathbf{E}$ in such a way as to produce a scalar.[26] Recalling that the magnetic field is a pseudovector, the most general possibility is $f_1 = A(V,B)\mathbf{V} \cdot \mathbf{E} + C(V,B)[\mathbf{V} \times \mathbf{B}] \cdot \mathbf{E}$, with A and C being scalar functions of the magnitudes of the speed and magnetic field. This result can be written as

$$f_1 = A_{ij}(V,\mathbf{B})V_iE_j, \qquad (8.76)$$

in terms of the unknown tensor $\mathbb{A}$, which now depends on the magnetic field vector and the magnitude of the speed. When computing $\frac{\partial f_1}{\partial \mathbf{V}}$ in (8.75), a term with the partial derivative of $\mathbb{A}$ appears, which gives a

[26]In this analysis, we use the tensorial concepts developed in Appendix B.

vector parallel to **V**, leading to a vanishing product in the Boltzmann equation and thereby simplifying the calculations. Substituting (8.76) into (8.75), we obtain $\Gamma_{ij}A_{jk} = -e\tau h_{\mathrm{FD}}(\varepsilon)\delta_{ik}$, where

$$\Gamma_{ij} = \delta_{ij} - e\tau\epsilon_{ikj}B_k/m^*. \tag{8.77}$$

Hence, the perturbation is $f_1 = -e\tau h_{\mathrm{FD}}(\varepsilon)(\Gamma^{-1})_{ij}V_iE_j$. The electric current is proportional to the electric field, allowing us to identify the conductivity tensor,

$$\sigma_{ij} = \frac{2e^2\tau}{\mathcal{V}}\sum_k V_iV_k(\Gamma^{-1})_{kj}h_{\mathrm{FD}}(\varepsilon), \tag{8.78}$$

$$= \frac{e^2\tau}{3}\int g(\varepsilon)V^2(\varepsilon)(\Gamma^{-1}(\varepsilon))_{ij}h_{\mathrm{FD}}(\varepsilon)\,\mathrm{d}\varepsilon, \tag{8.79}$$

where we have used the isotropy of the band structure in the parabolic approximation. At vanishing temperature, this result further simplifies to

$$\sigma_{ij} = \frac{e^2 g(\varepsilon_F)\tau(\varepsilon_F)V^2(\varepsilon_F)}{3}[\Gamma^{-1}(\varepsilon_F)]_{ij}. \tag{8.80}$$

In experiments, it is normally the resistivity tensor ρ that is measured, being defined by the relation $\mathbf{E} = \rho\mathbf{J}$. Hence,

$$\rho_{ij} = (\sigma^{-1})_{ij} = \frac{3\Gamma(\varepsilon_F)_{ij}}{e^2 g(\varepsilon_F)\tau(\varepsilon_F)V^2(\varepsilon_F)}. \tag{8.81}$$

8.8.3   Hall effect and the magnetoresistance

Having developed the theory of the electric response under magnetic fields, we now consider the geometry presented in Fig. 8.9, with $\mathbf{B} = B\hat{\mathbf{z}}$. The resistivity tensor becomes

$$\rho = \frac{1}{\sigma_0}\begin{pmatrix} 1 & \omega_c\tau & 0 \\ -\omega_c\tau & 1 & 0 \\ 0 & 0 & 1 \end{pmatrix}, \tag{8.82}$$

where $\sigma_0 = e^2 g(\varepsilon_F)\tau(\varepsilon_F)V^2(\varepsilon_F)/3$ is the electrical conductivity in the absence of the magnetic field and $\omega_c = eB/m^*$ is the cyclotron frequency of electrons.

The transversal magnetoresistance is defined as

$$\Delta\rho_{xx}(B) = \rho_{xx}(B) - \rho_{xx}(B = 0), \tag{8.83}$$

which evidently vanishes in the case under study. Indeed, even if we extend the model to a paraboloid model with $\varepsilon = \hbar^2(m^{*-1})_{ij}k_ik_j/2$, we again obtain a vanishing magnetoresistance. However, if the bands are not parabolic, finite magnetoresistance emerges. Also, when several bands are involved in the transport, nonvanishing values are obtained.[27]   [27] See Exercise 8.11.

The form of the resistivity matrix implies that the x and y components of the electric field are

$$E_x = J_x/\sigma_0 + \omega_c\tau J_y/\sigma_0, \tag{8.84}$$

$$E_y = -\omega_c\tau J_x/\sigma_0 + J_y/\sigma_0. \tag{8.85}$$

In the Hall geometry, current is allowed to flow along the x direction, but no current flows in the y direction. For this to take place, an electric field transverse to the imposed field in the x direction should develop,

$$E_y = -e\tau B E_x/m^* = -R_H B J_x,$$ (8.86)

where the last equality defines the Hall resistance R_H. In our case, it takes the value,

$$R_H = \frac{eB\tau}{m^*\sigma_0}.$$ (8.87)

Note that, in contrast to the conductivity, it depends on the sign of the electric carrier. In the case of semiconductors, which are studied in the next chapter, electricity is carried by both electrons and holes, with opposite charges. The sign of the Hall resistance will then depend on which carrier dominates. The sign of the Hall resistance is the same as the one obtained in Chapter 1, where a simple mean free path analysis of the classical transport was applied. However, the coefficients are different, and in particular, the emergence of the effective mass is well justified.

8.9   Thomas–Fermi screening

When external charges $\rho_{\text{ext}}(\mathbf{r})$ are introduced into a metal—for example substitutional ions—the free charges will rearrange to screen them, similarly to what happens in plasmas with Debye screening (Chapter 6). If the charge is positive, electrons will be attracted, creating a cloud that effectively reduces the charge of the ion, whereas if the added charge is negative, the region will be depleted, also reducing the effective charge. We will see that the screening is completely efficient and the cloud has exactly the same but opposite value as the added charge. We proceed by applying the approximation of Thomas and Fermi, who assumed that everywhere the distribution function is an equilibrium Fermi–Dirac distribution, but with density—or equivalently chemical potential—depending on space,

$$f(\mathbf{k},\mathbf{r}) = f_{\text{FD}}(\varepsilon(\mathbf{k}),\mu(\mathbf{r})) = \frac{1}{e^{(\varepsilon(\mathbf{k})-\mu(\mathbf{r}))/k_B T} + 1}.$$ (8.88)

Far from the added charge, the chemical potential μ_0 is fixed and the associated electron charge density cancels out the charge density of the crystal ions. Therefore, the net charge density is given by the imbalance of electrons,

$$\rho_e(\mathbf{r}) = -e\left[\int g(\varepsilon)f_{\text{FD}}(\varepsilon,\mu(\mathbf{r}))\,d\varepsilon - \int g(\varepsilon)f_{\text{FD}}(\varepsilon,\mu_0)\,d\varepsilon\right].$$ (8.89)

The total charge density generates an electric potential Φ,

$$\nabla^2\Phi = -\frac{\rho_e + \rho_{\text{ext}}}{\epsilon_0}.$$ (8.90)

Finally, the electron response to the electric field is described by the Boltzmann equation. As the distribution is a Fermi–Dirac distribution, the right-hand side of the equation cancels, and we are left only with the streaming and electric field terms,

$$\mathbf{V} \cdot \frac{\partial f_{\mathrm{FD}}}{\partial \mathbf{r}} + \frac{e}{\hbar} \nabla \Phi \cdot \frac{\partial f_{\mathrm{FD}}}{\partial \mathbf{k}} = 0. \tag{8.91}$$

The first term is transformed by noting that the spatial dependence is through the chemical potential $\frac{\partial f_{\mathrm{FD}}}{\partial \mathbf{r}} = \frac{\partial f_{\mathrm{FD}}}{\partial \mu} \nabla \mu = -\frac{\partial f_{\mathrm{FD}}}{\partial \varepsilon} \nabla \mu$, where in the last equality we have used the explicit form of the distribution. For the second term, the definition of the Bloch velocity is used. We finally obtain

$$\nabla \mu = e \nabla \Phi, \tag{8.92}$$

which combined with the Poisson equation (8.90) gives the Thomas–Fermi equation,

$$\nabla^2 \mu = \frac{e^2}{\epsilon_0} \left[\int g(\varepsilon) f_{\mathrm{FD}}(\varepsilon, \mu(\mathbf{r})) \, \mathrm{d}\varepsilon - \int g(\varepsilon) f_{\mathrm{FD}}(\varepsilon, \mu_0) \, \mathrm{d}\varepsilon \right] - \frac{e \rho_{\mathrm{ext}}}{\epsilon_0}. \tag{8.93}$$

This nonlinear equation for μ is the analogue to the Poisson–Boltzmann equation of Section 6.2.2, where here the Fermi–Dirac distribution is used. To solve it, we will consider small charges such that we can linearise the equation. The square bracket reduces to $\frac{\partial n}{\partial \mu} \delta \mu = g(\varepsilon_F) \delta \mu$, where the expression is evaluated in the low temperature limit. We can use the superposition principle and study specific external charge distributions, and for illustrative purposes we consider a localised charge $\rho_{\mathrm{ext}}(\mathbf{r}) = Q \delta(\mathbf{r})$,[28]

$$\nabla^2 \delta \mu = \kappa_{\mathrm{TF}}^2 \delta \mu - \frac{eQ}{\epsilon_0} \delta(\mathbf{r}). \tag{8.94}$$

The solution, written for the electric potential, has a Yukawa form,

$$\Phi = \frac{Q}{4\pi\epsilon_0} \frac{e^{-\kappa_{\mathrm{TF}} r}}{r}, \tag{8.95}$$

where the Thomas–Fermi wavevector is

$$\kappa_{\mathrm{TF}} = \sqrt{e^2 g(\varepsilon_F)/\epsilon_0}. \tag{8.96}$$

In metals, the density of states is large, and the screening length $\kappa_{\mathrm{TF}}^{-1}$ can be as small as a few angstroms. Conduction electrons are hence very efficient at screening any external charge.

8.10 Application: graphene

Since its first experimental realisation in 2002,[29] graphene has proven to be an extremely versatile material that presents unique properties. Graphene corresponds to a single two-dimensional layer of graphite, being first produced by simply peeling bulk graphite. This two-dimensional

[28] In Exercise 8.15, Fourier modes will be analysed to obtain the dielectric functions of metals.

[29] A. Geim and K. Novoselov won the Nobel Prize in Physics in 2010 for production of graphene and experimentation on this material, which shows fascinating properties.

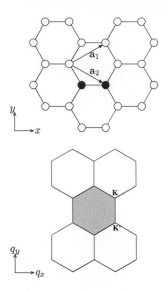

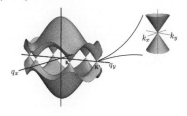

Fig. 8.10 Top: graphene lattice in real space. The atoms of a unit cell are presented in black. $\mathbf{a}_1$ and $\mathbf{a}_2$ are the two base vectors. Bottom: first Brillouin zone (in grey) for the graphene lattice, where the special points $\mathbf{K}$ and $\mathbf{K}'$ are indicated. Adapted from Abergel *et al.* (2010).

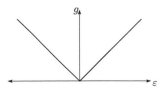

Fig. 8.11 Band structure of graphene. The conduction and valence bands meet at the points $\mathbf{K}$ and $\mathbf{K}'$. Zooming in on either of these points via the change of variables, $\mathbf{q} = \mathbf{K} + \mathbf{k}$ or $\mathbf{q} = \mathbf{K}' + \mathbf{k}$, the bands appear locally as cones. Adapted from Wilson (2006).

Fig. 8.12 Density of states of graphene near the Dirac point. The Fermi level is located at $\varepsilon_F = 0$.

material is a transparent conductor that is also mechanically stronger than steel. Under magnetic fields it presents the anomalous quantum Hall effect, becoming a playground for solid-state physics. From a fundamental point of view, it shows some analogies with particle physics, presenting exotic quantum properties. In terms of applications, graphene can be combined with other nanostructures to build transistors and other semiconductor devices.

Graphene owes its unusual electronic properties to the particular two-dimensional arrangement of the atoms. Here, carbon atoms sit on a honeycomb lattice, forming a two-dimensional crystal where each cell has two atoms, with the base vectors presented in Fig. 8.10. The electronic quantum states are confined in the transverse direction, while they are extended in the plane, labelled by their spin and a two-dimensional wavevector $\mathbf{q} = (q_x, q_y)$. The first Brillouin zone also has a hexagonal structure, and two points in reciprocal space, $\mathbf{K}$ and $\mathbf{K}'$, are particularly relevant. Indeed, when the band structure for graphene is obtained, it turns out that the valence and conduction bands meet at these points and the Fermi level for perfect graphene is located precisely at these intersection points. As we have learnt that the electronic transport properties depend on the Fermi surface and its vicinity, it is reasonable to look closely at the band structure near the Fermi level, as shown in Fig. 8.11. There, the dispersion takes the form of two cones. Changing variables to $\mathbf{q} = \mathbf{K} + \mathbf{k}$ or $\mathbf{q} = \mathbf{K}' + \mathbf{k}$, one finds that

$$\varepsilon = \pm \hbar V_F |\mathbf{k}|, \tag{8.97}$$

where V_F is the Bloch velocity, which in this case is independent of energy. The upper sign is for the conduction band and the lower sign for the valence band. The energy reference level has been adjusted to have $\varepsilon_F = 0$. Note that this dispersion relation is analogous to that of relativistic massless Dirac fermions in two dimensions.

To compute the density of states, we first obtain the number of states below ε,

$$G(\epsilon) = 4 \int_{\hbar V_F |\mathbf{k}| < \varepsilon} \frac{\mathrm{d}^2 k}{(2\pi/L)^2} = L^2 \frac{\varepsilon^2}{\pi \hbar^2 V_F^2}, \tag{8.98}$$

where the factor of 4 accounts for the spin degeneracy and that associated with choosing either K or K', called the valley degeneracy. Then, $g = L^{-2} \, \mathrm{d}G/\mathrm{d}\varepsilon$, i.e.

$$g(\varepsilon) = \frac{2|\varepsilon|}{\pi \hbar^2 V_F^2} = \frac{2|k|}{\pi \hbar V_F}. \tag{8.99}$$

This density of states is very unusual. It vanishes at the Fermi level, where the two bands meet, called the Dirac point (Fig. 8.12). However, it barely does so, and a small amount of impurities or an external potential would displace the Fermi level, leading to unusual transport properties. Naïvely, one would expect that at the Dirac point the electrical conductivity would vanish at zero temperature, but should increase when the Fermi level is moved away from this point. Experimentally, the position of the Fermi level can be controlled by placing a graphene layer on top of an insulating substrate (typically SiO_2), below which there is a conductor at fixed voltage (the gate voltage V_g). If V_g is positive, electrons will be attracted to populate the conduction band and consequently move the Fermi level up (Fig. 8.13). A negative voltage produces the opposite effect. It is customary to parameterise the state of the graphene layer using the value of the Fermi level ε_F, the Fermi wavevector k_F, or the carrier density n, which is defined as the number of electrons per unit area on top of the Dirac point. From this definition, we have $n = G(\varepsilon_F)/L^2 = k_F^2/\pi$.

The Boltzmann theory of transport for a two-dimensional system indicates that the conductivity is

$$\sigma = \frac{1}{2}e^2 V_F^2 g(\varepsilon_F)\tau(\varepsilon_F) = \frac{e^2 V_F k_F \tau(\varepsilon_F)}{2\pi\hbar}, \qquad (8.100)$$

where the factor of 2 reflects the dimensionality and in the second expression we have used (8.99). In graphene, the Bloch velocity is constant, hence any dependence of σ on ε_F comes from the explicit factor k_F or from the relaxation time. Again $\tau = 1/(n_{\text{imp}}w_0)$, but now in two dimensions instead of (8.30) one has

$$w_0(k) = \frac{k}{\hbar^2 V(k)} \int \left|\widehat{U}(k, \hat{\mathbf{k}} \cdot \hat{\mathbf{k}}')\right|^2 (1 - \cos\theta)\left(\frac{1+\cos\theta}{2}\right) \mathrm{d}\theta, \quad (8.101)$$

where the dimensionality is reflected in the change of the k exponent. The additional factor $(1 + \cos\theta)/2$ appears when the shape of the wavefunctions in the unit cell is considered, reflecting the honeycomb lattice, and now $\widehat{U}$ is computed using plane waves. Notably, this new factor suppresses backscattering events ($\theta = \pi$), increasing the conductivity.

Short-range Dirac delta potentials, $U(\mathbf{r}) = U_0 a_0^2 \delta(\mathbf{r})$, imply $w_0 \sim k$. Hence, the relaxation time goes as $\tau \sim k^{-1}$, resulting in a conductivity that is independent of ε_F. However, experiments show that $\sigma \sim n \sim \varepsilon_F^2$, contradicting this simple prediction. Taking account of such experiments, the relaxation time should rather go as $\tau \sim k$. It was rapidly realised that this was possible if the electron scattering was caused by Coulomb impurities located in the substrate supporting the graphene layer. In this case, $U(\mathbf{r}) = -Ze^2/(4\pi\epsilon r)$, where Ze is the charge of the impurity and ϵ is the dielectric constant. The two-dimensional Fourier transform gives $\widehat{U} = -Ze^2/(2\pi\epsilon q)$, where $q = |\mathbf{k} - \mathbf{k}'| = k\sqrt{2 - 2\hat{\mathbf{k}} \cdot \hat{\mathbf{k}}'}$. In summary, considering randomly located charged impurities, the relaxation time goes as $\tau \sim k$, which now agrees with the experimental result, $\sigma \sim k^2$.

More detailed experiments have shown that, although for large n the conductivity is linear in the carrier density, close to the Dirac point it goes to a constant. A large debate emerged on the origin and universality of this minimum value and whether the semiclassical Boltzmann equation could be used in this case. Two arguments are advanced to indicate that the kinetic equation should fail. First, we have shown that $\tau \sim k$; therefore, close to the Dirac point, the electron mean free time

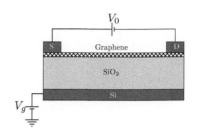

Fig. 8.13 Geometry used experimentally to change the position of the Fermi level. The gate voltage V_g imposed on the lower conductor attracts or repels electrons from the graphene layer. Source (S) and drain (D) conducting electrodes allow measurement of the conductivity and Hall effect. Adapted from Das Sarma *et al.* (2011).

would vanish. This invalidates one of the basic assumptions of the semi-classical quantum kinetic theory, namely that the mean free time should be longer than the period $2\pi\hbar/\varepsilon$ of the oscillation of the wavefunction phase or, alternatively, that the mean free path should be longer than the de Broglie wavelength. That is, close to the Dirac point, quantum coherence now prevails, invalidating the use of the random phase approximation in the derivation of the kinetic equation. The second objection is that, at low temperature, only electrons at the Fermi surface can respond to external perturbations. These electrons are treated statistically by means of the distribution function. However, at the Dirac point, the density of states vanishes; that is, there is a vanishing number of available electrons, which would make the coarse-graining procedure used to build the distribution function meaningless. More research is needed to settle this point, although it has been argued that charge fluctuations generate a Fermi level that depends on position, allowing use of the Boltzmann equation at every point. Using this approach, it was possible to compute the average conductivity, obtaining a minimum value for it in good agreement with experiment (Adam *et al.*, 2007).

Further reading

Electronic transport can be found in all books devoted to solid-state physics, where the concepts of the real and reciprocal lattices and the band structure of electrons are also described in detail. The book of Ashcroft and Mermin (1976) is already a classic on the subject. Also, (Ziman, 2001) is a classic text that describes in detail the properties of electrons and phonons. Kinetic theory of electrons and phonons is also presented in Callaway (1991), Jones and March (1973), Bruus and Flensberg (2004), and Vasko and Raichev (2006). Finally, Singleton (2001) is devoted entirely to the study of band theory and electronic properties.

A more profound discussion on the quantum treatment of nonequilibrium processes can be found in Jones and March (1973). The basis of the Landau theory of quantum liquids, which lies beyond the scope of this book, can be found in Lifshitz and Pitaevskii (1980) and Kadanoff and Baym (1962).

Graphene is a rapidly growing subject, and new articles and reviews appear periodically. Interested readers can go to the short introductory article by Wilson (2006). Detailed reviews on many aspects of graphene, including electronic transport, are given by Abergel *et al.* (2010) and Das Sarma *et al.* (2011). Finally, for a discussion on the Boltzmann approach and the conductivity problem, see Adam *et al.* (2007).

Exercises

(8.1) **Fermi level in insulators at low temperature.** The aim is to show that, in insulators, the Fermi level is located in the middle of the gap. At low temperatures, partially occupied levels occur only in a narrow region close to μ. We can hence consider just two bands and approximate the density of states using the van Hove expression at the band edges for the parabolic approximation (see Appendix D),

$$g(\varepsilon) = \begin{cases} g_V(\varepsilon_V - \varepsilon)^{1/2} & \varepsilon \leq \varepsilon_V, \\ 0 & \varepsilon_V < \varepsilon < \varepsilon_C, \\ g_C(\varepsilon - \varepsilon_C)^{1/2} & \varepsilon \geq \varepsilon_C, \end{cases}$$

where g_V and g_C characterise the bands and are, in general, different. Being an insulator, the valence band is completely occupied at zero temperature, which implies that the number of electrons is

$$N_e = \mathcal{V} \int_0^{\varepsilon_V} g(\varepsilon)\,\mathrm{d}\varepsilon.$$

At finite but small temperatures, the chemical potential is obtained from the equation,

$$N_e = \mathcal{V} \int_0^{\varepsilon_V} g(\varepsilon)f(\varepsilon)\,\mathrm{d}\varepsilon + \mathcal{V} \int_{\varepsilon_C}^{\infty} g(\varepsilon)f(\varepsilon)\,\mathrm{d}\varepsilon.$$

Using $k_\mathrm{B}T \ll \varepsilon_V, \varepsilon_C, (\varepsilon_C - \varepsilon_V)$, show that

$$\mu = \frac{\varepsilon_V + \varepsilon_C}{2} + \frac{k_\mathrm{B}T}{2}\log(g_V/g_C).$$

(8.2) **Conductivity tensor in an anisotropic metal.** Solids where the unit cell does not have cubic symmetry show anisotropic behaviour; for example, the dispersion relation is not isotropic. The parabolic approximation can be written as

$$\varepsilon = \frac{1}{2}\hbar^2 m_{ij}^{*-1}k_ik_j,$$

where m_{ij}^* is the effective mass matrix. Compute the conductivity tensor at vanishing temperature and show that it is also anisotropic.

(8.3) **Simple impurity model.** Point defects generate short-range potentials which depend on the specific properties of the defect. As a first approximation, we can model them as Dirac delta potentials, $U_i(\mathbf{r}) = U_0 a_0^3 \delta(\mathbf{r})$. Also, for the purpose of computations, the wavefunctions can be taken as plane waves, $\Psi_\mathbf{k} = e^{i\mathbf{k}\cdot\mathbf{r}}/\sqrt{\mathcal{V}}$. Compute the scattering amplitude $\widehat{U}_i$ and the relaxation time $\tau = 1/n_\mathrm{imp}w_0(k)$.

(8.4) **Ionic impurity.** A more detailed model is to consider impurities as ions of charge Ze. Thanks to Thomas–Fermi screening, they generate Yukawa-like potentials for the electrons,

$$U_i(\mathbf{r}) = -\frac{Ze^2 e^{-k_\mathrm{TF}r}}{4\pi\epsilon r},$$

where k_TF^{-1} is the Thomas–Fermi length and ϵ is the dielectric constant of the material. Compute the scattering amplitude $\widehat{U}_i$ and the relaxation time $\tau = 1/n_\mathrm{imp}w_0(k)$ using plane waves.

(8.5) **Low-temperature expansion of the Onsager transport integrals in metals.** In the calculation of the Onsager transport coefficients in metals, integrals of the form,

$$I_n = \int u(\varepsilon)(\varepsilon - \mu)^n h_\mathrm{FD}(\varepsilon)\,\mathrm{d}\varepsilon$$

appear. Changing variables to $\varepsilon = \mu + k_\mathrm{B}Tx$, show that, at low temperatures, the dominant contribution is

$$I_n = \begin{cases} (\pi k_\mathrm{B}T)^n J_n u(\mu) & ,n \text{ even} \\ (\pi k_\mathrm{B}T)^{n+1} J_{n+1} u'(\mu) & ,n \text{ odd} \end{cases},$$

where $J_0 = 1, J_2 = 1/3, J_4 = 7/15, \ldots$.

(8.6) **Onsager coefficients in metals.** Using the expansion described in Exercise 8.5, obtain the Onsager coefficients for a metal at low temperatures [eqns (8.62), (8.63), and (8.64)].

(8.7) **Cross Onsager coefficients in insulators.** Compute the cross Onsager coefficients $L_{12} = L_{21}$ for an insulator at low temperatures. Use the parabolic approximation for the dispersion relations of the conduction and valence bands.

(8.8) **Thermal conductivity of insulators.** Compute the thermal conductivity for an insulator at low temperatures. Use the parabolic approximation for the dispersion relations of the conduction and valence bands.

(8.9) **Modification of the Wiedemann–Franz law in the classical Lorentz model.** Consider the classical Lorentz model for electron transport (Chapter 3). Show that, in the hard sphere model, where energy is conserved at collisions, the conductivities are not related by the Wiedemann–Franz law. Find the value of $\kappa/\sigma T$.

(8.10) **The thermocouple.** A thermocouple is a device that can measure temperature differences electrically, thanks to the Seebeck effect. Consider two metals A and B, arranged as shown in Fig. 8.14. In one end the materials are in electrical contact, while in the other they are isolated from each other and therefore no electric current circulates. Placing the ends in contact with substances at different temperatures, show that the voltage measured between the open ends is

$$V = V_A - V_B = (S_B - S_A)(T_2 - T_1),$$

where $S_{A/B}$ are the Seebeck coefficients of the materials.

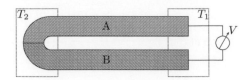

Fig. 8.14 Thermocouple configuration.

(8.11) **Two-band metals.** Some materials, for example aluminium, are metallic because two bands overlap at the Fermi level, leaving the valence band partially unfilled while the conduction band is partially filled, as shown in Fig. 8.15. Locally, close to the Fermi level, the bands can be approximated by parabolic dispersion relations:

$$\varepsilon_1(\mathbf{k}) = \varepsilon_{10} - \hbar^2(\mathbf{k} - \mathbf{k}_1)^2/2m_1^*,$$
$$\varepsilon_2(\mathbf{k}) = \varepsilon_{20} + \hbar^2(\mathbf{k} - \mathbf{k}_2)^2/2m_2^*,$$

with $\varepsilon_{10} > \varepsilon_{20}$.

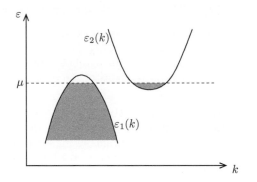

Fig. 8.15 Band structure for a metal with two bands overlapping.

Consider the relaxation time approximation, with characteristic times τ_1 and τ_2. First show that, if there are no interband transitions, the total conductivity is $\sigma = \sigma_1 + \sigma_2$, implying that the two bands work as resistances in parallel (the electrons can conduct in either of the channels). Now, in the presence of a magnetic field, show that this system does present a finite magnetoresistance.

(8.12) **Phonon kinetic equation.** Write down the collision term in the phonon kinetic equation that accounts for phonon–electron collisions.

(8.13) **Bloch–Grüneisen law: part 1.** Derive eqn (8.40) for the electrical conductivity when only one term in the expansion of Ψ is retained.

(8.14) **Bloch–Grüneisen law: part 2.** The objective is to compute explicitly I_2 [eqn (8.40)] to finally obtain the conductivity σ (8.43). First, transform the three-dimensional integrals to integrals over the magnitude and the unitary vector (e.g. $d^3k = dk\, d^2\hat{\mathbf{k}}k^2$). Then, using that phonons have low energy, in all factors except in f_{FD} and h_{FD}, substitute $k' = k$, $\varepsilon' = \varepsilon$, and $V' = V$. Show that, owing to $q \ll k$, the factors $\delta(\mathbf{k} - \mathbf{k}' \pm \mathbf{q})$ imply that the angle θ between $\hat{\mathbf{k}}$ and $\hat{\mathbf{k}}'$ is fixed to $\sin\theta = q/k$. Substitute then $\mathbf{V} \cdot \mathbf{V}' = V^2 \cos\theta \approx V^2(1 - q^2/2k^2)$. Show or use that the geometric integral $\int d^2\hat{\mathbf{k}}\, d^2\hat{\mathbf{k}}'\, d^2\hat{\mathbf{q}}\, \delta(\mathbf{k} - \mathbf{k}' \pm \mathbf{q}) = 8\pi^2/(k^2q)$ to leave I_2 as an integral over q, k, and k', where the last two can be transformed into integrals over ε and ε'. One of these can be carried out directly thanks to the Dirac delta functions, while the second can be computed by replacing the explicit expressions for the Fermi–Dirac and Planck distributions. Finally, making the change of variable $x = \hbar c q/k_{\mathrm{B}}T$, obtain the Bloch–Grüneisen expression (8.43).

(8.15) **Dielectric function of metals.** Using the Thomas–Fermi linearised equation, analyse the response of a metal to a static charge distribution $\rho_{\mathrm{ext}}(\mathbf{r}) = \rho_{\mathbf{k}} e^{i\mathbf{k}\cdot\mathbf{r}}$. Proceed similarly to Section 6.2.5 and derive the dielectric function of metals. The resulting expression, known as the plasma model, diverges for vanishing wavevector in accordance with macroscopic electromagnetism, which states that metals have infinite dielectric constant.

(8.16) **Finite-temperature effects on graphene.** Consider graphene at $\varepsilon_F = 0$. At finite temperature, not only must the Dirac point be considered, but also levels $k_{\mathrm{B}}T$ apart, giving rise to finite transport processes. Consider the model of scattering at Coulomb impurities and obtain the conductivity as a function of temperature.

(8.17) **Hall effect in graphene.** Obtain the Hall resistance at zero temperature as a function of the Fermi level. Use the Coulomb impurity model.

Semiconductors and interband transitions

9.1 Charge carriers: electrons and holes

In this chapter we analyse electronic transport when interband processes take place. For this purpose, we take semiconductors as a case study. Semiconductors are insulators where the bandgap is not extremely large compared with k_BT and, therefore, near the edges the valence or conduction band is not completely filled or empty, respectively. In the conduction band there is a small but finite concentration of electrons, while in the valence band some electrons are missing. It is customary for such states in the valence band where an electron is missing to be called holes. Both the electrons in the conduction band and the holes in the valence band are charge carriers and are responsible for the electronic properties of semiconductors.

Because the total number of electrons is fixed and the system is electrically neutral, the number of electrons in the conduction band equals the number of holes in the valence band. Indeed, in an intrinsic semiconductor,[1] electrons and holes are created in pairs when an electron in the valence band is excited to the conduction band. This generation process is compensated in equilibrium by recombination, where the electron goes down from the conduction to the valence band. This generation–recombination process involves energy exchange, for which there are many possible mechanisms. In semiconductors such as GaAs the band structure presents a direct bandgap (see Fig. 9.1), where the maximum of the valence band and the minimum of the conduction band occur at the same crystal momentum. In this case, the generation–recombination processes only involve energy exchange, with no momentum exchange. This can be achieved by absorbing or emitting a photon of appropriate energy, because the associated momentum of the photon, $p_\gamma = \hbar\omega/c$, is small, due to the large value of the light speed c. In Si and other semiconductors, the gap is indirect, with the two band extremes being located at different crystal momenta. Generation–recombination processes then also require the exchange of momentum, which cannot be achieved with a single photon. In this case, one or several phonons are involved. Other processes are possible; for example, thermal phonons are responsible for generating the equilibrium distribution of electrons and holes.

[1]Intrinsic semiconductors are pure materials, for example Si or GaAs. Extrinsic semiconductors, which are doped by other materials, will be described in Section 9.2.

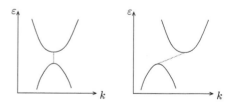

Fig. 9.1 Direct (left) and indirect (right) bandgaps in semiconductors.

Kinetic Theory and Transport Phenomena, First Edition, Rodrigo Soto.
© Rodrigo Soto 2016. Published in 2016 by Oxford University Press.

In equilibrium, the electrons follow the Fermi–Dirac distribution, which allows us to compute the total density of carriers in the conduction band (electrons) n and in the valence band (holes) p as

$$n = \int_{\varepsilon_C} g(\varepsilon) f_{\mathrm{FD}}(\varepsilon)\,\mathrm{d}\varepsilon \approx \int_{\varepsilon_C}^{\infty} g(\varepsilon) e^{-(\varepsilon-\mu)/k_{\mathrm{B}}T}\,\mathrm{d}\varepsilon, \qquad (9.1)$$

$$p = \int^{\varepsilon_V} g(\varepsilon)[1 - f_{\mathrm{FD}}(\varepsilon)]\,\mathrm{d}\varepsilon \approx \int_{-\infty}^{\varepsilon_V} g(\varepsilon) e^{-(\mu-\varepsilon)/k_{\mathrm{B}}T}\,\mathrm{d}\varepsilon, \qquad (9.2)$$

where we have used that μ is far from the band edges to approximate the Fermi–Dirac distributions f and $1 - f$ by the Boltzmann distribution. Therefore, from a statistical point of view, electrons and holes behave classically. At low temperatures, only energies close to the edges are relevant in the integrals. Hence, it is appropriate to make a parabolic approximation of the bands (Fig. 9.2),

$$\varepsilon_C(\mathbf{k}) = \varepsilon_C + \frac{\hbar^2(\mathbf{k} - \mathbf{k}_C)^2}{2m_C^*}, \qquad \varepsilon_V(\mathbf{k}) = \varepsilon_V - \frac{\hbar^2(\mathbf{k} - \mathbf{k}_V)^2}{2m_V^*}, \qquad (9.3)$$

where $m_{C/V}^*$ are called the effective mass of electrons and holes, and $\mathbf{k}_{C/V}$ are the locations of the extrema of the bands, which coincide for a direct gap. The densities of states are

$$g_C(\varepsilon) = \frac{1}{2\pi^2}\left(\frac{2m_C^*}{\hbar^2}\right)^{3/2}\sqrt{\varepsilon - \varepsilon_C}, \quad g_V(\varepsilon) = \frac{1}{2\pi^2}\left(\frac{2m_V^*}{\hbar^2}\right)^{3/2}\sqrt{\varepsilon_V - \varepsilon}, \qquad (9.4)$$

finally allowing one to obtain the carrier densities as

$$n = \frac{\sqrt{2}}{2}\left(\frac{k_{\mathrm{B}}T m_C^*}{\pi\hbar^2}\right)^{3/2} e^{-(\varepsilon_C-\mu)/k_{\mathrm{B}}T}, \qquad (9.5)$$

$$p = \frac{\sqrt{2}}{2}\left(\frac{k_{\mathrm{B}}T m_V^*}{\pi\hbar^2}\right)^{3/2} e^{-(\mu-\varepsilon_V)/k_{\mathrm{B}}T}, \qquad (9.6)$$

implying that electrons and holes are exponentially rare at low temperatures.

The position of the chemical potential can be obtained by imposing that, in an intrinsic semiconductor, the concentration of holes must equal that of electrons, resulting in[2]

$$\mu_i = \frac{\varepsilon_V + \varepsilon_C}{2} + \frac{3k_{\mathrm{B}}T}{4}\log(m_V^*/m_C^*). \qquad (9.7)$$

At low temperatures compared with the gap, the chemical potential of intrinsic semiconductors is exactly located in the middle position.

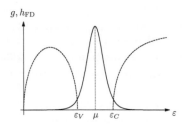

g, h_{FD}

Fig. 9.2 At finite temperature, the factor $h_{\mathrm{FD}}(\varepsilon)$ (solid line) overlaps with the valence and conduction bands, which are represented via the density of states g (dashed line).

[2]See Exercise 9.1.

9.2   Doped materials and extrinsic semiconductors

For use in microelectronics applications, semiconductors are usually doped to move the chemical potential and consequently alter the conduction properties. To be concrete, we will consider silicon (Si), which has an intrinsic bandgap of $\Delta = 1.12\,\text{eV}$. Each Si atom contributes four valence electrons. If some (few) silicon atoms are substituted by phosphorus atoms (P), which has five valence electrons, there will be an excess of electrons, which are placed in the conduction band, moving the chemical potential to higher energy. Note that the substitutional atoms are neutral, so there is no net charge added to the system; it is the number of electrons available for filling the bands that is changed, but the number of available states in the bands does not change (see discussion in Section 8.2). Conversely, if the substituted atoms are boron (B), which has three valence electrons, there will be a deficit of electrons and, consequently, the chemical potential will move to lower energy.[3] The phosphorus atoms are called donors, and the doped crystal is called an n-type extrinsic semiconductor. Boron is an acceptor, and silicon crystal doped with boron is called a p-type extrinsic semiconductor (see Fig. 9.3). Finally, the crystal is characterised by the density of donor N_D and acceptor N_A atoms.

To calculate the new position of the chemical potential, we use that the system remains electrically neutral, a condition that is expressed as $n - p = \delta n = N_D - N_A$. Using the expressions (9.5) and (9.6), it is found that $\mu = \mu_i + \delta\mu$, where

$$\sinh(\delta\mu/k_{\rm B}T) = \frac{\delta n}{\sqrt{2}(m_C^* m_V^*)^{3/4}(k_{\rm B}T/\pi\hbar^2)^{3/2}} e^{\Delta/2k_{\rm B}T}. \qquad (9.8)$$

The modification of the chemical potential can be positive or negative, but in either case the conductivity (8.36), which has factors $e^{-\Delta_{C/V}/k_{\rm B}T}$, is increased because either one or the other exponential factor will increase.

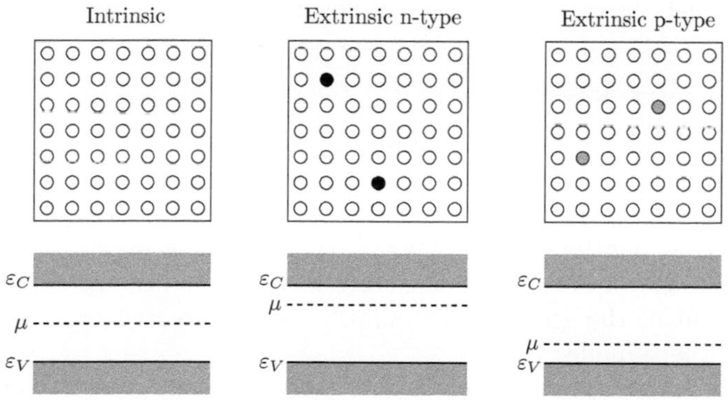

[3]Alternatively, the situation can be understood by noting that each atom forms covalent bonds with four neighbours. To create the bonds, four electrons are needed. Silicon thus uses all its valence electrons in these bonds. Phosphorus has an additional electron that is released to the conduction band. Boron, on the other hand, in order to create the bond, picks an electron from those available in the valence band, creating a hole.

Fig. 9.3 Position of the chemical potential in intrinsic and extrinsic semiconductors. The latter are obtained by substituting some silicon atoms by phosphorus to produce an n-type semiconductor or by boron to produce a p-type material. In the intrinsic case, the chemical potential lies at the middle of the bandgap, moving to higher energy for n-type semiconductors and the opposite for p-type semiconductors.

We have obtained that, when silicon is doped by phosphorus (n-type semiconductor), the valence band becomes more saturated and electrons are placed in the conduction band, and we say that the conduction is driven mainly by electrons (majority carriers). In the other case, when the lattice is doped with boron (p-type semiconductor), the conduction band is depleted and empty states appear in the valence band, allowing for electronic transitions in this region. In this case, it is said that the conduction is driven by holes.[4]

[4]Doping a material not only moves the position of the chemical potential, but also introduces discrete levels, which are located inside the bandgap. These levels, which do not constitute a band because the donors and acceptors are placed randomly, should be considered in detailed calculations of semiconductor properties. Here, we will not consider them because their presence does not alter the picture we will describe.

Note that the product np is independent on the location of μ. Hence, it is equal to the value for the intrinsic semiconductor. This result, $np = n_i p_i = n_i^2 = p_i^2$, is called the law of mass action.

9.3   Kinetic equation

The dynamical properties of semiconductors are completely determined by the motion of the charge carriers: electrons and holes. Although it is possible to work only with the distribution function of electrons, where we would need to indicate the band label, $n = C, V$, it is simpler to work with the distributions of electrons $f = f_C$ and holes $h = 1 - f_V$. Under many operational conditions, these distributions take values much smaller than one, being far from the degenerate case. Therefore, in equilibrium, they follow the Boltzmann distribution and the Pauli exclusion term is normally not needed when writing the kinetic equations.

Holes are not only defined as the absence of electrons, but rather they will be quasiparticles with specific electronic properties that we derive using phenomenological arguments.[5] Consider a photon with wavevector $\mathbf{q}$ and frequency ω that hits the semiconductor exciting an electron from the valence to the conduction band. The process is represented symbolically as $e_V + \gamma \rightarrow e_C$. The initial and final crystal momentum and energy of the electron are $(\mathbf{k}_v, \varepsilon_V(\mathbf{k}_v))$ and $(\mathbf{k}_c, \varepsilon_C(\mathbf{k}_c))$, respectively. Crystal momentum and energy conservation require that

$$\mathbf{k}_v + \mathbf{q} = \mathbf{k}_c, \qquad\qquad \varepsilon_V(\mathbf{k}_v) + \hbar\omega = \varepsilon_C(\mathbf{k}_c). \qquad (9.9)$$

Now, let us move the terms that refer to the valence band to the right-hand side. If we define $\mathbf{k}_h = -\mathbf{k}_v$ and $\varepsilon_h(\mathbf{k}_h) = -\varepsilon_V(\mathbf{k}_v)$, the conservation equations can be read as the generation of an electron–hole pair $(\gamma \rightarrow e + h)$,

$$\mathbf{q} = \mathbf{k}_e + \mathbf{k}_h, \qquad\qquad \hbar\omega = \varepsilon_e(\mathbf{k}_e) + \varepsilon_h(\mathbf{k}_h), \qquad (9.10)$$

where the hole crystal momentum and energy are precisely $\mathbf{k}_h$ and ε_h. Note that the valence–conduction band and the electron–hole representations are equivalent but should never be used simultaneously, otherwise double counting effects appear. To distinguish in which representation we are working, when speaking of holes, the energy and crystal momentum of the electron will simply be represented by ε_e and $\mathbf{k}_e$, without the subindex C.

We have found that for the crystal momentum $\mathbf{k}_h = -\mathbf{k}_v$. Hence, in presence of an electric field $\hbar\dot{\mathbf{k}}_h = -\hbar\dot{\mathbf{k}}_v = -(-e)\mathbf{E}$, where we used the equation of motion of electrons. That is, holes behave dynamically as having a positive charge. This can be seen pictorially in Fig. 9.4. In the presence of the field, electrons migrate in $\mathbf{k}$-space in the direction signalled by $-e\mathbf{E}$. Accordingly, holes migrate in the opposite direction, implying that they obey the equation $\hbar\dot{\mathbf{k}} = e\mathbf{E}$.

The Bloch velocity for holes is

$$\mathbf{V}_h = \frac{1}{\hbar}\frac{\partial \varepsilon_h}{\partial \mathbf{k}_h} = \frac{1}{\hbar}\frac{\partial (-\varepsilon_V)}{\partial (-\mathbf{k}_v)} = \mathbf{V}_v, \qquad (9.11)$$

[5]It is also possible to use Landau theory for Fermi liquids to characterise holes in a rigorous way.

Fig. 9.4 Top: When an electric field is applied, electrons migrate in $\mathbf{k}$-space in the same direction as $-e\mathbf{E}$ (pointing to the right in the figure). In the hole complementary figure (bottom), the migration gives rise to the opposite migration; that is, holes move in the same direction as $e\mathbf{E}$, allowing one to interpret them as having a positive charge, $q = +e$.

the same as electrons in the valence band. This is so because the Bloch velocity is a property of the quantum state and not if the state is occupied (valence electron) or empty (hole). In both cases the wavepacket moves with the same speed and direction.

Finally, when $h(\mathbf{k}_h) = 1 - f(\mathbf{k}_v)$ is replaced in the kinetic equation, the same collision term appears. In summary, the kinetic equation for holes is[6]

[6]See Exercise 9.2.

$$\frac{\partial h}{\partial t} + \mathbf{V}_h \cdot \frac{\partial h}{\partial \mathbf{r}} + \frac{e}{\hbar}\mathbf{E} \cdot \frac{\partial h}{\partial \mathbf{k}} = -\mathbb{W}[h](\mathbf{k}), \qquad (9.12)$$

where the dispersion relation,

$$\epsilon_h(\mathbf{k}) = -\varepsilon_V + \frac{\hbar^2(\mathbf{k} + \mathbf{k}_V)^2}{2m_V^*}, \qquad (9.13)$$

must be used.

9.3.1   Generation–recombination

The presence of two bands makes it possible to have interband transitions. Equivalently, in the language of holes, two charged species are present, electrons and holes, and reactions can take place. In the previous section, the generation process, when a photon creates an electron–hole pair was discussed. The reverse, recombination process, can also take place, with a photon being emitted.

Associated with this and other generation–recombination processes, new collision terms must be included in the kinetic equations, as usual consisting of gain and loss terms. Their explicit form depends on the transition rates specific to each process and on the distribution function of photons, phonons, or other agents that can trigger the processes. A simple case is obtained by considering thermal phonons at equilibrium, which are not dealt with explicitly but rather through their effect. For the holes, we write,

$$\left(\frac{\partial h(\mathbf{k})}{\partial t}\right)_{\text{g–r}} = \int \frac{\mathrm{d}^3 k'}{(2\pi)^3}\left[G(\mathbf{k}, \mathbf{k}') - R(\mathbf{k}, \mathbf{k}')f(\mathbf{k}')h(\mathbf{k})\right], \qquad (9.14)$$

which must be added to the right-hand side of eqn (9.12). Here, $G(\mathbf{k}, \mathbf{k}')$ is the spontaneous generation rate of an electron with crystal momentum $\mathbf{k}'$ and a hole with $\mathbf{k}$, and $R(\mathbf{k}, \mathbf{k}')$ is the recombination rate. A similar term exists for electrons, changing the roles of $\mathbf{k}$ and $\mathbf{k}'$. If the processes are driven by thermal phonons, they should satisfy the detailed balance condition for the equilibrium distributions: $G(\mathbf{k}, \mathbf{k}') - R(\mathbf{k}, \mathbf{k}')f_{\text{eq}}(\mathbf{k}')h_{\text{eq}}(\mathbf{k})$, which reads,

$$G(\mathbf{k}, \mathbf{k}') = e^{-\Delta/k_\mathrm{B}T}R(\mathbf{k}, \mathbf{k}'), \qquad (9.15)$$

where the low-temperature approximation is used. This relation reflects, for example, that many phonons are required to generate an electron–hole pair, reducing the generation transition rate by the Boltzmann factor, $e^{-\Delta/k_\mathrm{B}T}$.

Note that, according to the Fermi–Dirac distribution, the carrier populations do not depend on whether the semiconductor has a direct or indirect bandgap. The difference appears in the generation–recombination transition rates, which will generally be smaller for the latter case.

9.4   Hydrodynamic approximation

If we neglect the generation and recombination processes, the numbers of holes and electrons are conserved, and therefore the associated densities obey conservation equations with particle currents $\mathbf{J}_{e/h}$. For homogeneous systems, these currents obey Ohm's law, $\mathbf{J}_e = -e\mu_e n\mathbf{E}$ and $\mathbf{J}_h = e\mu_h p\mathbf{E}$, where the mobilities $\mu_{e/h}$ are defined by this proportionality.[7] The total electric current is $\mathbf{J} = e(\mathbf{J}_h - \mathbf{J}_e)$, resulting in a conductivity $\sigma = e^2(n\mu_e + p\mu_h)$. When the carrier densities depend on space, relaxation processes take place. In the long-wavelength limit, these relaxation processes are diffusive by symmetry. In summary,

$$\mathbf{J}_e = -D_e\nabla n - e\mu_e n\mathbf{E}, \qquad \mathbf{J}_h = -D_h\nabla p + e\mu_h p\mathbf{E}. \tag{9.16}$$

Similarly to what happens for Brownian particles, the mobilities and diffusion coefficients obey the fluctuation–dissipation relation. This is a consequence of the fact that the carriers follow the Boltzmann distribution and that the mobilities and diffusion coefficients are both limited by the same scattering mechanisms. Indeed, we can proceed as in Section 5.5.3 with the Einstein method to derive this relation. To be concrete, consider electrons first, placed in a constant electric field that pushes them in the negative z direction, while they are otherwise confined to $z \geq 0$. The system is described by the potential energy, $U = eEz$, and reaches a steady state with an electron density given by $n = n_0 e^{-eEz/k_B T}$, where Boltzmann statistics has been used. However, at the same time, the electron current (9.16) must vanish, leading to

$$D_e = k_B T\mu_e, \qquad\qquad D_h = k_B T\mu_h, \tag{9.17}$$

where in the second equality a similar reasoning for holes has been employed.

The generation–recombination contribution to the hydrodynamic equation is obtained by integrating the kinetic term (9.14) over wavevectors, resulting in a term, $\text{RG} = B(n_0 p_0 - np)$, where B quantifies the transition rates and n_0 and p_0 are the equilibrium concentrations.[8] This finally leads to the hydrodynamic equations for electrons and holes,

$$\frac{\partial n}{\partial t} = -\nabla \cdot \mathbf{J}_e + \text{RG}, \qquad \frac{\partial p}{\partial t} = -\nabla \cdot \mathbf{J}_p + \text{RG}. \tag{9.18}$$

9.5   Photoconductivity

If a semiconductor is irradiated by light with photons having energy greater than the gap, many electron–hole pairs are created. The carrier

[7]Recall that electrons and holes behave statistically as classical particles. Hence, the conductivity is proportional to the carrier density, as in the Lorentz model; see Exercise 9.4. Quantum effects appear in the scattering transition rates.

[8]In GaAs, $B \approx 10^{-10}\,\text{cm}^3\,\text{s}^{-1}$, while in Si, $B \approx 10^{-14}\,\text{cm}^3\,\text{s}^{-1}$. The larger value in GaAs is because it is a direct-bandgap semiconductor.

population increases, thereby augmenting the conductivity of the device. Using the hydrodynamic approach, the generation–recombination processes must cancel in the steady state, i.e.

$$B(n_i^2 - np) + I = 0, \tag{9.19}$$

where the law of mass action has been used and the last term accounts for the generation by photons, with a rate I that is proportional to the light intensity. At low temperatures we can neglect the thermal generation, giving

$$n = p = \sqrt{I/B}. \tag{9.20}$$

The conductivity that results after illumination is then

$$\sigma = e^2(n\mu_e + p\mu_h) = e^2(\mu_e + \mu_h)\sqrt{I/B}. \tag{9.21}$$

When the material is connected to a circuit, it can be used to measure light intensity. In this context, it is called a photoresistor.

The photoconductivity problem can also be analysed using kinetic theory, with the same result if we neglect processes where an electron receives two photons.[9]

[9]See Exercise 9.12.

9.6   Application: the diode or p–n junction

Semiconductors reach their full application potential when organised in heterogeneous structures, with the diode and transistor being the most well-known examples. A diode is a junction between p- and n-type semiconductors, while a transistor can be either a n–p–n or a p–n–p junction. Here, we will consider the diode or p–n junction; for other semiconductor devices, refer to the books indicated at the end of the chapter. Our aim is to understand the outstanding rectifying property of such a junction, where electricity can be transmitted in only one sense, as shown in Fig. 9.5.

Let us first consider the junction in equilibrium. When the p and n materials are placed in contact, the difference in chemical potential between the two materials will induce electrical transport to finally equilibrate the chemical potential. This reallocation of electrons will generate charge imbalances that, in turn, will produce internal electric fields. In the presence of an electric potential $\phi(\mathbf{r})$, the energy of an electron in the state $|\mathbf{k}, n\rangle$ is $\tilde{\varepsilon} = \varepsilon_n(\mathbf{k}) - e\phi(\mathbf{r})$. Therefore, in equilibrium, the distribution of electrons is

$$f_n(\mathbf{k}, \mathbf{r}) = \frac{1}{e^{(\varepsilon_n(\mathbf{k}) - e\phi(\mathbf{r}) - \mu)/k_\mathrm{B}T} + 1} = f(\varepsilon, \mathbf{r}). \tag{9.22}$$

It is convenient to define the electrochemical potential, $\psi(\mathbf{r}) = \mu + e\phi(\mathbf{r})$, which indicates the inflection point of the distribution. The electron and hole densities now depend on space. Using the Boltzmann approximation, valid if ψ is far from the band edges, we recover eqns (9.5) and (9.6) but with μ replaced by $\psi(\mathbf{r})$. The total charge density is

$$\rho(\mathbf{r}) = e[N_D(\mathbf{r}) - N_A(\mathbf{r}) + p(\mathbf{r}) - n(\mathbf{r})], \tag{9.23}$$

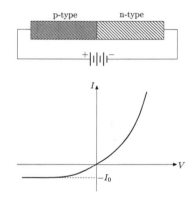

Fig. 9.5 *I–V* characteristic of a diode. When a potential difference V is applied to a p–n junction, the electric current I follows the law, $I = I_0(e^{V/k_\mathrm{B}T} - 1)$, where I_0 is an intrinsic property of the diode. The large asymmetry in the response between forward (positive) compared with reverse (negative) voltages can be modelled as a rectifier, where no current flows for reverse voltages.

which is finally responsible for generating the electric potential,

$$\nabla^2 \phi = -\frac{\rho}{\epsilon}, \tag{9.24}$$

where ϵ is the dielectric constant of the material. These constitute a coupled set of equations analogous to the Poison–Boltzmann equation in plasmas when computing Debye screening.

Let us consider a diode with an abrupt junction, where the donor and acceptor concentrations have a one-dimensional dependence,

$$N_D(x) = \begin{cases} N_D & x < 0 \\ 0 & x \geq 0 \end{cases}, \qquad N_A(x) = \begin{cases} 0 & x < 0 \\ N_A & x \geq 0 \end{cases}. \tag{9.25}$$

The distribution (9.22) shows that the reference values of μ and ϕ are arbitrary but must be defined consistently to generate the correct value for ψ. For simplicity, we choose to take $\mu = (\varepsilon_V + \varepsilon_C)/2$, and the Poisson equation reads,

$$\phi''(x) = -\frac{e}{\epsilon}\left[N_D(x) - N_A(x) - 2n_i \sinh\left(\frac{e\phi(x)}{k_B T}\right)\right]. \tag{9.26}$$

The boundary conditions for ϕ are such that, far from the junction, there is charge neutrality; that is, the right-hand side of the previous equation must vanish. Explicitly,

$$\sinh\left[\frac{e\phi(-\infty)}{k_B T}\right] = -\frac{N_D}{2n_i}, \qquad \sinh\left[\frac{e\phi(+\infty)}{k_B T}\right] = \frac{N_A}{2n_i}. \tag{9.27}$$

Similarly to Debye screening (see Section 6.2.2), a characteristic length λ emerges, where the effects of the junction are present. Beyond this region, the semiconductors are electrically neutral, while at distances on the order of λ from the junction, depleted regions are created. There, the potential varies and the materials are not electrically neutral: the p region is depleted of holes, being negatively charged, and the opposite happens in the n region. The final equilibrium configuration is shown in Figs. 9.6 and 9.7, with an electrochemical potential that joins the equilibrium values of the p and n parts in a transition region of length λ. For small doping ($N_A, N_D \ll n_i$) eqn (9.26) can be linearised giving, $\lambda = \sqrt{\frac{\epsilon k_B T}{2e^2 n_i}}$. The opposite case, when doping is large, is the most interesting for applications. Here, the solution of the nonlinear eqn (9.26) shows that there is an almost complete depletion near the junction. Approximating that the region is fully depleted yields for each region,

$$\lambda_{n,p} = \sqrt{\frac{2\epsilon[\phi(+\infty) - \phi(-\infty)](N_A/N_D)^{\pm 1}}{e(N_A + N_D)}}.$$

When the p–n junction is placed in an external field, in contact with metals, an electric current is established. Calculation of the resulting current is quite complex, and several approximations should be made in order to render the problem tractable. This was first achieved by Shockley in 1949. He noted that, in the depletion region, there is a deficit of carriers (both electrons and holes), as shown in Fig. 9.7. Therefore,

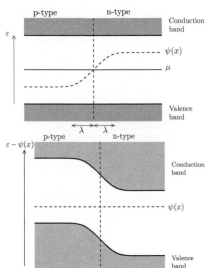

Fig. 9.6 Band diagram near a p–n junction. Top: the energies ε of the valence and conduction bands as well as the chemical potential are homogeneous. The development of an electric field gives rise to an inhomogeneous electrochemical potential $\psi(x)$. In equilibrium, electrons follow a Fermi–Dirac distribution that is centred at $\psi(x)$. Therefore, depletion of holes in the valence band develops in the p region near the junction. Similarly, electrons are depleted in the conduction band near the junction in the n region. Bottom: an alternative representation, showing the bands relative to $\psi(x)$, which is now horizontal, with the Fermi–Dirac distribution centred about it. The bands are now bent, with the same implications for the populations of holes and electrons as for the top representation.

this region has a higher resistivity than the rest of the system, and the imposed voltage will be dropped essentially over this region. Far from the depletion regions, the electric field is weak, and again deep in the p and n regions, the material is neutral. Second, we approximate that, in the depletion region, no recombination processes take place, only occurring far from the junction. These and other technical details allow one to obtain the electric current I when a voltage V is imposed as

$$I = I_0 \left(e^{V/k_{\mathrm{B}}T} - 1 \right), \qquad (9.28)$$

where I_0 is an intrinsic property of the diode. This relation is called the *I–V* characteristic of the diode. The origin of the large difference between the forward (positive) and reverse (negative) voltages can be understood by recalling that the deficits of the carrier densities create a rise of the electric potential in the depletion region, which is proportional to the depletion region size. Now, if a forward potential is externally imposed, the potential rise at the depletion region should be smaller, decreasing its extension. As the depletion region is the most resistive part, the net effect is a reduction of the resistivity and increase of the current. In contrast, for a reverse potential, the depletion region increases its size, and consequently the current is drastically reduced.

A complete solution of the diode response is complex because the base state is inhomogeneous. Numerical methods applied to the kinetic equation nevertheless give very accurate predictions.

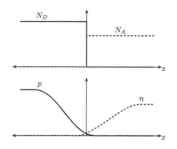

Fig. 9.7 Top: concentrations of donors N_D and acceptors N_A. Bottom: resulting concentration of holes p and electrons n. A depletion region near the junction is formed, creating a net electric charge.

Further reading

Semiconductors are described in most solid-state physics books, for example Ashcroft and Mermin (1976), Jones and March (1973), or Singleton (2001). There are also books especially devoted to semiconductor devices that present a kinetic approach (Pulfrey, 2010; Jacoboni, 2010).

Exercises

(9.1) **Chemical potential.** Derive eqn (9.7) imposing that $n = p$ for an intrinsic semiconductor.

(9.2) **Kinetic equation for holes.** Define the hole distribution function, $h(\mathbf{r}, \mathbf{k}_h, t) = 1 - f(\mathbf{r}, \mathbf{k}_v, t)$, where $\mathbf{k}_h = -\mathbf{k}_v$. Substitute it in the kinetic equation for electrons (8.7) to finally obtain the kinetic equation for holes (9.12).

(9.3) **Hall effect.** Analyse the Hall effect in semiconductors. Consider an extrinsic semiconductor and show that the sign of the Hall resistance depends on the relative population of carriers.

(9.4) **Electrical conductivity of extrinsic semiconductors.** Compute the electrical conductivity of extrinsic semiconductors. Use the relaxation time approximation. At low temperatures all the electrons and holes have nearly the same energy, and we can use a single relaxation time for each species: τ_e and τ_h. Show that the electric current for electrons can be written as $\mathbf{J}_e = -e\mu_e n\mathbf{E}$, and similarly for holes. Give an expression for the mobilities $\mu_{e/h}$.

(9.5) **Thermal conductivity of extrinsic semiconductors.** Compute the thermal conductivity of extrinsic semiconductors. Express the result in terms of the intrinsic thermal conductivity and the excess density of carriers δn.

(9.6) **Density of carriers in extrinsic semiconductors.** Consider an extrinsic semiconductor with chemical potential μ and band edges ε_V and ε_C. At low temperatures compared with the gap, $\Delta = \varepsilon_C - \varepsilon_V$, we can use the parabolic approximation for the bands, with respective effective masses $m_{V/C}^*$. Compute the total density of carriers in each band (eqns (9.5) and (9.6)), defined as the number of electrons in the conduction band n or the number of holes in the valence band p.

(9.7) **Wiedemann–Franz law.** Show that, in a semiconductor with low concentration of carriers, that is, when they obey the Boltzmann distribution, the Wiedemann–Franz law is not satisfied.

(9.8) **Thomas–Fermi screening.** Obtain the screening length in the Thomas–Fermi approximation for extrinsic semiconductors. Use the Boltzmann approximation for the electron and hole distributions.

(9.9) **Auger recombination.** Besides the direct electron–hole recombination described in the text, in some materials, especially in solar cells and light-emitting diodes (LEDs), there is another process that takes place, called Auger recombination. Here an electron and a hole recombine but the extra energy, instead of being transferred to a photon, is given to another electron in the conduction band. Also, the energy can be transferred to a hole. Write down the corresponding terms in the kinetic and hydrodynamic equations.

(9.10) **Poisson equation in the p–n junction.** Derive eqn (9.26).

(9.11) **Gated semiconductors.** When a semiconductor film is placed in the geometry shown in Fig. 9.8, the gate voltage V_g changes the position of the chemical potential. Indeed, a positive voltage will attract electrons to the semiconductor, increasing μ, and the opposite happens when V_g is negative. With the gate voltage fixed, a current is imposed between the source and drain contacts. Show that, at low temperatures, the electrical conductivity as a function of the gate voltage $\sigma(V_g)$ gives information on the density of states $g(\epsilon)$.

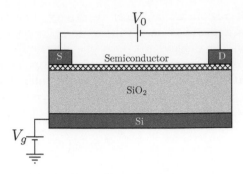

Fig. 9.8 Geometry used experimentally to change the position of the chemical potential. The gate voltage V_g imposed in the lower conductor attracts or repels electrons from the semiconductor film. Source (S) and drain (D) conducting electrodes allow the measurement of the conductivity.

(9.12) **Kinetic description of photoconductivity.** Consider a direct-bandgap semiconductor illuminated by monochromatic light of frequency ω, with $\hbar\omega > \Delta$. Assume that the transition rate is independent of the wavevectors, with the sole condition that energy is conserved. At vanishing temperature, there are no holes or free electrons and the effect of the photons is to create electron–hole pairs, where the carriers are in the region of size $\hbar\omega - \Delta$ close to the band edges. The final assumption is that no secondary processes take place, where photons can excite an electron to the newly emptied states in the valence band or photons can excite the electrons in the conduction band to even higher energies. Under these conditions, solve the kinetic equation and compute the resulting conductivity using the formula (8.34). Note that discontinuities appear in the distribution function.

(9.13) **Diode laser.** When a current circulates in a diode, holes and electrons cross the depletion region and recombine in the quasineutral regions. In a direct-gap semiconductor such recombination generates light, which if properly confined by mirrors, can give rise to laser emission. Instead of dealing with the full diode geometry, consider a homogeneous extrinsic semiconductor that by some mechanism pumps electrons into the conduction band. Write down the kinetic equation considering stimulated recombination. Assuming that only a small fraction γ of photons escape from the mirrors, show that laser light is generated (see Section 7.8 for the operation of lasers).

Numerical and semianalytical methods

10

10.1 Direct approach

The various kinetic equations that we have studied in this book, and also others that describe different phenomena, are integrodifferential equations for one or several distributions, with important nonlinearities. Besides, as we have shown when studying their spectra, they include both diffusive and wave phenomena, associated with the conserved fields, together with relaxational dynamics. In the language of partial differential equations, this means that such kinetic equations are both hyperbolic and parabolic, needing sophisticated methods to solve them numerically. Furthermore, the integral terms, for example in the classical Boltzmann equation,[1] are nonlocal in velocity space, thus requiring the performance of extensive velocity integrations. Finally, and this is a critical issue, the distribution functions usually depend on many variables: position, velocity or wavevector, and time. It is therefore extremely demanding in terms of computer time and memory to solve kinetic equations by direct numerical methods. Only in situations where there are important symmetries in configurational and velocity space, allowing a reduction of the dimensionality of the problem, is a direct approach feasible.

Discarding in general the direct solution, there are two approaches to solve kinetic equations. These do not aim to find the solution for the distribution function, which we are normally not interested in, but rather, the methods we will describe look to extract relevant statistical information from the distribution functions. The first method consists in finding a set of coupled equations for fields that depend on position and time, but not on velocities, thereby reducing the dimensionality of the problem. These equations will be further solved numerically using standard methods for partial differential equations. The second method uses a discrete but large number of particles, which follow trajectories designed to statistically sample at each time the distribution function.

10.2 Method of moments

Consider the classical Boltzmann equation for a monocomponent gas,

$$\frac{\partial f}{\partial t} + \mathbf{c} \cdot \nabla f + \frac{\mathbf{F}}{m} \cdot \frac{\partial f}{\partial \mathbf{c}} = J[f, f], \qquad (10.1)$$

[1]In this chapter, to avoid overloading the notation, we will use the language of classical kinetic theories, but the results are not limited to those and examples will be given for quantum kinetic models.

Kinetic Theory and Transport Phenomena, First Edition, Rodrigo Soto.
© Rodrigo Soto 2016. Published in 2016 by Oxford University Press.

where J is the collision operator. We know that, at long times, the distribution will be a local Maxwellian plus corrections, which will decay in time. Inspired by this, we propose solutions,

$$f(\mathbf{r}, \mathbf{c}, t) = f_{\mathrm{MB}}(\mathbf{c}; n, \mathbf{v}, T) \sum_n a_n P_n(\mathbf{c}), \qquad (10.2)$$

where n, $\mathbf{v}$, T, and a_n depend on space and time and we do not require the coefficients a_n to be necessarily small. The functions P_n constitute a basis set for functions on the velocity space, and polynomials are normally used.

We first demand that n, $\mathbf{v}$, and T have their canonical meanings of density, mean velocity, and temperature; that is, for example, for the density, it is required that $\int f \, \mathrm{d}^3 c = n$, which restricts one degree of freedom. A simple solution for this condition is that

$$P_0 = 1, \quad a_0 = 1, \qquad (10.3)$$

$$\int f_{\mathrm{MB}}(\mathbf{c}) P_n(\mathbf{c}) \, \mathrm{d}^3 c = 0, \quad \forall n \neq 0. \qquad (10.4)$$

Similar conditions are imposed for the mean velocity and temperature,

[2]For simplicity, we group in $n = 1$ the three components of the velocity.

resulting in[2]

$$P_1 = \mathbf{c}, \quad a_1 = 0, \qquad (10.5)$$

$$P_2 = mc^2/2 - 3k_{\mathrm{B}}T/2, \quad a_2 = 0, \qquad (10.6)$$

$$\int f_{\mathrm{MB}}(\mathbf{c}) \begin{Bmatrix} 1 \\ \mathbf{c} \\ c^2 \end{Bmatrix} P_n(\mathbf{c}) \, \mathrm{d}^3 c = 0, \quad \forall n \geq 3. \qquad (10.7)$$

These conditions can easily be fulfilled if orthogonal polynomials are used. As the base distribution is the Maxwellian, these turn out to be the Hermite polynomials in the reduced velocity, $\mathbf{C} = \mathbf{c} - \mathbf{v}$.

The aim now is to find dynamical equations for n, $\mathbf{v}$, T, and a_n, $n \geq 3$. To do so, the Boltzmann equation (10.1) is projected onto a set of functions $Q_n(\mathbf{c})$, which again are normally polynomials that could be the same as P_n; that is, we compute

$$\int Q_n(\mathbf{c}) (\text{Boltzmann eqn}) \, \mathrm{d}^3 c. \qquad (10.8)$$

For this purpose, we recall the development of Section 4.3.1,

$$\frac{\partial}{\partial t} (n \langle Q_n \rangle) + \nabla \cdot (n \langle \mathbf{c} Q_n \rangle) - n\mathbf{F} \cdot \langle \nabla_{\mathbf{c}} Q_n \rangle = I[Q_n], \qquad (10.9)$$

where $I[Q_n]$ vanishes for the collisional invariants $(Q = 1, \mathbf{c}, c^2)$. The result is a set of partial differential equations in time and space, which are notably simpler than the original kinetic equation.

The moment equations (10.9) are equivalent to the kinetic equation if an infinite set of functions is used in (10.2) and also infinite moment equations are used. In practice, of course, finite numbers of functions are used and the same number of equations are obtained. The choice of the functions P_n and Q_n is based on physical arguments, demanding that they express the most important phenomena that we are aiming to describe. Also, analysis of the spectrum of the linear equation indicates which are the slowest modes, dominating the dynamics.

For the Boltzmann equation, two choices are made. The first uses only the local equilibrium function, whereas the second, called the Grad method, includes irreversible fluxes.

10.2.1 Local equilibrium moment method

Here, the distribution function is simply approximated to

$$f(\mathbf{r}, \mathbf{c}, t) = f_{\mathrm{MB}}(\mathbf{c}; n, \mathbf{v}, T), \tag{10.10}$$

and the Boltzmann equation is projected into $Q_n = \{1, m\mathbf{c}, mc^2/2\}$, generating precisely the equations for the mass density, velocity, and temperature. The polynomials Q_n are collisional invariants, resulting in the vanishing of the right-hand side of eqn (10.9). After performing several simplifications, the equations read,

$$\frac{\partial n}{\partial t} + n\nabla \cdot \mathbf{v} = 0, \tag{10.11}$$

$$mn\frac{\mathrm{D}\mathbf{v}}{\mathrm{D}t} = -\nabla p, \tag{10.12}$$

$$nc_v\frac{\mathrm{D}T}{\mathrm{D}t} = -nk_{\mathrm{B}}T\,\nabla \cdot \mathbf{v}, \tag{10.13}$$

where we have defined the convective derivative operator, $\frac{\mathrm{D}}{\mathrm{D}t} = \frac{\partial}{\partial t} + (\mathbf{v} \cdot \nabla)$, to simplify the notation. The result is the Euler fluid equations for an ideal gas with equation of state $p = nk_{\mathrm{B}}T$ and heat capacity per particle $c_v = dk_{\mathrm{B}}/2$, where $d = 2, 3$ is the spatial dimensionality.[3]

[3]The use of the Maxwell–Boltzmann distribution gives only reversible fluxes, $\mathbb{P} = p\mathbb{I}$ and $\mathbf{q} = 0$. Hence the Euler equations are obtained.

10.2.2 Grad's method

The Euler equations are reversible in time because they do not include nonequilibrium fluxes. Grad noted in 1949 that it is possible to include momentum and heat fluxes by using a limited number of Hermite polynomials in the reduced velocity, $\mathbf{C} = \mathbf{c} - \mathbf{v}$. In three dimensions,

$$f(\mathbf{c}) = f_{\mathrm{MB}}(\mathbf{c}; n, \mathbf{v}, T)\left[1 + \frac{m^2}{5nk_{\mathrm{B}}^3 T^3}\left(C^2 - \frac{5k_{\mathrm{B}}T}{m}\right)\mathbf{C} \cdot \mathbf{q}\right.$$
$$\left. + \frac{m}{2nk_{\mathrm{B}}^2 T^2}p_{ij}\mathbf{C}_i\mathbf{C}_j\right], \tag{10.14}$$

where the numerical factors have been chosen such that $P_{ij} = nk_{\mathrm{B}}T\delta_{ij} + p_{ij}$ is the stress tensor and $\mathbf{q}$ is the heat flux. The tensor p_{ij} gives the nonequilibrium contribution to the stress, which is symmetric and has vanishing trace, resulting in five degrees of freedom. In summary, the model considers 13 moments as dynamical variables: n, $\mathbf{v}$, T, $\mathbb{p}$, and

[4]In two dimensions, eight moments are obtained.

$\mathbf{q}$.[4] The equations of motion are obtained by projecting the Boltzmann equation into $Q_n = \{1, m\mathbf{c}, mC^2/2, mC_xC_y, mC_yC_z, mC_xC_z, m(C_x^2 - C_y^2), m(C_y^2 - C_z^2), mC^2\mathbf{C}/2\}$. The equations for Grad's 13-moment method are

$$\frac{\mathrm{D}n}{\mathrm{D}t} + n\nabla \cdot \mathbf{v} = 0,$$

$$mn\frac{\mathrm{D}\mathbf{v}}{\mathrm{D}t} + \nabla \cdot \mathbb{P} = 0,$$

$$nc_v\frac{\mathrm{D}T}{\mathrm{D}t} + \nabla \cdot \mathbf{q} + \mathbb{P} : \nabla \mathbf{v} = 0,$$

$$\frac{\mathrm{D}\mathbb{p}}{\mathrm{D}t} + \mathbb{p}\nabla \cdot \mathbf{v} + \frac{2}{d+2}[\nabla \mathbf{q}] + [\mathbb{p} \cdot \nabla \mathbf{v}] + p[\nabla \mathbf{v}] = -\frac{\mathbb{p}}{\tau},$$

$$\frac{\mathrm{D}\mathbf{q}}{\mathrm{D}t} + \frac{d+4}{d+2}(\mathbf{q}\nabla \cdot \mathbf{v} + \mathbf{q} \cdot \nabla \mathbf{v}) + \frac{2}{d+2}q_k\nabla V_k + \frac{k_BT}{m}\nabla \cdot \mathbb{p}$$

$$+ \frac{(d+4)k_B}{m}\mathbb{p} \cdot \nabla T - \frac{1}{mn}\overbrace{(\nabla \cdot \mathbb{P})\mathbb{p}} + \frac{(d+2)k_B}{2m}p\nabla T = -\frac{(d-1)\mathbf{q}}{d\tau},$$

where the equations for the stress tensor and heat flux have a collisional contribution. The collisional terms have been evaluated for hard spheres of diameter D and depend on the characteristic time,

$$\tau = \frac{d+2}{2^{d+1}}\frac{1}{nD^{d-1}}\sqrt{\frac{m}{\pi k_BT}}. \tag{10.15}$$

The square bracket is used to indicate the symmetric traceless part, namely for a tensor, $[\mathbb{A}] = A_{ij} + A_{ji} - \frac{2}{d}\delta_{ij}A_{kk}$, and the overbrace term must be read as $(\partial P_{jk}/\partial x_k)p_{ij}$.

Note that, in Grad's method, the choice of the polynomials is based on physical grounds. In particular, noting that the equations for the first three fields depend on the stress tensor and the heat flux, it is then natural to add these fields as new variables. As a consequence, the method does include all polynomials of order zero, one, and two, but not all polynomials of order three are included; for example, $C_xC_yC_z$ does not take part in the expansion.

Finally, it must be remarked that such distributions with a finite number of polynomials are not guaranteed to be everywhere positive. Therefore, one should not try to extract more information from the distribution functions than the moments that define them. Only the moments must be considered as valid information.

The partial differential equations generated by Grad's methods are not properly hydrodynamic equations because not all fields are conserved quantities.[5] Rather, they constitute what is called a generalised hydrodynamic model.

[5]To be fair, besides conserved fields, hydrodynamic equations can also include other slow variables such as order parameters near a critical second-order transition or the fields that describe spontaneous breaking of continuous symmetries, for example the director in crystal liquids (Forster, 1995; Chaikin and Lubensky, 2000).

10.3 Particle-based methods

10.3.1 Sampling

The kinetic equations can be studied numerically by noting that the distribution functions are statistical objects. Indeed, they give the coarse-grained number of particles in a given configuration, which is an average of the real number of particles over small boxes in phase space and time. So, instead of studying the distribution function f directly, we can simulate a large number of particles $\widetilde{N}$, which evolve in phase space, such that an estimator $\widehat{f}$ of the distribution function is given by the number density of particles in phase-space boxes. On most occasions, we will not be interested directly in the distribution function but rather in averages that are computed directly from the sampling particles. The sampling particles must not be confused with the real ones, and their dynamics should be designed to reproduce the evolution of the distribution function and not necessarily the motion of real particles. The number $\widetilde{N}$ must be large to have good sampling and avoid unphysical fluctuations, but nevertheless much smaller than the real number of particles N (of the order of the Avogadro number).[6] Each sampling particle represents $\Lambda = N/\widetilde{N}$ real particles.

The kinetic equations that we have studied take into account several forms of microscopic dynamics that can be classified mainly into five categories: (i) streaming motion of independent particles, where each particle follows deterministic equations of motion; (ii) Brownian motion, where many contributions sum up to a random fluctuating force; (iii) long-range forces, where many distant particles produce a net effect on other particles, an effect that can be modelled by a mean field force; (iv) collisions, where a few—normally one or two—particles collide at instantaneous events; and finally (v) quantum effects that either suppress (in the case of fermions) or enhance (in the case of bosons) the different transition rates. Generally, these mechanisms combine and two or more of them are present in a single kinetic model.

10.3.2   Random numbers

When evolving the sampling particles, several actions will be stochastic, for example the random selection of the particles that will suffer scattering events. In all computer languages and programming systems, it is possible to generate random numbers which are uniformly distributed in the interval $[0, 1]$.[7] Such algorithms guarantee that, if a sequence of such numbers is generated, they will be statistically independent. Using these numbers, all sorts of distributions can be generated. In some cases, simple changes of variable are enough; for example, if x is uniform in $(0, 1]$, then $y = a + (b - a)x$ will be uniform in $(a, b]$ or $z = -\tau \ln(x)$ will follow an exponential distribution with characteristic scale τ. Also, if x_1 and x_2 are uniformly distributed in $(0, 1]$, then

$$\eta_1 = \sqrt{-2\ln x_1}\cos(2\pi x_2), \qquad \eta_2 = \sqrt{-2\ln x_1}\sin(2\pi x_2) \qquad (10.16)$$

are two independent normally distributed numbers; that is, they follow a Gaussian distribution with null average and unit variance.[8]

When it is not possible to make a change of variable, numbers can be drawn from a given probability density $P(x)$ with $a \le x \le b$, using the acceptance–rejection method. Let $P_{\max}$ be an upper bound of P.

[6]Experience has shown that, for the Boltzmann equation, for example, 20 particles per cubic mean free path is sufficient.

[7]Normally, 1 is excluded, and in some cases, zero is excluded as well. In most cases, the presence or not of these excluded numbers is not relevant as they occur rarely, but eventually they can generate problems when some nonanalytic functions are applied to them.

[8]See Exercise 10.3 for demonstration of the Box–Muller transformation.

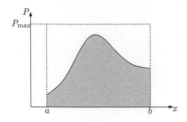

Fig. 10.1 Acceptance–rejection method to generate numbers according to a probability density $P(x)$. Pairs of numbers are uniformly generated in the rectangle $[a, b] \times [0, P_{max}]$, being selected if they fall below the curve $P(x)$ in grey. The chance of selecting a number in the interval $[x, x + dx]$ will then be proportional to the area $P(x)\,dx$, which is precisely the definition of a probability density.

Pairs of numbers x and y are drawn from uniform distributions in $[a, b]$ and $[0, P_{max}]$, respectively. If $y < P(x)$ then x is accepted, otherwise another pair is generated until the condition is fulfilled. The numbers x generated by this procedure follow the desired distribution, as can be seen geometrically in Fig. 10.1.

On many occasions, one needs to generate random unitary vectors. The simplest way is an acceptance–rejection method. Three numbers x_1, x_2, and x_3 are drawn at random, uniformly in $[-1, 1]$. If $x_1^2 + x_2^2 + x_3^2 \leq 1$, the point is selected and normalised, otherwise a new triplet is generated until success.

Finally, if an action is to be performed with a probability p, with $0 \leq p \leq 1$ (for example, to perform a collision or not), this is easily achieved by the Monte Carlo method. When such a decision must be taken, a random number s is drawn from a uniform distribution in $[0, 1]$. The action is performed if $s < p$. The interpretation is direct: there is a probability p for $s < p$ and $1 - p$ for $s > p$, which is precisely the desired result.

10.3.3 Streaming motion

Streaming is the simplest motion because particles do not interact and each one follows the deterministic equation of motion. The kinetic equation is linear,

$$\frac{\partial f}{\partial t} + \frac{\partial}{\partial \Gamma}(\Phi(\Gamma)f) = 0, \qquad (10.17)$$

which is associated with the equation of motion, $\dot{\Gamma} = \Phi(\Gamma)$. These equations of motion should be solved for all the sampling particles. The $\tilde{N}$ uncoupled ordinary differential equations can be integrated in time steps Δt using any of the various methods available in the literature [see, for example Press *et al.* (2007)]. If an explicit—Euler-like—method is used, the evolution will be $\Gamma_a(t + \Delta t) = \Gamma_a(t) + \Phi(\Gamma_a(t))\Delta t$, with $a = 1, 2, \ldots, \tilde{N}$. However, the Euler method is not a good choice because of its instability and low precision, and higher-order methods should be used, for example Runge–Kutta, Verlet, leapfrog, predictor–corrector, or implicit methods, where the election depends on the specific properties of the differential equations. For example, for Hamiltonian equations, $\dot{p}_a = F_a$, $\dot{x}_a = p_a/m$, the leapfrog method is particularly well suited,

$$\begin{aligned} p_a(t + \Delta t/2) &= p_a(t - \Delta t/2) + F_a[x(t)]\Delta t, \\ x_a(t + \Delta t) &= x_a(t) + p_a(t + \Delta t/2)\Delta t/m, \end{aligned} \qquad (10.18)$$

which makes an error $O(\Delta t^4)$ at each time step.

10.3.4 Brownian motion

Consider a general Fokker–Planck equation for the distribution function $f(\Gamma, t)$, where Γ is a multidimensional variable,

$$\frac{\partial f}{\partial t} = -\frac{\partial}{\partial \Gamma}[\Phi(\Gamma)f] + D\frac{\partial^2 f}{\partial \Gamma^2}. \qquad (10.19)$$

Assume that, at time t, the distribution function is known: $f(\Gamma, t)$. For a short time interval Δt, the solution of the Fokker–Planck equation is

$$f(\Gamma, \Delta t) \approx \int G_{\Delta t}(\Gamma, \Gamma') f(\Gamma', t) \, \mathrm{d}\Gamma', \qquad (10.20)$$

where the Green function is

$$G_{\Delta t}(\Gamma, \Gamma') = \frac{1}{\sqrt{2\pi D \Delta t}} e^{-[\Gamma - (\Gamma' + \Phi(\Gamma')\Delta t)]^2 / 2D\Delta t}. \qquad (10.21)$$

This represents the deterministic streaming from Γ' to $\Gamma' + \Phi(\Gamma')\Delta t$, plus the spreading associated with diffusion. The new distribution is obtained by deterministically displacing the sampling particles and adding on top a random displacement that is Gaussian distributed. Take, for each particle, a normal random number η with a Gaussian distribution of unit variance and vanishing average, and evolve the sampling particles with

$$\Gamma_a(t + \Delta t) = \Gamma_a(t) + \Phi(\Gamma_a(t))\Delta t + \sqrt{2D\Delta t} \, \eta_a. \qquad (10.22)$$

By construction, they describe the requested distribution.

We have obtained a rule, for finite time steps, to evolve the sampling particles. This is an Euler-like method, which is precise only for small time steps with an error per time step, $\mathcal{O}(\Delta t^{3/2})$. More precise methods can be obtained using a strategy similar to the Runge–Kutta method; for example, Heun's method, which is still of order $\mathcal{O}(\Delta t^{3/2})$ but improves the deterministic part, generating first an intermediate predictor value $\overline{\Gamma}$, which is later corrected,

$$\begin{aligned} \overline{\Gamma}_a &= \Gamma_a(t) + \Phi(\Gamma_a(t))\Delta t + \sqrt{2D\Delta t} \, \eta_a, \\ \Gamma_a(t + \Delta t) &= \Gamma_a(t) + \left[\Phi(\Gamma_a(t)) + \Phi(\overline{\Gamma}_a) \right] \Delta t/2 + \sqrt{2D\Delta t} \, \eta_a, \end{aligned} \qquad (10.23)$$

where the same random numbers η_a must be used in the predictor and corrector steps. This process can then be repeated iteratively to advance in time using either the Euler or Heun rules, where for each time step, new independent random numbers must be generated.

Consider the rule (10.22). It can be cast into a differential equation by dividing by Δt and taking the limit of vanishing time step. However, the random term becomes singular, diverging as $1/\sqrt{\Delta t}$. This divergence, however, does not lead to divergent trajectories, because when $\Delta t \to 0$ the number of generated random numbers grows, producing a small net effect because, being independent, they mutually cancel. This singular equation is normally written as

$$\dot{\Gamma} = \Phi(\Gamma) + \sqrt{2D}\xi(t), \qquad (10.24)$$

where ξ, called a noise, is a stochastic process[9] with a Gaussian distribution, vanishing average, and correlation,

$$\langle \xi(t)\xi(t') \rangle = \delta(t - t'); \qquad (10.25)$$

That is, the noises at different instants are independent, and the variance diverges. Equation (10.24) with the prescription (10.25) is called a Langevin equation for white noise, being the most well-known example of a stochastic differential equation.[10]

[9]Stochastic processes are random variables that depend on time.

[10]If the diffusion coefficient depends on Γ, the singularity of the limiting process generates an ambiguity regarding how the noise should be interpreted; either the intensity of the noise should be evaluated with the value of Γ before its action—called the Itô interpretation—or with the value during its action—the Stratonovich interpretation. In this chapter we will not deal with this problem, and interested readers are directed for example to the book by van Kampen (2007). At the Fokker–Planck level, however, no ambiguity is present.

10.3.5　Long-range forces

When we studied particles interacting with long-range forces in Chapter 6, we justified the use of the Vlasov–Poisson equation by noting that, although the microscopic Hamiltonian includes pair interactions, in practice particles interact with all of the others simultaneously, allowing us to approximate this interaction by a mean field. This mean field, which is a smooth function of space, does not include the details of the interactions with nearby particles, but it was shown that these give only small corrections to the dominant dynamics. We will build the particle-based numerical method on the same principles; that is, particles will not interact in pairs but only through a mean field.

Consider a plasma composed of several species, $\alpha = 1, 2, \ldots$, with masses m_α and charges q_α. The number of particles of each species N_α is macroscopic, and for the numerical method we will sample the plasma using $\widetilde{N}_\alpha$ particles. Note that $\Lambda_\alpha = N_\alpha/\widetilde{N}_\alpha$ can differ from species to species. Each of the sampling particles must move as a representative real particle and must respond to the field in the same way. Consequently, the inertial mass and the charge of the sampling particles are the same as for the real ones. Their equation of motion is

$$m_a \ddot{\mathbf{r}}_a = -q_a \nabla \left[\Phi_{\mathrm{mf}}(\mathbf{r}_a) + \Phi_{\mathrm{ext}}(\mathbf{r}_a) \right], \quad a = 1, \ldots, \widetilde{N}. \tag{10.26}$$

However, to generate the same field, we have to assign them larger-scaled charges, $\widetilde{q}_\alpha = \Lambda_\alpha q_\alpha$,[11]

$$\nabla^2 \Phi_{\mathrm{mf}}(\mathbf{r}) = -\sum_\alpha \frac{\widetilde{q}_\alpha n_\alpha(\mathbf{r})}{\epsilon_0} = -\sum_\alpha \frac{\Lambda_\alpha q_\alpha n_\alpha(\mathbf{r})}{\epsilon_0}. \tag{10.27}$$

The Newton equation (10.26) is solved for the sampling particles using the leapfrog method (10.18). To obtain the field, several methods exist, with different levels of approximations that have been designed to better describe specific situations. Here we describe the particle-in-cell method and the projection onto a basis set.

Particle-in-cell method (PIC)

To compute the mean field potential, space is divided into cells to which the particles are allocated according to their coordinates. Summing the total charge, the charge density ρ can be computed for each cell. It is now possible to solve the Poisson equation using either the fast Fourier transform or Gauss–Seidel method. The electric field is then obtained by computing the gradient by finite differences or directly by using the Fourier transform. Finally, the force on each particle is obtained by using interpolation methods from the electric field in the cells to the particle positions (see Fig. 10.2).

When particles are allocated to the cells, their charge is assigned to the geometric centre of the cell. This procedure leads to a truncation error in the charge density (and thus in the fields) of the order of the cell size. The cells, on the other hand, cannot be made too small because, to have

[11]If we were studying self-gravitating systems, the inertial mass—which appears in Newton's second law—remains the same, but the gravitational mass, which generates the field and appears in the Poisson equation, must be larger.

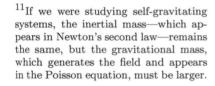

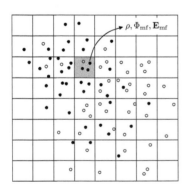

Fig. 10.2 Illustration of the particle-in-cell method. Charges are assigned to cells, in which the charge density, mean field potential, and electric fields are obtained.

a sufficiently smooth potential, many particles must be in each cell on average. Besides, when a particle moves from one cell to a neighbouring one, a discontinuous change is produced in the density. To overcome these problems, charges are spread over lengths of the order of the cell size. A coarse-grained density profile is used,

$$\rho(\mathbf{r}) = \sum_a q_a S(\mathbf{r} - \mathbf{r}_a), \qquad (10.28)$$

where S is a cloud function with the conditions of being short ranged—of the order of one cell size—and being normalised to unity, $\int S(\mathbf{x}) \, d^d x = 1$, with d the dimension of the cell division of space. Charges are therefore allocated fractionally to cells, according to the partial integration of S in the cell (see Fig. 10.3). Using this procedure, each cell has an allocated charge density ρ_c, from which the potential can be computed by solving the Poisson equation.

In the PIC method, we can take advantage of symmetries present in the system to gain precision and time. For example, if the system is uniform along the y and z directions, cells can be built only along the relevant x direction and the Poisson equation is solved in 1D only.

Projection onto a basis set

Imagine that we want to study the evolution of an elliptical galaxy.[12] The stars interact via the mean field, which has almost spherical symmetry. It is then natural to use basis functions for the mass density that reflect this symmetry and take into account the possible deviations from perfect sphericity,

$$\rho(\mathbf{r}) = \sum_a m_a \delta(\mathbf{r} - \mathbf{r}_a) = \sum_{nlm} \rho_{nlm} R_{nl}(r) Y_{lm}(\Omega), \qquad (10.29)$$

where Y_{lm} are the spherical harmonics. The choice of the radial functions R_{nl} is not arbitrary, but rather must satisfy three conditions. First, that the first terms take good account of the mass density of the galaxy during its evolution, allowing the truncation of the expansion at a finite number of terms. Second, we require that the associated Poisson equation, $\nabla^2[P_{nl}(r) Y_{lm}(\Omega)] = R_{nl}(r) Y_{lm}(\Omega)$, with radial part,

$$\frac{1}{r} \frac{d^2}{dr^2} [r P_{nl}(r)] - \frac{l(l+1)}{r^2} P_{nl}(r) = R_{nl}(r), \qquad (10.30)$$

must have simple—ideally analytic—solutions. Finally, the following biorthogonality condition is required:

$$\int P_{nl}(r) Y_{lm}^*(\Omega) R_{n'l'}(r) Y_{l'm'}(\Omega) \, d^3 r = \delta_{n,n'} \delta_{l,l'} \delta_{m,m'}. \qquad (10.31)$$

Hence, the coefficients of the expansion (10.29) are[13]

$$\rho_{nlm} = \sum_a m_a P_{nl}(r_a) Y_{lm}^*(\Omega_a). \qquad (10.32)$$

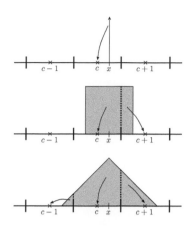

Fig. 10.3 Charge assignment in the PIC method using different cloud functions S for a charge located at x. Top: for the Dirac delta function, charges are assigned to the closest cell centre. Middle: for a rectangular cloud, the charge is distributed between the two closest cells, with a fraction proportional to the area in each cell. Bottom: in the case of a triangular cloud, charges are distributed between three cells. Adapted from Hockney and Eastwood (1988).

[12] Here, we use the example of a self-gravitating system, but the method can equally be applied to plasmas, for example.

[13] Alternatively, one can require that the basis functions R_{nl} be orthogonal, in which case the coefficients ρ_{nlm} would be obtained in terms of R_{nl} evaluated at the particle positions. However, in elliptical galaxies, there is a known weak divergence of the density at the origin that makes this approach inappropriate. The potential, however, is regular at the origin.

Then, the mean field potential, solution of $\nabla^2\Phi_{\mathrm{mf}} = 4\pi G\rho$, is

$$\Phi_{\mathrm{mf}}(\mathbf{r}) = 4\pi G\sum_{nlm}\rho_{nlm}P_{nl}(r)Y_{lm}(\Omega). \qquad (10.33)$$

The evaluation of the potential and the force field, which is obtained by analytically differentiating the basis functions, has computational cost $\mathcal{O}(\widetilde{N}M)$, where M is the total number of functions used in the expansion.

Quad- and oct-tree

Finally, there is another method, proposed by Barnes and Hut in 1986 to model self-gravitating systems, which is properly not a solution for the Poisson equation, but uses hypotheses similar to those of the Vlasov equation. It aims to compute the gravitational force on each of the sampling particles,

$$\mathbf{F}_a = \sum_b \frac{G\Lambda m_a m_b}{r_{ab}^2}\hat{\mathbf{r}}_{ab}, \qquad (10.34)$$

where $\mathbf{r}_{ab} = \mathbf{r}_b - \mathbf{r}_a = r_{ab}\hat{\mathbf{r}}_{ab}$ is the relative distance and we have added the scaling factor for the gravitational masses. Even though there are many fewer sampling particles than the real number of stars in a galaxy, they must nevertheless be numerous to achieve numerical accuracy. Thus, direct calculation of the forces becomes unaffordable with a computational cost that scales as $\mathcal{O}(\widetilde{N}^2)$. The idea of Barnes and Hut is that, because the forces are long ranged, distant particles need not be summed pairwise but can be grouped into a single effective particle with mass equal to the sum of the masses and located at their centre of mass. In this way, we reduce the number of forces to be computed. Particles, though, are not grouped statically as in the PIC method, but rather dynamically depending on their distance from the particle a for which we are computing the force: if they are far away, large groups can be made, but if they are close, the groups must be smaller.

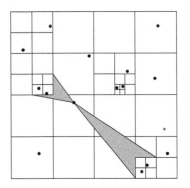

Fig. 10.4 Example of a quad-tree implementation in two dimensions.

This dynamic grouping is achieved using a hierarchical tree that we describe in two dimensions for simplicity, although the extension to three dimensions is direct. Once the particles are placed in the area, which we take as a square, it is divided into four equal squares (into eight equal cubes in three dimensions), and we proceed recursively until each square contains either one or no particles. At each level of division, each square stores the number of particles within itself, their total mass, and the position of the centre of mass. The cost of constructing this structure is $\mathcal{O}(\widetilde{N}\log_4\widetilde{N})$. Now, when computing the force on each particle, we start from the top level and descend the levels until either there is one particle in the square or the encompassing angle from the particle to the square is smaller than a certain threshold θ_c. The force is then computed from the grouped particles inside that square. In the example shown in Fig. 10.4, the angle on the left is larger than θ_c and for the force evaluation one must descend another level, whereas the angle presented on the right is smaller than θ_c and the force can be computed from this group without going to lower levels. Note that, for the particle represented by an asterisk, the opposite situation is obtained, showing how the grouping is dynamically generated. The computational cost for the evaluation of the force on a single particle is $\mathcal{O}(\log_4 \widetilde{N})$, with a prefactor that grows with decreasing θ_c.

10.3.6 Collisions

In gases, electrons in a metal, and other examples, particles suffer instantaneous collisions or scattering events, which change their velocities. Collisions can involve one particle, as in the case of electrons in a metal scattering at impurities, two particles as in a classical gas or when electrons scatter with phonons, or even three particles as when quartic-order anharmonicities are considered in phonon dynamics.

One-particle collisions are the simplest case because the sampling particles must follow the same dynamics as the real particles. Consider a classical particle with a transition rate $W(\mathbf{c}, \mathbf{c}')$, such that $W \mathrm{d}t$ is the probability that it changes from velocity $\mathbf{c}$ to velocity $\mathbf{c}'$ in the time interval $\mathrm{d}t$. The total transition rate is

$$\widehat{W}(\mathbf{c}) = \int W(\mathbf{c}, \mathbf{c}') \, \mathrm{d}^3 c'. \qquad (10.35)$$

Then, at each time step Δt, each particle will suffer a collision with probability $\widehat{W} \Delta t$, a decision that is made using the Monte Carlo method. If a particle is selected to collide, the resulting velocity will be distributed with probability $W(\mathbf{c}, \mathbf{c}')/\widehat{W}(\mathbf{c})$. To obtain the velocity with the correct distribution, velocities are first generated with an a priori distribution $P(\mathbf{c}')$ using some algorithm.[14] As the generation mechanism is biased, the attempted velocities must be selected with a probability proportional to $W(\mathbf{c}, \mathbf{c}')/[\widehat{W}(\mathbf{c}) P(\mathbf{c}')]$, and the acceptance–rejection method described in Section 10.3.2 is applied.

For two-particle collisions, we first consider deterministic cases as in classical gases where, given the incoming velocities and some geometrical factors (for example, the impact parameter and azimuthal angle), the outgoing velocities are completely determined. These collisions are dealt with by using the direct simulation Monte Carlo (DSMC) method, originally proposed in the context of aerodynamics and later extended in use to other areas of physics and chemistry. The key point is to recall that here we do not intend to track the real collisions between particles but to sample them statistically as they take place in a gas described by the Boltzmann equation. First, the volume is divided into boxes, which can reflect the symmetries of the problem under study, and the particles are allocated to the boxes according to their positions. In each of these boxes of volume $\mathcal{V}_c$, $\widetilde{N}_c$ particles sample the local distribution function $f(\mathbf{r}, \mathbf{c})$. In a time step Δt, the total number of collisions that should take place for the sampling particles is $N_{\mathrm{coll}} = \widetilde{N}_c \nu \Delta t/2$, where

$$\nu = \frac{\sigma}{n} \int f(\mathbf{c}_1) f(\mathbf{c}_2) |\mathbf{c}_2 - \mathbf{c}_1| \, \mathrm{d}^3 c_1 \, \mathrm{d}^3 c_2 \qquad (10.36)$$

[14] For example, the direction of $\mathbf{c}'$ can be uniformly distributed in the unit sphere in the case of collisions that conserve energy, or $\mathbf{c}$ can be drawn from a Maxwellian distribution in other general cases.

is the collision frequency and the factor of 1/2 avoids double counting. We recall that σ is the total cross section, where we consider spherical particles for simplicity. Considering that each sampling particle represents $\Lambda = N/\widetilde{N}$ real particles,

$$N_{\text{coll}} = \widetilde{N}_c^2 \Lambda \sigma \langle v_{\text{r}} \rangle \Delta t / 2 \mathcal{V}_c, \tag{10.37}$$

where $\langle v_{\text{r}} \rangle$ is the average relative velocity between particles in the cell. As also reflected in the expression for the collision frequency, the probability for the collision of a pair of sampling particles is proportional to their relative velocity,

$$P_{\text{coll}}(i, j) = \frac{|\mathbf{c}_i - \mathbf{c}_j|}{\sum_{k=1}^{\widetilde{N}_c} \sum_{l<k} |\mathbf{c}_k - \mathbf{c}_l|}. \tag{10.38}$$

The evaluations of this probability and the average relative velocity in (10.37) are costly, as all pairs must be considered. However, there is a method that efficiently and simply draws the colliding particles. Let us take a good estimate of the maximum relative velocity $v_{\text{r,max}}$, with the sole condition that it is larger than all relative velocities; for example, we can take $v_{\text{r,max}} = 2 \max_i |\mathbf{c}_i|$. Now, select any two particles i and j at random in the cell. The pair is accepted with probability $|\mathbf{c}_i - \mathbf{c}_j|/v_{\text{r,max}}$ and the collision is performed as discussed below. This process is repeated, generating in total N_{cand} candidate pairs. By construction, the ratio of the number of collisions performed to the number of candidates is $N_{\text{coll}}/N_{\text{cand}} = \langle v_{\text{r}} \rangle / v_{\text{r,max}}$. Using (10.37) we obtain that the number of candidates should be

$$N_{\text{cand}} = \widetilde{N}_c^2 \Lambda \sigma v_{\text{r,max}} \Delta t / 2 \mathcal{V}_c, \tag{10.39}$$

which is now simple to evaluate. This expression is not necessarily an integer number. If we truncate it, fewer collisions than expected would be performed, affecting all transport properties. In practice, this number is truncated to an integer number, but the fractional part is added to the next time step, allowing the correct number of collisions to be generated on average.

For each pair of accepted particles, a hard sphere collision should be performed. The velocity of the centre of mass is conserved, and by energy conservation, the magnitude of the relative velocity is also preserved. As discussed in Section 4.2.2, for particles that collide randomly in three dimensions, following the molecular chaos hypothesis, the relative velocity is rotated uniformly in the sphere. Therefore, new velocities are taken from the collision rule,

$$\mathbf{c}_i' = (\mathbf{c}_i + \mathbf{c}_j)/2 + |\mathbf{c}_i - \mathbf{c}_j|\hat{\mathbf{n}}'/2, \tag{10.40}$$
$$\mathbf{c}_j' = (\mathbf{c}_i + \mathbf{c}_j)/2 - |\mathbf{c}_i - \mathbf{c}_j|\hat{\mathbf{n}}'/2, \tag{10.41}$$

where $\hat{\mathbf{n}}$ is uniformly distributed in the unit sphere.

The scheme for the DSMC method can be summarised as follows. At each time step (i) the particles are streamed freely, (ii) they are allocated to cells, (iii) for each cell, N_{cand} is evaluated, and (iv) in each cell, N_{cand} pairs are generated and collisions are performed with probability $|\mathbf{c}_i - \mathbf{c}_j|/v_{r,max}$.

We now consider the probabilistic case; for example, in the dynamics of phonons. Collisions in general are characterised by transition rates $W_{\mathbf{c}_1,\mathbf{c}_2 \to \mathbf{c}_1',\mathbf{c}_2'}$. Here we can proceed by two different methods, depending on whether the transition rates respect conservation laws or not. First, if there are no conservations, we can draw at random a pair of particles i and j and, with some a priori distribution $P(\mathbf{c})$, the two outgoing velocities $\mathbf{c}_1'$ and $\mathbf{c}_2'$. The transition is then selected with probability $[W_{\mathbf{c}_1,\mathbf{c}_2 \to \mathbf{c}_1',\mathbf{c}_2'}/P(\mathbf{c}_1')P(\mathbf{c}_2')]/W_{max}$, where W_{max} is an upper bound of the numerator. Following reasoning similar to that applied in the deterministic case, the number of collisions that should be attempted is

$$N_{cand} = \widetilde{N}_c^2 \Lambda W_{max} \Delta t / 2 \mathcal{V}_c. \tag{10.42}$$

This method, however, fails when strict conservations apply. First, the Dirac delta functions make the transition rates unbounded, making the definition of W_{max} impossible, and second, with probability one the selected outgoing velocities will not satisfy the conservations. Under this condition, the total transition rates,

$$\widehat{W}_{\mathbf{c}_1,\mathbf{c}_2} = \int W_{\mathbf{c}_1,\mathbf{c}_2 \to \mathbf{c}_1',\mathbf{c}_2'} \, d^3 c_1' \, d^3 c_2', \tag{10.43}$$

are first computed, being bounded from above by $\widehat{W}_{max}$. Again, a pair of particles i and j are drawn at random, and the collision is performed with probability $\widehat{W}_{\mathbf{c}_1,\mathbf{c}_2}/\widehat{W}_{max}$. Once selected, the outgoing velocities are sorted as described for one-particle collisions. The total number of collisions that should be attempted is given by eqn (10.42) with W changed to $\widehat{W}$.

10.3.7 Quantum effects

In the case of collisions of quantum gases, the transition rates are multiplied by factors $(1 \pm f')$, evaluated at the final states. The upper sign is used for bosons, representing stimulated emission, and the lower case for fermions, reflecting the Pauli exclusion principle. In both cases it is therefore necessary to estimate the distribution function on the final states. We recall the discussion in Section 7.4 where the distribution function was introduced, corresponding to a coarse-grained average of the occupation number distribution. The coarse-graining procedure, to be effective, was performed over wavevector windows larger than the spacing between states, $2\pi/L$; that is, a number $s \gg 1$ must be chosen to define the averaging window size $\Delta = 2\pi s/L$. On the other hand, s cannot be excessively large because Δ must be much smaller than the reciprocal wavevectors.

In the methods we have described in this chapter, each sampling particle represents Λ real particles. Hence, the mean distance between particles in $\mathbf{k}$-space is increased by a factor $\Lambda^{1/d}$, where $d = 1, 2, 3$ is the relevant spatial dimensionality. Here d can be lower than the dimensionality of the underlying physical problem if there is some symmetry; for example, in isotropic cases, the distribution function depends only on $k = |\mathbf{k}|$. Consequently, to obtain smooth estimations of the distribution function using the sampling particles, the coarse-grained average should be made over windows of size $\widetilde{\Delta} = 2\pi s \Lambda^{1/d}/L$. To perform the coarse graining, the $\mathbf{k}$-space is divided into these small volumes. For a volume centred around $\mathbf{k}_i$, the number of sampling particles in the volume $\widetilde{N}_i$ is obtained, resulting in the following estimation for the distribution function:

$$f_i = \frac{\Lambda \widetilde{N}_i}{\text{Number of states in the volume}} = \frac{\widetilde{N}_i}{s^d}, \qquad (10.44)$$

where we have used that the number of states in the volume is Λs^d.

Because of fluctuations associated with the finite number of sampling particles, it may happen that for fermions the estimated distribution is not strictly smaller than one. In such cases, the transition rate factor $1 - f'$ is simply replaced by zero, keeping track of the fraction of collision events where such correction was performed. If it turns out that this fraction is large, either the number of sampling particles $\widetilde{N}$ or the window size factor s must be enlarged.

10.3.8   Boundary conditions

These are directly implemented on the sampling particles, following the same rules as for real ones. In a time step, after the streaming phase, the particles that have reached a boundary are identified. Then, they are reflected back either using a specular law or with a transition probability, as described in Section 4.6. In the latter case, acceptance–rejection methods are used to obtain the post-collisional velocity.

Further reading

The moment method was originally presented in Grad (1949) and later reviewed in Grad (1958). There are many books on numerical methods. Press *et al.* (2007) presents a detailed description of general methods, where ordinary and partial differential equations and random numbers are particularly suited for our purpose. Different numerical methods to solve stochastic processes are presented in San Miguel and Toral (2000). The DSMC method is described in the original book by Bird (1994) and also in the more general book by Garcia (2000). Hockney and Eastwood (1988) is a classical text for plasma physics, with detailed descriptions of the PIC and other methods. Finally, the Octree method is described in detail in the original article (Barnes and Hut, 1986).

Ratchets, for which an example is presented in Exercise 10.6, are described in detail in Jülicher *et al.* (1997), Reimann (2002), and Astumian and Hänggi (2002).

Exercises

(10.1) **Grad's equation for a uniform shear flow.** Solve Grad's equation for a gas in a uniform shear flow, $n = n_0$, $\mathbf{v} = \dot{\gamma}y\hat{\mathbf{x}}$, and $T = T_0$, where the shear rate $\dot{\gamma}$ is assumed to be small compared with the collision frequency. Obtain that $P_{xy} = \eta\dot{\gamma}$ and derive the viscosity. Compare it with the mean free path approximation and the full result of the Boltzmann equation.

(10.2) **Moment equations for semiconductors.** The objective is to derive the moment equations for semiconductors. Here, because the only conserved quantity is the particle density, the base distribution function is a Maxwellian with a fixed temperature given by the lattice and a vanishing velocity, with only the density as a relevant field. Hence, the polynomials that perturb the reference distribution can include $\mathbf{k}$ and k^2. Therefore, as a first model, we can work with a five-moment approximation,

$$f(\mathbf{r}, \mathbf{k}, t) = n \left(\frac{2\pi m^*}{k_B T}\right)^{3/2} e^{-\frac{\hbar^2 k^2}{2m^* k_B T}}$$
$$\times \left[1 + \mathbf{a} \cdot \mathbf{k} + b\left(k^2 - \frac{3k_B m^* T}{\hbar^2}\right)\right],$$

where the dynamical fields are n, $\mathbf{a}$, and b. Interpret these fields. Obtain their equations of motion for a single-band semiconductor (no holes) and no generation–recombination.

(10.3) **Box–Muller transformation.** Show that η_1 and η_2 given in eqn (10.16) are distributed with

$$P(\eta_1, \eta_2) = \frac{1}{2\pi} e^{-(\eta_1^2 + \eta_2^2)/2};$$

that is, they are independent numbers with a Gaussian distribution of unit variance and vanishing average.

(10.4) **Numerical method for the Langevin equation.** Consider the Langevin equation with white noise [eqns (10.24) and (10.25)] and define the discrete instants, $t_n = n\Delta t$, with $\Gamma_n = \Gamma(t_n)$. Integrating the Langevin equation between t_n and t_{n+1} gives

$$\Gamma_{n+1} - \Gamma_n = A_n + B_n,$$

with

$$A_n = \int_{t_n}^{t_n + \Delta t} F[\Gamma(t)]\,\mathrm{d}t, \quad B_n = \int_{t_n}^{t_n + \Delta t} \xi(t)\,\mathrm{d}t.$$

The deterministic term can be approximated using explicit or implicit methods,

$$A_n^{\text{explicit}} = F(\Gamma_n)\Delta t,$$
$$A_n^{\text{implicit}} = [F(\Gamma_n) + F(\Gamma_{n+1})]\Delta t/2.$$

For the stochastic term show that:

(a) B_n are Gaussian random variables,
(b) $\langle B_n \rangle = 0$,
(c) $\langle B_n^2 \rangle = \Delta t$, and
(d) $\langle B_n B_m \rangle = 0$, for $n \neq m$.

Conclude that the iteration rule (10.22) solves numerically the Langevin equation.

(10.5) **Diffusion of particles.** Simulate the motion of inertial Brownian particles described by the Fokker–Planck equation in three dimensions,

$$\frac{\partial f}{\partial t} + \mathbf{v} \cdot \nabla f = \frac{\gamma}{M} \frac{\partial}{\partial \mathbf{v}} \cdot (\mathbf{v}f) + \frac{\gamma k_B T}{M^2} \frac{\partial^2 f}{\partial v^2}.$$

Initially, particles are located at the origin and their velocities obey an equilibrium Maxwellian distribution. Compute the mean square displacement from the origin as a function of time. Show that for short times ballistic motion is obtained, $\langle \Delta r^2 \rangle \sim t^2$, and that for large times diffusion appears, $\langle \Delta r^2 \rangle \sim t$. Obtain the diffusion coefficient and determine the crossover time between the two regimes.

(10.6) **Ratchet motion of Brownian particles.** Consider overdamped Brownian particles moving in an asymmetric external one-dimensional potential, which is periodically turned on and off,

$$U(x) = U_0[\sin(x) + \alpha \sin(2x)]g(t),$$

with g being periodic with period τ, given by

$$g(t) = \begin{cases} 1 & 0 \le t < \tau/2 \\ 0 & \tau/2 \le t < \tau \end{cases}$$

and α measuring the asymmetry of the potential. Show that, when $k_B T \sim U_0$, rectified motion is obtained characterised by a nonvanishing average velocity. Show that the rectification is suppressed if the potential is permanently switched on.

(10.7) **Fokker–Planck solution via the Langevin equation.** In some cases, solution of the Fokker–Planck equation can easily be found by the use of the associated Langevin equation. Consider, for example, the case of lateral diffusion of a light ray, discussed in Section 5.8. For a ray emitted along the $\hat{\mathbf{z}}$ direction from $\mathbf{r} = 0$, the light intensity f that has slightly deviated to point in the direction, $\hat{\mathbf{n}} = n_x\hat{\mathbf{x}} + \hat{\mathbf{z}}$, with $|n_x| \ll 1$, at $\mathbf{r} = x\hat{\mathbf{x}} + z\hat{\mathbf{z}}$, is given by the Fokker–Planck equation [eqn (5.57)],

$$\frac{\partial f}{\partial z} = -n_x\frac{\partial f}{\partial x} + \frac{1}{L}\frac{\partial^2 f}{\partial n_x^2},$$

where $L = c/D_r$, c is the speed of light, and D_r is the coefficient of rotational diffusion. Here, z plays the role of time and the associated Langevin equation reads,

$$\frac{dx}{dz} = n_x, \qquad \frac{dn_x}{dz} = \xi(z),$$

with ξ a white noise with correlation,

$$\langle\xi(z)\xi(z')\rangle = \frac{2}{L}\delta(z - z').$$

Show, by direct integration, that

$$n_x(z) = \int_0^z \xi(z')\,dz', \quad x(z) = \int_0^z (z - z')\xi(z')\,dz'.$$

As n_x and x are linear in the noise, which is a Gaussian random variable, they will also be Gaussian. It can be directly verified that their mean values vanish, and therefore, to obtain the full distribution, it only remains to compute the second moments, $\langle n_x(z)^2\rangle$, $\langle x(z)^2\rangle$, and $\langle n_x(z)x(z)\rangle$. Using the explicit integral expressions for n_x and x and the noise correlation derive the results of Section 5.8.

(10.8) **Brownian charged particles.** Consider an ensemble of Brownian overdamped charged particles, with equal number of particles having positive and negative charge, $\pm q$. The particles are placed between two oppositely charged walls, which generate a uniform electric field, $\mathbf{E} = E_0\hat{\mathbf{x}}$, which must be added to the field generated by the particles. The system is confined along this direction, while being periodic and homogeneous in the other directions. Simulate the system using the PIC method where the Poisson equation is solved by a tridiagonal matrix method. Compute the density profiles when varying the temperature.

(10.9) **Two-stream instability in plasmas.** Consider a plasma composed of charged particles of equal masses and charges $\pm q$, with equal numbers of particles of both species. Initially both species are uniform with a Maxwellian distribution of velocities, but one species has an extra imposed velocity $V\hat{\mathbf{x}}$.

Consider that the plasma remains homogeneous in y and z and only develops inhomogeneities along x. Simulate the plasma using the PIC method and solve the Poisson equation using the tridiagonal method (Press *et al.*, 2007), considering periodic boundary conditions. Show that density instabilities develop.

(10.10) **Thermalisation of a uniform gas.** Perform a DSMC simulation of a hard sphere gas. The gas is homogeneous, so a single cell is sufficient. Take an initial condition with all particles having the same speed but random directions. Design a routine that measures the H function, binning velocities in coarse cells. Show that H decreases and that it reaches the stationary value after a few mean collision times. Finally, show that the resulting distribution function is a Maxwellian.

(10.11) **DSMC simulation of a uniform shear flow.** To study the viscous response of a gas subject to a uniform shear flow, $V' = dv_x/dy$, it is possible to perform DSMC simulations with the so-called Lees–Edwards boundary condition. The system is periodic in all directions, but when particles emerge from the bottom or top layers they are reinjected at the opposite side with an additional velocity, $\pm V'L$, where L is the length of the box. The system is uniform in x and z, allowing cells to be defined only along y. Simulate a system under this condition and compute the stress tensor. Show that for small shear rates V' the response is viscous and extract the viscosity.

(10.12) **Radiative cooling of an electron gas.** Consider a homogeneous electron gas which is initially at a high temperature, with electrons following a Maxwell–Boltzmann distribution. The electron-electron collisions are slightly inelastic due to a small amount of energy which is radiated and lost. An inelastic collision rule that mimics the inelastic hard sphere gas, introduced in Section 4.9, can be written for electrons obeying a parabolic dispersion relation, $\varepsilon = \hbar^2 k^2/2m$,

$$\mathbf{k}'_{1/2} = \mathbf{k}_{1/2} \pm (1 - 2q)\left[(\mathbf{k}_2 - \mathbf{k}_1)\cdot\hat{\mathbf{n}}\right]\hat{\mathbf{n}},$$

where $\hat{\mathbf{n}}$ is a uniformly distributed random unit vector and $q \ll 1$ measures the amount of energy that is lost in each collision.

(a) Show that this collision rule conserves momentum and compute the energy loss in each collision.

(b) Run a simulation with a simple transition rate independent of the incoming velocities. Show that at long times the Fermi–Dirac distribution is reached.

Mathematical complements

A.1 Fourier transform

For a function $h(\mathbf{r})$ that depends on space, the convention used in this book for the Fourier transform is the following:

$$\widetilde{h}(\mathbf{k}) = (2\pi)^{-3} \int h(\mathbf{r})e^{-i\mathbf{k}\cdot\mathbf{r}} \, \mathrm{d}^3 r, \tag{A.1}$$

and, correspondingly, the antitransform is

$$h(\mathbf{r}) = \int \widetilde{h}(\mathbf{k})e^{i\mathbf{k}\cdot\mathbf{r}} \, \mathrm{d}^3 k. \tag{A.2}$$

The convention used for functions $h(t)$ that depend on time is

$$\widetilde{h}(\omega) = (2\pi)^{-1} \int h(t)e^{i\omega t} \, \mathrm{d}t, \tag{A.3}$$

with the antitransform being

$$h(t) = \int \widetilde{h}(\omega)e^{-i\omega t} \, \mathrm{d}\omega. \tag{A.4}$$

For functions that have spatiotemporal dependence, the complete Fourier transform and antitransform are

$$h(\mathbf{k}, \omega) = (2\pi)^{-4} \int \widetilde{h}(\mathbf{r}, t)e^{-i(\mathbf{k}\cdot\mathbf{r}-\omega t)} \, \mathrm{d}^3 r \, \mathrm{d}t, \tag{A.5}$$

$$h(\mathbf{r}, t) = \int \widetilde{h}(\mathbf{k}, \omega)e^{i(\mathbf{k}\cdot\mathbf{r}-\omega t)} \, \mathrm{d}^3 k \, \mathrm{d}\omega. \tag{A.6}$$

Finally, it is direct to show that the Fourier transform of the convolution, $h(\mathbf{r}) = \int f(\mathbf{r}')g(\mathbf{r} - \mathbf{r}') \, \mathrm{d}^3 r'$ is $\widetilde{h}(\mathbf{k}) = (2\pi)^3 \widetilde{f}(\mathbf{k})\widetilde{g}(\mathbf{k})$.

A.2 Dirac delta distributions

The Dirac delta distribution[1] $\delta(x)$ is defined to be concentrated around $x = 0$ such that, for any continuous test function f, the following property is fulfilled:

$$\int f(y)\delta(y - x) \, \mathrm{d}y = f(x). \tag{A.7}$$

[1] Formally, they are not functions but rather distributions, although sometimes in informal contexts they are called Dirac delta functions.

It is possible to build the Dirac delta distributions as a limiting process of simple functions; for example,

$$\delta(x) = \lim_{\epsilon \to 0} \begin{cases} 0 & x < -\epsilon/2 \\ 1/\epsilon & -\epsilon/2 \le x \le \epsilon/2 \\ 0 & x > \epsilon/2 \end{cases} . \tag{A.8}$$

Other functions used in this book that converge to the Dirac delta distribution are[2]

$$\delta(x) = \lim_{\epsilon \to 0} \frac{1}{\sqrt{2\pi\epsilon}} e^{-x^2/2\epsilon},$$

$$\delta(x) = \lim_{\epsilon \to 0} -\frac{\partial}{\partial \epsilon} \frac{1}{e^{x/\epsilon}+1}, \tag{A.9}$$

$$\delta(x) = \lim_{t \to \infty} \frac{1}{\pi t} \left[\frac{\sin(xt)}{x} \right]^2 .$$

[2]See Exercise A.1.

In the case of distributions obtained as a limiting process, it is sufficient to demand that they be concentrated around the origin and that

$$\int \delta(x)\,\mathrm{d}x = 1. \tag{A.10}$$

Finally, Fourier transforming the Dirac delta distribution and then performing the antitransform, we have[3]

[3]See Exercise A.2.

$$\delta(x) = \frac{1}{2\pi} \int e^{ikx}\,\mathrm{d}k = \frac{1}{2\pi} \int e^{-ikx}\,\mathrm{d}k. \tag{A.11}$$

It is easy to prove that $x\delta(x) = 0$. Indeed, for any continuous test function f,

$$\int f(y)(y-x)\delta(y-x)\,\mathrm{d}y = \left[f(y)(y-x) \right]_{y=x} = 0. \tag{A.12}$$

Hence, $x\delta(x)$ behaves as a null distribution.

Using the change of variables rule, one can verify that

$$\delta(ax) = \frac{1}{|a|}\delta(x) \tag{A.13}$$

and that

$$\delta(h(x)) = \sum_n \frac{1}{|h'(x_n)|} \delta(x - x_n), \tag{A.14}$$

where x_n are all the roots of the equation $h(x) = 0$.

The Dirac delta derivative $\delta'(x)$ is defined by the following relation:

$$\int f(y)\delta'(y-x)\,\mathrm{d}y = -f'(x), \tag{A.15}$$

where f is any test function with continuous derivative. With this definition, it is possible to use the integration by parts method whenever Dirac deltas are involved in the integrand.

Finally, distributions allow the extension of the possible solutions of the equation, $(x - x_0)f(x) = 0$ to $f(x) = A\delta(x - x_0)$, for any constant A.

A.3 Eigenvalues of a perturbed operator

A.3.1 Statement of the problem

Consider a matrix $\mathbb{M}$ that can be written as

$$\mathbb{M} = \mathbb{M}_0 + \epsilon \mathbb{M}_1, \tag{A.16}$$

where ϵ is a small parameter. We wish to find the eigenvalues and eigenvectors of the matrix in a perturbative manner, using ϵ as a power ordering parameter. The eigenvalue problem is written as

$$\mathbb{M}\mathbf{v}^i = \lambda^i \mathbf{v}^i. \tag{A.17}$$

We propose a series expansion for both the eigenvalues and eigenvectors,

$$\mathbf{v}^i = \mathbf{v}_0^i + \epsilon \mathbf{v}_1^i + \epsilon^2 \mathbf{v}_2^i + \dots, \tag{A.18}$$
$$\lambda^i = \lambda_0^i + \epsilon \lambda_1^i + \epsilon^2 \lambda_2^i + \dots. \tag{A.19}$$

Substituting (A.16), (A.18), and (A.19) into (A.17) and ordering by powers of ϵ, we obtain the following hierarchy of equations:

$$\mathcal{O}(\epsilon^0) : \mathbb{M}_0 \mathbf{v}_0^i = \lambda_0^i \mathbf{v}_0^i, \tag{A.20}$$

$$\mathcal{O}(\epsilon^n) : \mathbb{M}_0 \mathbf{v}_n^i + \mathbb{M}_1 \mathbf{v}_{n-1}^i = \sum_{k=0}^{n} \lambda_k^i \mathbf{v}_{n-k}^i; \quad n \geq 1. \tag{A.21}$$

We now proceed to solve these equations order by order, which as we will see, provides both the eigenvalues and the eigenvectors.

A.3.2 Order $\mathcal{O}(\epsilon^0)$

Equation (A.20) is simply the eigenvalue problem associated with the unperturbed matrix $\mathbb{M}_0$. Each eigenvalue can be nondegenerate or degenerate.

A.3.3 Order $\mathcal{O}(\epsilon^1)$

Equation (A.21) can be reordered as

$$\left(\mathbb{M}_0 - \lambda_0^i\right) \mathbf{v}_1^i = \left(\lambda_1^i - \mathbb{M}_1\right) \mathbf{v}_0^i, \tag{A.22}$$

which has the form, $\mathbb{A}\mathbf{x} = \mathbf{b}$, where $\mathbb{A}$ is a noninvertible matrix because λ_0^i is an eigenvalue of $\mathbb{M}_0$. Then, the equation has a solution only if the right-hand side belongs to the image of $\mathbb{A}$. This condition can be stated in a simpler form, which we will analyse separately for the degenerate and nondegenerate cases.

Nondegenerate case

We first consider that the eigenvalue is nondegenerate. Let $\mathbf{u}^i$ be a solution of

$$\left(\mathbb{M}_0 - \lambda_0^i\right)^T \mathbf{u}^i = 0, \tag{A.23}$$

where T indicates the transpose of the matrix. That is, $\mathbf{u}^i$ is the i-th left eigenvalue, which is known to exist and is unique except for a scalar factor. Transposing this expression and multiplying it on the right by $\mathbf{v}_1^i$, we obtain $\mathbf{u}^{iT}\left(\mathbb{M}_0 - \lambda_0^i\right)\mathbf{v}_1^i = 0$. Then, multiplying (A.22) on the left by $\mathbf{u}^{iT}$, we have the condition,

$$\mathbf{u}^{iT}\left(\lambda_1^i - \mathbb{M}_1\right)\mathbf{v}_0^i = 0. \tag{A.24}$$

It can be shown that this is a necessary and sufficient condition for eqn (A.22) to have a solution. The expression (A.24) is a linear equation for the single unknown λ_1^i, which can be solved directly, obtaining the next order of the expansion of the eigenvalue. For its solution, we first need to find $\mathbf{u}^i$, which corresponds to the left eigenvector of $\mathbb{M}_0$ associated with the relevant eigenvalue.

Once λ_1^i has been determined, $\mathbf{v}_1^i$ is obtained by solving the linear equation (A.22).

Degenerate case

If the eigenvalue λ_0^i is degenerate, with degeneracy g, there are two possibilities: Either we have g linear independent eigenvectors ($\mathbf{v}_0^{i,j}$; $j = 1, 2, \ldots, g$), or there are fewer eigenvectors and we have a Jordan block. The first case is simpler. There, it can be shown that there are also g left eigenvectors,

$$\left(\mathbb{M}_0 - \lambda_0^i\right)^T \mathbf{u}^{i,j} = 0; \quad j = 1, 2, \ldots, g. \tag{A.25}$$

The role of the first-order correction is to break the degeneracy; that is, the g eigenvalues will differ to this order, as shown in Fig. A.1. Also, the eigenvectors that to zeroth order can be any linear combination of the g eigenvectors, to first order will adopt definitive values. To determine this degeneracy break, we multiply (A.22) on the left by the projector, $\mathbb{P}_i = \sum_j \mathbf{u}^{i,j}\mathbf{u}^{i,jT}$. Noting that $\mathbf{u}^{i,j}$ are left eigenvalues, the left-hand side is zero, giving therefore

$$\mathbb{P}_i \mathbb{M}_1 \mathbf{v}_0^i = \lambda_1^i \mathbb{P}_i \mathbf{v}_0^i. \tag{A.26}$$

This is a generalised eigenvalue equation that gives the first-order corrections to the eigenvalue λ_1^i and the zeroth-order eigenvectors, with the degeneracy broken. Having determined these quantities, the corrections to the eigenvectors $\mathbf{v}_1^i$ are obtained by solving the linear equation (A.22).

In the cases studied in Chapters 3, 4, 5 of the transport coefficients, the degeneration is not broken at order ϵ and the analysis must be continued in a completely analogous way to order ϵ^2.

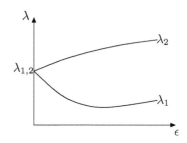

Fig. A.1 Evolution of a degenerate eigenvalue with increasing ϵ. There is a first linear correction that breaks the degeneracy, as described in the text. Later, quadratic and higher-order corrections occur.

When there are fewer than g eigenvectors, the analysis is notably more complex; for example, the eigenvalues are nonanalytic in ϵ and fractional powers appear. In this book we will not encounter these cases, and we defer study of this case.[4]

[4]See, for example, Exercise A.6.

Exercises

(A.1) **Functions that converge to the Dirac delta distribution.** Show that the functions in eqn (A.9) converge to Dirac delta distributions. That is, they are normalised to unity and that, in the limit, they become concentrated around the origin.

(A.2) **Fourier representation of the Dirac delta distribution.** Show that the integral,

$$\int_{-L}^{L} \frac{dk}{2\pi} e^{ikx},$$

in the limit $L \to \infty$ is normalised to unity and becomes concentrated around the origin.

(A.3) **Nondegenerate numerical case.** Consider the matrix,

$$\mathbb{M} = \begin{pmatrix} 1 & 0 & 0 \\ 0 & 2 & 0 \\ 0 & 0 & 3 \end{pmatrix} + \begin{pmatrix} 0 & 0.1 & 0 \\ 0 & 0 & 0 \\ 0 & 0 & 0 \end{pmatrix}.$$

Compute the eigenvalues to first order in the correction.

(A.4) **Degenerate numerical case.** Consider the matrix,

$$\mathbb{M} = \begin{pmatrix} 1 & 0 & 0 \\ 0 & 1 & 0 \\ 0 & 0 & 1 \end{pmatrix} + \begin{pmatrix} 0 & 0.1 & 0 \\ 0 & 0 & 0 \\ 0 & 0 & 0 \end{pmatrix}.$$

Compute the eigenvalues to first order in the correction.

(A.5) **Functional case.** Consider the operator,

$$\mathcal{O} = \frac{d^2}{dx^2} + \epsilon V \frac{d}{dx}$$

acting on the space of periodic square-integrable functions on the interval $[0, L]$. Considering the standard internal product,

$$(f, g) = \int_0^L f^*(x) g(x)\, dx,$$

compute the eigenvalues and eigenfunctions of $\mathcal{O}$ to first order in ϵ. Note that, in this example, the unperturbed operator, $\mathcal{O}_0 = \frac{d^2}{dx^2}$, is Hermitian. Hence, the left and right eigenvectors are the same, simplifying the analysis.

(A.6) **Jordan block case.** Consider the matrix,

$$\mathbb{M} = \begin{pmatrix} 1 & 1 \\ 0 & 1 \end{pmatrix} + \epsilon \begin{pmatrix} 1 & 1 \\ 1 & 1 \end{pmatrix}.$$

Compute analytically the eigenvalues. Expand them for small ϵ and show that fractional powers appear.

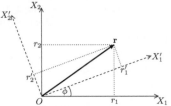

B   Tensor analysis

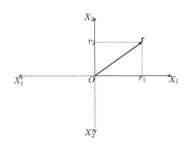

Fig. B.1 Transformation of the Cartesian components of a vector $\mathbf{r}$ when the coordinate system is rotated by an angle ϕ.

B.1   Basic definitions

Many of the calculations in kinetic theory are greatly simplified by using tensor analysis. Here, we will consider tensors in three dimensions in the context of the rotation group. In three dimensions, there are symmetries of space implying that, if we perform some coordinate transformations, physical laws should remain unchanged. Two symmetry operations will be considered: rotation and parity, which are represented in Figs. B.1 and B.2.

The basic object under study is a vector. In terms of the coordinate system, it is represented by its three Cartesian coordinates, $\mathbf{r} = (r_1, r_2, r_3)$. When we rotate the coordinate system as shown in Fig. B.1, the vector is represented by a new set of Cartesian coordinates r_i' that are related to the previous ones by

$$\text{Rotation: } r_i' = R_{ik} r_k, \tag{B.1}$$

where R is the rotation matrix. In two and three dimensions, it is given by

$$\mathbb{R}_{2D} = \begin{pmatrix} \cos\phi & \sin\phi \\ -\sin\phi & \cos\phi \end{pmatrix}, \tag{B.2}$$

$$\mathbb{R}_{3D} = \begin{pmatrix} \cos\theta\cos\psi & \cos\phi\sin\psi + \sin\phi\sin\theta\cos\psi & \sin\phi\sin\psi - \cos\phi\sin\theta\cos\psi \\ -\cos\theta\sin\psi & \cos\phi\cos\psi - \sin\phi\sin\theta\sin\psi & \sin\phi\cos\psi + \cos\phi\sin\theta\sin\psi \\ \sin\theta & -\sin\phi\cos\theta & \cos\phi\cos\theta \end{pmatrix}, \tag{B.3}$$

where the first one is written in terms of the angle represented in Fig. B.1 and the second is written in terms of the Euler angles ϕ, θ, and ψ. Representation of the rotation matrices in terms of other parameters is also possible (for example, quaternions are particularly used in the context of computer graphics). Independently of the particular representation, the rotation transformation is required to keep the length of vectors unchanged; that is, we demand that $|\mathbf{r}'|^2 = |\mathbf{r}|^2$. In terms of the rotation matrix, as this relation should be valid for any vector, we obtain

$$R_{ik} R_{il} = \delta_{kl}, \tag{B.4}$$

where δ is the identity or Kronecker delta tensor,

$$\delta_{ik} = \begin{cases} 1 & i = k, \\ 0 & i \neq k, \end{cases} \tag{B.5}$$

Fig. B.2 Transformation of the Cartesian components of a vector $\mathbf{r}$ when the coordinate system is reflected through the origin (parity transformation).

which is symmetric under exchange of the index. The Kronecker delta is also called the identity tensor $\mathbb{I}$. In summary, the matrix times its transpose gives the identity, so

$$\mathbb{R}^{-1} = \mathbb{R}^T, \tag{B.6}$$

and we say that the rotation matrices are unitary.

The parity transformation consists in reflecting the axis, and therefore the components simply change sign,

$$\text{Parity: } r_i' = -r_i. \tag{B.7}$$

Under these transformations, most objects in physics can be classified as follows:

Scalars: these are represented by a single variable ψ whose value does not change under rotation and parity transformations,

$$\text{Rotation: } \psi' = \psi, \tag{B.8}$$

$$\text{Parity: } \psi' = \psi. \tag{B.9}$$

As examples we have the mass and charge. We can also define scalar fields that are scalar quantities which depend on position and time, for example the mass density field.

Vectors: these are represented by three components r_i whose values change under rotation and parity as

$$\text{Rotation: } r_i' = R_{ik} r_k, \tag{B.10}$$

$$\text{Parity: } r_i' = -r_i. \tag{B.11}$$

The best known examples are the position, velocity, acceleration, and force vectors in mechanics. It is possible to have vector fields, such as the electric field.

Tensors: tensors are characterised by their rank n and are represented by 3^n components $T_{i_1, i_2, \ldots, i_n}$, with $i_k = 1, 2, 3$. If $n = 2$ they are represented by a matrix. The components transform under rotation and parity transformations as

$$\text{Rotation: } T_{i_1, i_2, \ldots, i_n}' = R_{i_1 k_1} R_{i_2 k_2} \ldots R_{i_n k_n} T_{k_1, k_2, \ldots, k_n}, \tag{B.12}$$

$$\text{Parity: } T_{i_1, i_2, \ldots, i_n}' = (-1)^n T_{k_1, k_2, \ldots, k_n}. \tag{B.13}$$

Note that, according to this definition, scalars and vectors are tensors of rank 0 and 1, respectively. Depending on the context, it is possible that the reader may have come across different examples of tensors. In mechanics, the inertia tensor is of rank 2, as is the dielectric tensor in electromagnetism. In fluid dynamics and elasticity, the stress tensor is a tensor field of rank 2.[1] Finally, in elasticity of materials, the elastic constants constitute a tensor of rank 4 (Landau and Lifshitz, 1986).

[1]See Exercise B.10.

Other physical quantities behave similarly to tensors with respect to the rotation transformation, but have a different sign under parity transformation. These are called pseudoscalars, pseudovectors, and pseudotensors. The best known example is the angular momentum, defined as $\mathbf{L} = \mathbf{r} \times \mathbf{p}$, where the cross product is defined using the right-hand rule. Under parity transformation, both $\mathbf{r}$ and $\mathbf{p}$ change sign and, therefore, $\mathbf{L}$ does not. We say that the angular momentum is a pseudovector. We thus define the following types:

Pseudoscalars: these are represented by a single variable ψ whose value does not change under rotation but changes sign under parity,

$$\text{Rotation: } \psi' = \psi, \tag{B.14}$$

$$\text{Parity: } \psi' = -\psi. \tag{B.15}$$

Pseudovectors: these are represented by three components r_i whose values change under rotation and parity as

$$\text{Rotation: } r_i' = R_{ik} r_k, \tag{B.16}$$

$$\text{Parity: } r_i' = r_i. \tag{B.17}$$

Pseudotensors: these are characterised by their rank n and are represented by 3^n components $T_{i_1, i_2, \ldots, i_n}$, with $i_k = 1, 2, 3$. The components transform under rotation and parity transformations as

$$\text{Rotation: } T_{i_1, i_2, \ldots, i_n}' = R_{i_1 k_1} R_{i_2 k_2} \ldots R_{i_n k_n} T_{k_1, k_2, \ldots, k_n}, \tag{B.18}$$

$$\text{Parity: } T_{i_1, i_2, \ldots, i_n}' = (-1)^{n+1} T_{k_1, k_2, \ldots, k_n}. \tag{B.19}$$

B.2   Isotropic tensors

Of special interest are tensors whose components remain invariant under rotation transformation. These special tensors are called isotropic, implying that they look the same from any direction. They do not depend on the coordinate system that is used and, hence, only depend on scalar values. It follows directly from the definition that any scalar is isotropic. Also, no vector is isotropic, as can be checked from the relation (B.10). This is because a vector points to a direction and hence cannot look the same from any direction; we say that there is a special direction pointed to by the vector.

In the case of tensors of rank 2, it can be proved that the only isotropic tensor is the Kronecker delta or any tensor proportional to it.

For rank 3, there are no isotropic tensors. The Levi–Civita symbol, which is defined as

$$\epsilon_{ijk} = \begin{cases} 1 & ijk = 123, 231 \text{ or } 312 \\ -1 & ijk = 321, 213 \text{ or } 132 \\ 0 & \text{if any index is repeated,} \end{cases} \tag{B.20}$$

is an isotropic pseudotensor of rank 3. It is also called the totally antisymmetric tensor.

There are three isotropic tensors of rank 4, which are written in terms of the Kronecker delta as

$$T^{[1]}_{ijkl} = \delta_{ij}\delta_{kl}, \tag{B.21}$$

$$T^{[2]}_{ijkl} = \delta_{ik}\delta_{jl}, \tag{B.22}$$

$$T^{[3]}_{ijkl} = \delta_{il}\delta_{jk}. \tag{B.23}$$

Linear combinations of these three tensors are also isotropic.

As a simple application, consider the rank 2 tensorial integral,

$$\int x_i x_j e^{-ax^2} \, \mathrm{d}^3x. \tag{B.24}$$

Evidently, the resulting tensor only depends on a, which is a scalar value. It does not have any preferred direction and must therefore be isotropic, allowing us to write,

$$\int x_i x_j e^{-ax^2} \, \mathrm{d}^3x = b\delta_{ij}, \tag{B.25}$$

where b is a scalar. To compute it, we take the trace of the previous expression by multiplying it by δ_{ij} and summing over repeated indices. Noting that $\sum_i \delta_{ii} = 3$, we have $b = \frac{1}{3} \int x^2 e^{-ax^2} \, \mathrm{d}^3x$. This single integral can easily be computed in Cartesian or spherical coordinates, and we have the final result,

$$\int x_i x_j e^{-ax^2} \, \mathrm{d}^3x = \frac{\pi^{3/2}}{2a^{5/2}}\delta_{ij}. \tag{B.26}$$

B.3   Tensor products, contractions, and Einstein notation

Tensors can be combined to produce tensors of higher or lower rank using the tensor product and contraction operations, respectively. Having two vectors **a** and **b**, it is possible to construct a rank 2 tensor as

$$T_{ik} = a_i b_k, \tag{B.27}$$

with the tensor product operation.[2] In an analogous manner, it is also possible to construct higher-rank tensors. We recall that here the language has not been used carelessly: one can directly verify that $\mathbb{T}$ is indeed a tensor.

With the same two vectors, we can compute the scalar product $\phi = \mathbf{a} \cdot \mathbf{b}$, which one can verify indeed transforms as a scalar. In terms of the components,

$$\phi = \sum_{i=1}^{3} a_i b_i = a_i b_i, \tag{B.28}$$

[2]This expression is called a dyadic and is represented as $\mathbb{T} = \mathbf{a} \otimes \mathbf{b}$, although we will not use this notation in the present book.

where in the second equation we have used Einstein notation, which simply means that any repeated index must be summed over. The contraction or summation of repeated indices is the usual operation of matrix–vector multiplication or matrix–matrix multiplication in linear algebra. Indeed, if $\mathbf{b} = \mathbb{A}\mathbf{x}$ then $b_i = A_{ik}x_k$, and if $\mathbb{C} = \mathbb{A}\mathbb{B}$ then $C_{ik} = A_{ij}B_{jk}$. Again, contraction of tensors produces valid tensors of the resulting rank.

If the tensor product or contraction operations involve pseudotensors, the result will be a tensor or a pseudotensor, depending on whether the number of pseudotensors is even or odd, respectively.

We note that the isotropic tensors appear in the construction of tensors; for example, the dot product can also be written as $\mathbf{a} \cdot \mathbf{b} = a_i \delta_{ik} b_k$. The cross product $\mathbf{c} = \mathbf{a} \times \mathbf{b}$ is obtained as $c_i = \epsilon_{ijk} a_j b_k$. Other properties of the Kronecker delta are $\delta_{ij} a_j = a_i$ and $A_{ij} \delta_{ij} = A_{ii}$. We finally have the useful relations,[3]

$$\epsilon_{ijk}\epsilon_{ilm} = \delta_{jl}\delta_{km} - \delta_{jm}\delta_{kl}, \qquad (B.29)$$

$$\epsilon_{ijk}\epsilon_{ijl} = \delta_{kl}. \qquad (B.30)$$

[3]For an application of these relations, see Exercise B.5.

B.4   Differential operators

The gradient operator is a vector with components given by

$$\nabla_i = \frac{\partial}{\partial x_i}. \qquad (B.31)$$

Contracting it with vector fields and with the Levi–Civita tensor results in the divergence and curl operators,

$$\nabla \cdot \mathbf{v} = \frac{\partial}{\partial x_i} v_i, \qquad \nabla \times \mathbf{v} = \epsilon_{ijk} \frac{\partial}{\partial x_j} v_k. \qquad (B.32)$$

B.5   Physical laws

The main application of tensor analysis comes from the observation that physical laws are tensorially consistent. In an equality, both sides should be tensors or pseudotensors of equal rank. Also, if the tensors are (anti)symmetric under the exchange of some indices on one side of the equation, the same must be true on the other side. Finally, the trace of a tensor is also preserved in equations.[4]

Using this principle, we can verify that the following laws are tensorially consistent:

[4]This can easily be verified, because if we have the equation $A_{ij} = B_{ij}$, then contracting with the Kronecker delta we obtain $A_{ii} = B_{ii}$. That is, the traces are equal.

$$m\mathbf{a} = \mathbf{F}, \qquad \nabla \cdot \mathbf{E} = \rho/\epsilon_0, \qquad (B.33)$$

$$\frac{\partial \rho}{\partial t} = D\nabla^2 \rho, \qquad \frac{\partial \mathbf{B}}{\partial t} + \nabla \times \mathbf{E} = 0, \qquad (B.34)$$

where we have to use that the magnetic field $\mathbf{B}$ is a pseudovector. Analogously, examples of expressions that are tensorially inconsistent are $m\mathbf{a} = \mathbf{L}$, $\nabla^2 \mathbf{E} = \mathbf{B}$, or $\frac{\partial \rho}{\partial t} = \mathbf{v} \cdot (\nabla \times \mathbf{v})$.

Exercises

(B.1) **Poisson equation.** Show that, if $\mathbf{E}(\mathbf{r})$ is a vector field, then $\rho = \epsilon_0 \nabla \cdot \mathbf{E}$ transforms as a scalar.

(B.2) **Quadratic forms.** Show that, if $\mathbf{a}$ and $\mathbf{b}$ are vectors and $\mathbb{C}$ is a rank 2 tensor, then $\phi = \mathbf{a} \cdot \mathbb{C} \cdot \mathbf{b} = a_i B_{ik} b_k$ transform as a scalar.

(B.3) **Pseudoscalars.** Consider the velocity vector $\mathbf{v}$ and the angular momentum pseudovector $\mathbf{L}$. Show that the quantity $\psi = \mathbf{v} \cdot \mathbf{L}$ is a pseudoscalar.

(B.4) **Isotropic tensors.** Verify that the isotropic tensors of rank 4 indeed do not change the values of their components under rotations.

(B.5) **Double cross product.** Using the relation (B.29) prove that

$$\mathbf{a} \times (\mathbf{b} \times \mathbf{c}) = (\mathbf{a} \cdot \mathbf{c})\mathbf{b} - (\mathbf{a} \cdot \mathbf{b})\mathbf{c}.$$

Analogously, if $\mathbf{v}$ is a vector field, show using the same relation that

$$\nabla^2 \mathbf{v} = \nabla(\nabla \cdot \mathbf{v}) - \nabla \times (\nabla \times \mathbf{v}),$$

an expression that is sometimes written as

$$\triangle \mathbf{v} = \mathrm{grad}(\mathrm{div}\,\mathbf{v}) - \mathrm{curl}(\mathrm{curl}\,\mathbf{v}).$$

(B.6) **Tensorial integrals.** Let $\phi_{\mathrm{MB}}(\mathbf{c})$ be the Maxwell distribution function. Consider the symmetric $\mathbb{S}$ and antisymmetric $\mathbb{A}$ rank 2 tensors. Compute

$$I = \int \phi_{\mathrm{MB}}(\mathbf{c})\, \mathbf{c} \cdot \mathbb{S} \cdot \mathbf{c}\, \mathrm{d}^3 c,$$

$$I = \int \phi_{\mathrm{MB}}(\mathbf{c})\, \mathbf{c} \cdot \mathbb{A} \cdot \mathbf{c}\, \mathrm{d}^3 c.$$

(B.7) **Some integrals.** Using tensor analysis compute the following integrals:

$$I = \int (\mathbf{x} \cdot \mathbf{a})^2 e^{-x^2/2}\, \mathrm{d}^3 x,$$

$$I = \int (\mathbf{x} \cdot \mathbf{a})(\mathbf{x} \cdot \mathbf{b}) e^{-x^2/2}\, \mathrm{d}^3 x,$$

where $\mathbf{a}$ and $\mathbf{b}$ are known vectors.

(B.8) **Scalar waves.** The wave equation for a scalar field ψ could be generalised to

$$\frac{\partial^2 \psi}{\partial t^2} - A_{ik} \frac{\partial^2 \psi}{\partial x_i \partial x_k} = 0,$$

with A_{ik} a rank 2 tensor. This generalisation comprises the cases,

$$\frac{\partial^2 \psi}{\partial t^2} - c^2 \frac{\partial^2 \psi}{\partial x \partial y} = 0,$$

$$\frac{\partial^2 \psi}{\partial t^2} - c^2 \left(\frac{\partial^2 \psi}{\partial x^2} - \frac{\partial^2 \psi}{\partial y^2} \right) = 0.$$

Show that, without loss of generality, $\mathbb{A}$ can be considered as a symmetric tensor. Show that, if the medium is isotropic, then we recover the usual wave equation.

(B.9) **Vectorial waves.** We wish to write the wave equation for a vectorial field $\boldsymbol{\psi}$ (for example, the fluid velocity or the electric field). By a wave equation we mean a partial differential equation with second-order derivatives in time and space. The most general equation of this type is

$$\frac{\partial^2}{\partial t^2}\psi_i - C_{ijkl} \frac{\partial}{\partial x_j} \frac{\partial}{\partial x_k} \psi_l = 0,$$

where $\mathbb{C}$ is a rank 4 tensor.

Knowing that the isotropic rank 4 tensors are of the form, $C_{ijkl} = C_1 \delta_{ij}\delta_{kl} + C_2 \delta_{ik}\delta_{jl} + C_3 \delta_{il}\delta_{jk}$, write the equation that results for an isotropic medium. Using the relation (B.29), write the equation using the vector operators $\nabla(\nabla \cdot \boldsymbol{\psi})$ and $\nabla \times (\nabla \times \boldsymbol{\psi})$. How many wave speeds exist for a vectorial field? What is the vectorial character of each of these waves? In the case of the electric field in vacuum there is additionally the Maxwell equation $\nabla \cdot \mathbf{E} = 0$. Verify that, in this case, there is only one speed for electromagnetic waves.

(B.10) **Viscous stress tensor in a fluid.** In a viscous fluid the stress tensor $\mathbb{P}$ with components P_{ik} is isotropic and has an isotropic contribution plus another that is proportional to the velocity gradients. Assuming that space is isotropic, show that the most general expression for the stress tensor can be written as

$$\mathbb{P} = p\mathbb{I} - \eta\left[(\nabla \mathbf{v}) + (\nabla \mathbf{v})^T - \frac{2}{3}(\nabla \cdot \mathbf{v})\mathbb{I} \right] - \zeta(\nabla \cdot \mathbf{v})\mathbb{I},$$

where p is called the hydrostatic pressure, η is the shear viscosity, and ζ is the bulk viscosity. In the above expression we use the notation $\mathbb{I}$ for the identity tensor, $\nabla \mathbf{v}$ is the gradient tensor (with derivatives in the rows and velocity components as columns), and the superscript T indicates matrix transpose.

(B.11) **Cubic symmetry.** Show that any rank 2 tensor that has cubic symmetry is automatically isotropic. By cubic symmetry we mean that it has parity symmetry and that any two directions can be exchanged without modifying the components of the tensor.

Scattering processes

C.1 Classical mechanics

C.1.1 Kinematics of binary collisions

Consider classical particles that interact through pairwise conservative forces that decay sufficiently rapidly at large distances. We are interested in describing the collision of a pair of particles that come from infinity with velocities $\mathbf{c}_1$ and $\mathbf{c}_2$. By energy conservation it is not possible for the particles to become captured in bound states. Therefore, the interaction lasts for a finite time, after which the particles escape again to infinity with velocities $\mathbf{c}_1'$ and $\mathbf{c}_2'$. For simplicity the case of particles of equal mass is considered; the extension to dissimilar masses is direct. Energy and momentum conservation imply that

$$
\begin{aligned}
c_1'^2 + c_2'^2 &= c_1^2 + c_2^2, \\
\mathbf{c}_1' + \mathbf{c}_2' &= \mathbf{c}_1 + \mathbf{c}_2.
\end{aligned}
\tag{C.1}
$$

Defining the pre- and postcollisional relative velocities as $\mathbf{g} = \mathbf{c}_2 - \mathbf{c}_1$ and $\mathbf{g}' = \mathbf{c}_2' - \mathbf{c}_1'$, respectively, the general solution of eqns (C.1) is[1]

$$
\mathbf{g}' = \mathbf{g} - (\mathbf{g} \cdot \hat{\mathbf{n}})\hat{\mathbf{n}},
\tag{C.2}
$$

where $\hat{\mathbf{n}}$ is a unit vector that will be determined later in terms of the interparticle potential and the relative velocity, but up to now is arbitrary. If the particles interact through a central potential, $\hat{\mathbf{n}}$ points to the centre of mass. Noting that $\mathbf{g}' \cdot \hat{\mathbf{n}} = -\mathbf{g} \cdot \hat{\mathbf{n}}$, one can directly invert the relation (C.2) as

$$
\mathbf{g} = \mathbf{g}' - (\mathbf{g}' \cdot \hat{\mathbf{n}})\hat{\mathbf{n}},
\tag{C.3}
$$

where we note that the same expression is recovered with the roles of $\mathbf{g}$ and $\mathbf{g}'$ exchanged.

Following Cercignani (2013), the transformations $\mathbf{g} \to \mathbf{g}'$ [eqn (C.2)] and $\mathbf{g}' \to \mathbf{g}$ [eqn (C.3)] are characterised by the same Jacobian $J_{\hat{\mathbf{n}}}$, which depends on the unit vector. The Jacobian of the combined transformation is hence $J_{\hat{\mathbf{n}}}^2$. However, as the combined transformation yields the initial relative velocity, the Jacobian must be one, from which we obtain that $|J_{\hat{\mathbf{n}}}| = 1$. Finally, using that the centre-of-mass velocity is conserved, we get that the Jacobian of the transformation from the initial to the final velocities is one; that is,

$$
\mathrm{d}^3 c_1' \, \mathrm{d}^3 c_2' = \mathrm{d}^3 c_1 \, \mathrm{d}^3 c_2,
\tag{C.4}
$$

an expression that is used in the derivation of the Lorentz and Boltzmann equations in Chapters 3 and 4.

[1] See Exercise C.1 to show that this relation also holds for particles with different masses.

C.1.2 Geometrical parameterisation

The unit vector $\hat{\mathbf{n}}$ is determined by the geometry of the collision. In the centre-of-mass reference frame, the impact parameter b is defined as the distance from the projected unscattered trajectory to the origin. By conservation of energy, $|\mathbf{g}'| = |\mathbf{g}|$, therefore angular momentum conservation implies that, for the postcollisional state, the impact parameter is also b. The deflection angle χ is related to ϕ_0, which points to the point of minimum distance, by $\chi = |\pi - 2\phi_0|$ (Fig. C.1). Furthermore, ϕ_0 is obtained by integrating the motion under the influence of the interparticle potential U (Landau and Lifshitz, 1976),

$$\phi_0 = \int_{r_{\min}}^{\infty} \frac{(b/r^2)\,\mathrm{d}r}{\sqrt{1 - (b/r)^2 - (2U/\mu g^2)}}, \tag{C.5}$$

where $\mu = m_1 m_2/(m_1 + m_2)$ is the reduced mass for particles of masses m_1 and m_2, which equals $\mu = m/2$ for particles of equal mass.

If there is a beam of particles that approach the centre, it is useful to define the differential cross section $\mathrm{d}\sigma$ as the number of particles that are scattered between χ and $\chi + \mathrm{d}\chi$ divided by the particle flux n, which is assumed to be uniform. If there is a one-to-one relation between b and χ, the particles that will be deflected in the referred range of angles have impact parameters between $b(\chi)$ and $b(\chi) + \mathrm{d}b(\chi)$. Now, by simple geometry, $\mathrm{d}\sigma = 2\pi b\,\mathrm{d}b$, which using the chain rule gives $\mathrm{d}\sigma = 2\pi b(\chi)\left|\frac{\partial b(\chi)}{\partial \chi}\right|\mathrm{d}\chi$. Finally, instead of the differential of the deflection angle, the cross section is usually referred to the element of solid angle, $\mathrm{d}\Omega = 2\pi \sin\chi\,\mathrm{d}\chi$, as

$$\mathrm{d}\sigma = \frac{b(\chi)}{\sin\chi}\left|\frac{\partial b(\chi)}{\partial\chi}\right|\mathrm{d}\Omega. \tag{C.6}$$

C.1.3 Scattering for hard sphere, Coulomb, and gravitational potentials

In the case of hard spheres of diameter D, the impact parameter and the deflection angle are related by $b = D\sin\phi_0 = D\sin((\pi - \chi)/2)$. Replacing this expression in (C.6) gives

$$\mathrm{d}\sigma_{\mathrm{hs}} = (D^2/4)\,\mathrm{d}\Omega; \tag{C.7}$$

that is, after the collision, particles are scattered isotropically. This result is used in the statistical modelling of hard sphere collisions, where the postcollisional relative velocity is selected from an isotropic distribution.

Coulomb and self-gravitating systems are characterised by a potential $U = \alpha/r$, where the sign of α depends on the signs of the interacting charges for the Coulomb case and is always negative for the gravitational interaction. Using eqn (C.5), one gets, after some trigonometric transformations, $\phi_0 = \arctan\left(\mu g^2 b/\alpha\right)$. The deflection angle therefore satisfies

$$\tan\left(\frac{\chi}{2}\right) = \frac{\alpha}{\mu g^2 b}, \tag{C.8}$$

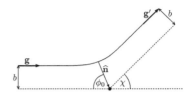

Fig. C.1 Geometrical parameterisation of the scattering process in the centre-of-mass reference frame. The precollisional relative velocity $\mathbf{g}$ and the postcollisional one, $\mathbf{g}'$ depend on the impact parameter b. The deflection angle is χ, and the unit vector $\hat{\mathbf{n}}$ points from the point of minimal distance to the origin, forming an angle ϕ_0 with the horizontal axis.

which allows one to compute the differential cross section as

$$d\sigma = \left(\frac{\alpha}{2\mu g^2 \sin^2 \frac{\chi}{2}}\right)^2 d\Omega. \tag{C.9}$$

Notably, this does not depend on the sign of α. This implies that, for Coulomb interactions, the statistical scattering properties do not depend on the signs of the charges.

For large separations, the deflection angle for gravitational interactions is $\chi \approx 2\alpha/(\mu g^2 b)$. After the collision, the relative velocity acquires a transversal component, $\delta g_\perp = g \sin \chi \approx -\frac{4Gm}{gb}$, where for particles of equal mass we have used that $\alpha = -Gmm$ and $\mu = m/2$. Finally, each of the colliding particles gets a transversal velocity component of

$$\delta c_\perp = \frac{g \sin \chi}{2} \approx -2Gm/gb, \tag{C.10}$$

where we have used that $\mathbf{c}_{2/1} = \mathbf{V} \pm \mathbf{g}/2$ and that the velocity of the centre of mass $\mathbf{V}$ is unchanged. This result is used to derive the dynamical friction properties for self-gravitating systems in Chapter 6.

C.2   Quantum mechanics

C.2.1   Time-dependent perturbation theory

The method used in this book to study transitions or scattering processes between quantum states is time-dependent perturbation theory (Messiah, 2014; Cohen-Tannoudji *et al.*, 1992). Consider a system described by an unperturbed Hamiltonian H_0 whose eigenstates $|n\rangle$ and energies ε_n are known,

$$H_0|n\rangle = \varepsilon_n|n\rangle. \tag{C.11}$$

Initially, the system is prepared in the pure state $|i\rangle$, and at time $t = 0$ a perturbation in the Hamiltonian is switched on, such that it now reads,

$$H = H_0 + V(t), \tag{C.12}$$

where we write $V(t) = \epsilon \widehat{V}(t)$ ($\epsilon \ll 1$) to indicate that the perturbation, which may depend on time, is small.

Our aim is to determine how the system evolves and the probability that, after some time, the system will be in a state $|k\rangle$, different from the initial one. We assume that the perturbation is weak enough that the Hilbert space is unchanged and that the state of the system at any time $|\psi(t)\rangle$ can be written in terms of the unperturbed states,

$$|\psi(t)\rangle = \sum_j c_j(t)|j\rangle. \tag{C.13}$$

Inserting this expansion into the Schrödinger equation, $i\hbar\frac{d}{dt}|\psi\rangle = H|\psi\rangle$, results in

$$i\hbar\frac{d}{dt}c_k = \varepsilon_k c_k + \epsilon \sum_j \langle k|\widehat{V}(t)|j\rangle c_j, \tag{C.14}$$

where we have used the orthogonality of the eigenstates, $\langle k|j\rangle = \delta_{jk}$.

We now look for perturbative solutions of the previous equation. We first note that, in the absence of the perturbation, the amplitudes are described simply by $c_k^{(0)}(t) = e^{-i\varepsilon_k t/\hbar}c_k^{(0)}(0)$, which suggests that it will be convenient to perform the change of variables,

$$c_k(t) = e^{-i\varepsilon_k t/\hbar}b_k(t), \tag{C.15}$$

and it is expected that, for $\epsilon \ll 1$, the amplitudes b_k will evolve slowly in time.[2] In this case, we can expand

$$b_k = b_k^{(0)} + \epsilon b_k^{(1)} + \dots, \tag{C.16}$$

where now the time derivatives of b_k are of order 1. Replacing this expansion in (C.14) gives

$$\frac{\mathrm{d}}{\mathrm{d}t}b_k^{(0)} = 0, \tag{C.17}$$

$$\frac{\mathrm{d}}{\mathrm{d}t}b_k^{(1)} = -\frac{i}{\hbar}\sum_j \langle k|\widehat{V}(t)|j\rangle e^{i(\varepsilon_k-\varepsilon_j)t/\hbar}b_j^{(0)}. \tag{C.18}$$

As expected, the zeroth-order amplitudes state that the initial condition is preserved, $b_k^{(0)} = \delta_{ik}$, where $|i\rangle$ is the state in which the system was initially prepared. At first order, therefore, we get

$$b_k^{(1)}(t) = -\frac{i}{\hbar}\int_0^t \langle k|\widehat{V}(s)|i\rangle e^{i(\varepsilon_k-\varepsilon_i)s/\hbar}\,\mathrm{d}s, \tag{C.19}$$

from which the probability of finding the system in the state $|k\rangle$, $|b_k|^2$, is readily obtained.

C.2.2   Fermi golden rule

Two cases are of particular interest. The first consists of a constant perturbation $\widehat{V}$. In this case,

$$b_k(t) = -\frac{i}{\hbar}\epsilon\langle k|\widehat{V}|i\rangle\int_0^t e^{i(\varepsilon_k-\varepsilon_i)s/\hbar}\,\mathrm{d}s = -\frac{i}{\hbar}\langle k|V|i\rangle\left[\frac{e^{i(\varepsilon_k-\varepsilon_i)t/\hbar}-1}{i(\varepsilon_k-\varepsilon_i)/\hbar}\right], \tag{C.20}$$

where we have replaced $\epsilon\widehat{V} = V$. The probability of finding the system in a state $|k\rangle$ at a time t after the perturbation was switched on is then

$$p_k(t) = |b_k(t)|^2 = \frac{|\langle k|V|i\rangle|^2}{\hbar^2}\left[\frac{\sin\left(\frac{\varepsilon_k-\varepsilon_i}{2\hbar}t\right)}{\left(\frac{\varepsilon_k-\varepsilon_i}{2\hbar}\right)}\right]^2. \tag{C.21}$$

When the final states are discrete, oscillations take place. However, the picture is completely different when they are continuous or quasi-continuous, as is the case for quantum gases studied in Chapters 7, 8, and 9. In order to continue working with normalised functions, we consider finite volumes where the quantum states are quasicontinuous. The

[2]Indeed, this can be formally proved using multiple time scale methods (Cohen-Tannoudji *et al.*, 1992).

probabilities p_k are factored as a term that depends explicitly on the final state times the function $F((\varepsilon_k - \varepsilon_i)/\hbar, t)$, which depends only on the energy of the final state. For quasicontinuous states we can label them as $|k\rangle = |\varepsilon, \mu\rangle$, where μ groups all the other quantum numbers for a given energy (for example, the direction of the momentum for free particles). Now, the probability of transit to a state in a window of ε and μ is

$$p(t) = \sum_{\mu} \int g(\mu, \varepsilon) \frac{|\langle \varepsilon, \mu | V | i \rangle|^2}{\hbar^2} F((\varepsilon - \varepsilon_i)/\hbar, t)\, \mathrm{d}\varepsilon, \qquad \text{(C.22)}$$

where g is the density of states and we have used the general rule, $\sum_k = \sum_{\mu} \int g(\mu, \varepsilon)\, \mathrm{d}\varepsilon$. For large times, the function F is peaked for ε close to ε_i, and indeed it becomes a Dirac delta function, $\lim_{t\to\infty} F((\varepsilon - \varepsilon_i)/\hbar, t) = 2\pi\hbar t \delta(\varepsilon - \varepsilon_i)$ (see Exercise C.3), which allows us to write,

$$p(t) = \sum_{\mu} \int g(\mu, \varepsilon) |\langle \varepsilon, \mu | V | i \rangle|^2 \frac{2\pi t}{\hbar} \delta(\varepsilon - \varepsilon_i)\, \mathrm{d}\varepsilon. \qquad \text{(C.23)}$$

The probability increases linearly with time, allowing us to define the probability transition rate from the initial state $|i\rangle$ to a quasicontinuous state $|k\rangle$ as

$$\overline{W}_{i \to k} = \frac{2\pi}{\hbar} |\langle k | V | i \rangle|^2\, \delta(\varepsilon_k - \varepsilon_i). \qquad \text{(C.24)}$$

This important result, which states that, under constant perturbations, transitions can only take place between states with equal energy, is known as the Fermi golden rule.

The second case of interest is that of an oscillatory perturbation. As we are considering up to linear order in perturbation theory, it suffices to deal with monochromatic perturbations, $V(t) = V_\omega \sin(\omega t)$ or $V(t) = V_\omega \cos(\omega t)$. Both cases give, for long times, the transition rate,[3]

$$\overline{W}_{i \to k} = \frac{\pi}{\hbar} |\langle k | V_\omega | i \rangle|^2 \left[\delta(\varepsilon_k - \varepsilon_i - \hbar\omega) + \delta(\varepsilon_k - \varepsilon_i - \hbar\omega) \right]. \qquad \text{(C.25)}$$

This states that, under oscillatory perturbations, transitions can occur only between states that differ in energy by one quantum $\hbar\omega$. The first term gives $\varepsilon_k = \varepsilon_i + \hbar\omega$, meaning that the system absorbs one quantum from the field, while for the second term, $\varepsilon_k = \varepsilon_i - \hbar\omega$, one quantum is emitted to the field. Note that both processes present equal transition rates.

---

[3]See Exercise C.2.

Exercises

(C.1) **Collision rule for different masses.** Show that, in a binary collision of particles with different masses m_1 and m_2, the collision rule (C.2) is also satisfied. To show this, it is convenient to write the energy and momentum conservation equations in terms of the relative and centre-of-mass velocities.

(C.2) **Oscillatory perturbation.** Derive eqn (C.25) for the perturbation $V(t) = V_\omega \sin(\omega t)$.

(C.3) **Dirac delta function.** Consider the function $f(x,t) = \frac{1}{\pi t} \left[\frac{\sin(xt)}{x} \right]^2$. Show that, as a function of x, it is normalised to unity and that when $t \to \infty$ it becomes concentrated around the origin, therefore becoming a Dirac delta distribution (see Fig. C.2).

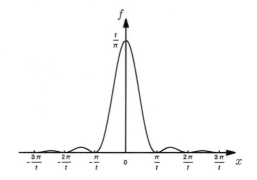

Fig. C.2 The function $f(x,t) = \frac{1}{\pi t} \left[\frac{\sin(xt)}{x} \right]^2$ converges to a Dirac delta function for large t.

D
Electronic structure in crystalline solids

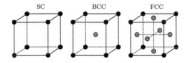

Fig. D.1 Examples of simple lattices with cubic geometry: simple cubic (SC) with atoms at the corners, body-centred cubic (BCC) with an additional atom in the centre, and face-centred cubic (FCC) with atoms at the centre of the faces.

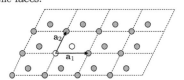

Fig. D.2 Example of a lattice in 2D. The base vectors $\mathbf{a}_1$ and $\mathbf{a}_2$ span a rhombic unit cell. Each unit cell has two atoms, which are displayed as empty circles for a specific cell.

D.1 Crystalline solids

Most solids present a crystalline structure, where atoms are organised regularly in cells that repeat periodically. The simplest example is the single cubic lattice, where the atoms sit at the corners of packed cubes. In practice, it is rare for atoms to adopt this simple structure, although polonium is one example. Rather, more complex structures appear in Nature. Some examples where atoms are arranged in periodic lattices are shown in Fig. D.1, including the body-centred cubic and face-centred cubic lattices. In Chapter 8 it is shown that the crystalline structure of solids is the main factor responsible for the particular electronic properties that they present.

In all cases, the lattice is first described by a unit cell that can have one or more atoms located at positions $\mathbf{r}_l$ relative to a reference point of the cell. Then cells are repeated periodically by displacing them by an integer number of some base vectors $\mathbf{a}_i$, such that the new cells are obtained by the translation vectors (see Fig. D.2),

$$\mathbf{R} = n_1\mathbf{a}_1 + n_2\mathbf{a}_2 + n_3\mathbf{a}_3, \quad n_i \in \mathbb{Z}. \tag{D.1}$$

In three dimensions, there are three base vectors. They do not need to be orthogonal, but only linearly independent such that they can span the whole volume. The volume of the unit cell is given by the scalar triple product $v = (\mathbf{a}_1 \times \mathbf{a}_2) \cdot \mathbf{a}_3$.

Associated with these base vectors, we define the reciprocal base vectors $\mathbf{b}_j$ as those satisfying the biorthogonality condition $\mathbf{a}_i \cdot \mathbf{b}_j = 2\pi\delta_{ij}$. These vectors have dimensions of reciprocal distance, and their integer combinations generate the reciprocal vectors,

$$\mathbf{G} = m_1\mathbf{b}_1 + m_2\mathbf{b}_2 + m_3\mathbf{b}_3, \quad m_i \in \mathbb{Z}. \tag{D.2}$$

D.2 Band structure

The electrons in the system can be classified into two types. The core electrons are in deep energy levels and will remain bound to each specific atom. The exterior or valence electrons, in contrast, are not completely bound to a single atom but rather are subject to the attraction of the neighbouring atoms as well. These electrons will become delocalised and are the main subject of study in electronic transport. We thus separate this problem into two: the valence electrons and the ions, the latter being composed of the nucleus and core electrons. The ions are arranged periodically and, as such, generate a periodic potential V for the valence electrons, satisfying $U(\mathbf{r}+\mathbf{R}) = U(\mathbf{r})$ for any translation vector (D.1). It is easy to verify that any periodic function can be written as a Fourier series in the reciprocal vectors,[1]

$$U(\mathbf{r}) = \sum_{\mathbf{G}} U_{\mathbf{G}} e^{i\mathbf{G}\cdot\mathbf{r}}. \tag{D.3}$$

Consider now a macroscopic solid, composed of many unit cells. For simplicity, although it is not strictly necessary to do so, we will consider that the solid is a parallelepiped, aligned with the base vectors $\mathbf{a}_i$, with N_i cells on each side. By macroscopic we understand that the total number of cells $N = N_1 N_2 N_3$ is on the order of the Avogadro number. We will also assume that the box is periodic for the purpose of computation of the electron wavefunctions Ψ,[2] which are then written as Fourier series,

$$\Psi(\mathbf{r}) = \sum_{\mathbf{q}} \Psi_{\mathbf{q}} e^{i\mathbf{q}\cdot\mathbf{r}}, \tag{D.4}$$

where $\mathbf{q} = m_1\mathbf{b}_1/N_1 + m_2\mathbf{b}_2/N_2 + m_3\mathbf{b}_3/N_3$, with $m_i \in \mathbb{Z}$, are vectors in reciprocal space. Note that Ψ is not necessarily periodic in the cell.

It is convenient to define the first Brillouin zone (FBZ), also called the Wigner–Seitz cell, as the set of vectors $\mathbf{q}$ that are closer to $\mathbf{G} = 0$ than any other reciprocal vector. Figure D.3 presents the FBZ for some lattices. Then, any vector in the reciprocal space can be uniquely decomposed as $\mathbf{q} = \mathbf{k} + \mathbf{G}$, where $\mathbf{k}$ belongs to the FBZ.

D.2.1 Bloch theorem

Let us now analyse the Schrödinger equation for the valence electrons,

$$\left[-\frac{\hbar^2}{2m}\nabla^2 + U(\mathbf{r}) \right] \Psi(\mathbf{r}) = \varepsilon\Psi(\mathbf{r}), \tag{D.5}$$

where m is the electron mass,[3] U is the periodic potential generated by the ions, and ε are the energy levels. Using the representations (D.3) and (D.4), the Schrödinger equation reads,

$$\sum_{\mathbf{q}} \frac{\hbar^2 q^2}{2m} \Psi_{\mathbf{q}} e^{i\mathbf{q}\cdot\mathbf{r}} + \sum_{\mathbf{G}}\sum_{\mathbf{q}} U_{\mathbf{G}} \Psi_{\mathbf{q}} e^{i(\mathbf{G}+\mathbf{q})\cdot\mathbf{r}} = \varepsilon \sum_{\mathbf{q}} \Psi_{\mathbf{q}} e^{i\mathbf{q}\cdot\mathbf{r}}. \tag{D.6}$$

The summed vector $\mathbf{G}+\mathbf{q}$ also belongs to the reciprocal space and therefore we can relabel it as $\mathbf{q}$, such that now the equation adopts the simpler form,

$$\sum_{\mathbf{q}} \left(\frac{\hbar^2 q^2}{2m} \Psi_{\mathbf{q}} + \sum_{\mathbf{G}} U_{\mathbf{G}} \Psi_{\mathbf{q}-\mathbf{G}} \right) e^{i\mathbf{q}\cdot\mathbf{r}} = \varepsilon \sum_{\mathbf{q}} \Psi_{\mathbf{q}} e^{i\mathbf{q}\cdot\mathbf{r}}. \tag{D.7}$$

[1] Indeed, using this representation, we have $U(\mathbf{r}+\mathbf{R}) = \sum_{\mathbf{G}} U_{\mathbf{G}} e^{i\mathbf{G}\cdot(\mathbf{r}+\mathbf{R})}$. However, $\mathbf{G}\cdot\mathbf{R} = 2\pi n$, with $n \in \mathbb{Z}$ by the orthogonality condition, proving that U is periodic.

[2] Again, this is not strictly necessary, but it simplifies the calculation enormously. This will allow us to use exponential Fourier functions instead of sine and cosine functions. The final results are nevertheless the same if fixed boundary conditions are used.

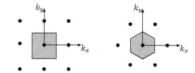

Fig. D.3 First Brillouin zone (FBZ) in grey for a square (left) and hexagonal (right) lattice. Dots represent the reciprocal vectors of the lattice.

[3] This is the bare electron mass $m = 9.1\times10^{-31}$ kg. Later we will see that, in some cases, electrons move in a solid as if they have an effective mass, different from this value.

As the Fourier functions are linearly independent, we end up with the following set of equations:

$$\frac{\hbar^2 q^2}{2m}\Psi_{\mathbf{q}} + \sum_{\mathbf{G}} U_{\mathbf{G}}\Psi_{\mathbf{q}-\mathbf{G}} = \varepsilon\Psi_{\mathbf{q}}. \tag{D.8}$$

The equations for $\mathbf{q}$ and $\mathbf{q} + \mathbf{G}$ are coupled and constitute a set of infinite size. Each set of equations can be labelled by a unique vector $\mathbf{k}$ in the FBZ, and we note that the sets of equations for different vectors in the FBZ do not couple (see Fig. D.4). For each $\mathbf{k} \in$ FBZ there are infinite coupled equations (one for each reciprocal vector $\mathbf{G}$), and as such, an infinite set of discrete energy levels is obtained, which we label as $\varepsilon_{n,\mathbf{k}}$, where $n \in \mathbb{N}$ is an integer index. The wavefunctions are labelled accordingly as $\Psi_{n,\mathbf{k}}$.

Fig. D.4 A one-dimensional case with reciprocal vector b where the FBZ spans the range $[-b/2, b/2]$, shown in grey. In eqn (D.8) each vector $\mathbf{k}$ in the first Brillouin zone (FBZ) couples to an infinite number of vectors $\mathbf{k} + m\mathbf{b}$ outside the FBZ but not to other vectors inside the FBZ.

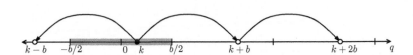

Note that, because the equations are coupled, the election of choosing the FBZ to label the levels is arbitrary and other options are possible. This also implies that the levels are periodic in the reciprocal space, $\varepsilon_{\mathbf{k}} = \varepsilon_{\mathbf{k}+\mathbf{G}}$. Finally, the vectors in the reciprocal space used to label the states $\mathbf{k}$ belong to the FBZ, with a distance between successive vectors equal to $\Delta k = |\mathbf{b}_i|/N_i$, which for macroscopic systems is extremely small. Consequently, the vectors $\mathbf{k}$ are quasicontinuous, densely filling the FBZ, and we will use the notation $\varepsilon(\mathbf{k})$ for the dispersion relation. This property will allow us to use differential calculus when manipulating them.

Before continuing with the energy bands, there is an additional property of the Schrödinger equation that deserves some attention. Because different vectors in the FBZ do not couple in the solution of eqn (D.8), the eigenfunctions are built only by contributions of vectors that differ in reciprocal vectors $\mathbf{G}$. Then, for the eigenfunctions, the sum in (D.4) is restricted to

$$\Psi_{\mathbf{k},n}(\mathbf{r}) = \sum_{\mathbf{G}} \Psi^{(n)}_{\mathbf{k}+\mathbf{G}} e^{i(\mathbf{k}+\mathbf{G})\cdot\mathbf{r}}. \tag{D.9}$$

By factoring out $e^{i\mathbf{k}\cdot\mathbf{r}}$, we obtain the representation of a periodic function in the unit cell. We have obtained the Bloch theorem, which states that the wavefunctions for a periodic potential can be written as

$$\Psi_{\mathbf{k},n}(\mathbf{r}) = e^{i\mathbf{k}\cdot\mathbf{r}} u_{\mathbf{k},n}(\mathbf{r})/\sqrt{\mathcal{V}}, \tag{D.10}$$

where $u_{\mathbf{k},n}$ is periodic in the unit cell and the normalisation factor has been put explicitly such that

$$\int |\Psi|^2 \, \mathrm{d}^3 r = 1. \tag{D.11}$$

Using the transformation, $\nabla(e^{i\mathbf{k}\cdot\mathbf{r}} u(\mathbf{r})) = e^{i\mathbf{k}\cdot\mathbf{r}}(\nabla + i\mathbf{k})u(\mathbf{r})$, the Schrödinger equation can be written,

$$H_{\mathbf{k}} u_{\mathbf{k}} = \varepsilon_{\mathbf{k}} u_{\mathbf{k}}, \tag{D.12}$$

where now the domain for $u_{\mathbf{k}}$ is the unit cell and the effective Hamiltonian is

$$H_{\mathbf{k}} = -\frac{\hbar^2}{2m}(\nabla + i\mathbf{k})^2 + U. \tag{D.13}$$

D.2.2   Energy bands

In the previous sections it was shown that the energy levels in a periodic system can be labelled as $\varepsilon_n(\mathbf{k})$, where $\mathbf{k}$ is a quasicontinuous vector in the FBZ and $n = 1, 2, 3, \ldots$ is an integer index that labels the bands. Also, we showed that the energy levels for each $\mathbf{k}$ can be obtained from a Schrödinger-like equation in the unit cell, for a Hamiltonian that depends explicitly on $\mathbf{k}$ (D.13). It is therefore expected that, except for isolated accidents, the energy levels will be continuous functions of $\mathbf{k}$ and correspond to hypersurfaces in the FBZ.

To simplify the visualisation, consider first a 1D case where the bands are simple functions of $k = |\mathbf{k}|$. Figure D.5 presents a scheme of the band structure. Looking at the energy axis, two types of regions are clearly differentiated. Associated with each band, there is an energy interval such that, for each energy in the interval, there are one or various electronic states with this energy value.[4] We will call these intervals the energy bands. Complementary to them, there are empty regions. In these energy gaps there are no electronic states; these energy levels are therefore forbidden.

In three dimensions the structure of the energy hypersurfaces in the FBZ is more complex, but again, when these levels are projected onto the energy axis, energy bands and gaps appear. Finally, it is also possible that two energy bands overlap, as in Exercise 8.11.

D.2.3   Bloch velocity and crystal momentum equation

Electrons in a Bloch state do not have, as we have seen, a well-defined momentum or wavevector. We can, however, compute their velocity as the average of the velocity operator,

$$\mathbf{V}_n(\mathbf{k}) = \langle \Psi_{n,\mathbf{k}}| -\frac{i\nabla}{m\hbar} |\Psi_{n,\mathbf{k}}\rangle. \tag{D.14}$$

To do so, we start by computing the energy level near a given wavevector $\mathbf{k}$. Consider a small wavevector $\boldsymbol{\Delta}$. The energy level $\varepsilon_n(\mathbf{k} + \boldsymbol{\Delta})$ can

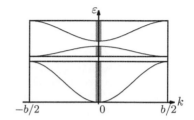

Fig. D.5 Simple representation of the energy bands in one dimension. The FBZ lies between $-b/2$ and $b/2$, where b is the reciprocal vector. Gaps appear between successive bands.

[4]To be precise, since the vectors are quasicontinuous, a small energy tolerance $\delta\varepsilon \sim L^{-1}$ must be accepted; that is, for each energy value in a band, there will be an electronic state with an energy sufficiently close to it.

be expanded as

$$\varepsilon_n(\mathbf{k} + \mathbf{\Delta}) \approx \varepsilon_n(\mathbf{k}) + \frac{\partial \varepsilon_n(\mathbf{k})}{\partial \mathbf{k}} \cdot \mathbf{\Delta}. \qquad (D.15)$$

The energy difference can also be obtained by noting that, when using the u eigenfunctions, the energies are associated with the Hamiltonians,

$$H_{\mathbf{k}} = -\frac{\hbar^2}{2m}(\nabla + i\mathbf{k})^2 + U, \qquad (D.16)$$

$$H_{\mathbf{k}+\mathbf{\Delta}} = -\frac{\hbar^2}{2m}(\nabla + i\mathbf{k} + i\mathbf{\Delta})^2 + U \approx H_{\mathbf{k}} - \frac{\hbar^2}{m}(\nabla + i\mathbf{k}) \cdot i\mathbf{\Delta}. \quad (D.17)$$

Then, according to time-independent perturbation theory,[5]

$$\varepsilon_n(\mathbf{k} + \mathbf{\Delta}) - \varepsilon_n(\mathbf{k}) \approx -\langle u_{n,\mathbf{k}} | \frac{\hbar^2}{m}(\nabla + i\mathbf{k}) \cdot i\mathbf{\Delta} | u_{n,\mathbf{k}} \rangle$$

$$\approx -\langle \Psi_{n,\mathbf{k}} | \frac{\hbar^2}{m} i\nabla | \Psi_{n,\mathbf{k}} \rangle \cdot \mathbf{\Delta}, \qquad (D.18)$$

where in second line we have applied the same transformation used to obtain (D.12). Comparing (D.15) and (D.18) and recalling the definition of the mean velocity (D.14), we obtain that

$$\mathbf{V}_n(\mathbf{k}) = \frac{1}{\hbar} \frac{\partial \varepsilon_n(\mathbf{k})}{\partial \mathbf{k}}. \qquad (D.19)$$

Note that, if the electrons were free, $\varepsilon = \hbar^2 k^2/2m$, the Bloch velocity would be $\mathbf{V} = \hbar\mathbf{k}/m$, the known result from quantum mechanics.

Consider a weak external force $\mathbf{F}$ acting on the electrons. The work done by the force in an interval δt on an electron in a Bloch state is $\delta W = \mathbf{F} \cdot \mathbf{V}(\mathbf{k})\delta t$. This work modifies the energy of the electron, which can be written as a change in $\mathbf{k}$, $\delta W = \delta\varepsilon = \frac{\partial\varepsilon}{\partial\mathbf{k}} \cdot \delta\mathbf{k}$. Equating both expressions and recalling the definition of the Bloch velocity, we find the semiclassical equation of motion,

$$\hbar\dot{\mathbf{k}} = \mathbf{F}, \qquad (D.20)$$

which is a Newton-like equation for the crystal momentum $\hbar\mathbf{k}$. Note that it is not the momentum of the particle because the Bloch states are not eigenstates of the momentum operator.

D.2.4   Self-consistent potential

In the derivation of the band structure, we consider a single electron moving in the periodic potential. However, in solids, there are many electrons that interact with each other via the Coulomb potential, which is quite intense (as large as 13.6 eV for electrons separated by 1.06 Å). However, we note that the charge density of an electron in a Bloch state is $\rho = e|\Psi|^2 = e|u|^2$, which is periodic. Then, the band theory makes the following first (Hartree) approximation. The periodic potential that should be considered for computing the band structure is

[5]Take a Hamiltonian H_0 with normalised states satisfying $H_0|n\rangle = \varepsilon_n^0|n\rangle$. For the perturbed Hamiltonian, $H = H_0 + V$, the energy levels are

$$\varepsilon_n = \varepsilon_n^0 + \langle n|V|n\rangle$$

to first order in V (Messiah, 2014; Cohen-Tannoudji *et al.*, 1992).

not only the one produced by the ions and core electrons, but also the self-consistent potential generated by the charge density of the valence electrons. This approximation renders the problem nonlinear because the potential that enters in the Schrödinger equation is quadratic in the wavefunctions. The Hartree approximation was further refined by Fock by imposing that the total wavefunction should reflect that electrons are fermions and, therefore, must be antisymmetric under the exchange of any pair of electrons. Under this Hartree–Fock approximation, there are no direct electron–electron interactions, but rather they interact via the self-consistent potential, similarly to what happens in the Vlasov model for plasmas, studied in Chapter 6.

The self-consistent approach, although very successful, is still an approximation and does not fully consider the quantum correlations of the interacting electrons. A more formal approach is the theory of Fermi liquids developed by Landau. There, the relevant objects of study, instead of the bare electrons, are the excitations that develop in the complete system. These excitations behave as quasiparticles that follow Fermi–Dirac statistics and carry charge, but have finite lifetime. We will not use this approach here, as it lies beyond the scope of this book. Interested readers are directed to the books by Lifshitz and Pitaevskii (1980) and Kadanoff and Baym (1962).

D.3 Density of states

In the thermodynamic limit, the vectors in reciprocal space become dense, and instead of treating them separately it is useful to group them according to their energies. The density of states $g(\varepsilon)$ is defined such that $g(\varepsilon)\, d\varepsilon$ is the number of quantum states between ε and $\varepsilon + d\varepsilon$, divided by the volume $\mathcal{V}$. It is an intensive quantity that provides information on how the bands are arranged.

The density of states depends on the band structure, which in turn depends on the ionic potential and the geometry of the unit cell. In the next paragraphs, some examples are given to illustrate the general features of g.

D.3.1 Free electron gas

The first model to consider is the free electron gas, where there is no potential and the energy levels are

$$\varepsilon = \frac{\hbar^2 k^2}{2m}, \tag{D.21}$$

where the wavevectors are given by $\mathbf{k} = 2\pi \mathbf{n}/L$. Here, it is convenient to first obtain $G(\varepsilon)$, equal to the number of states with energy equal to or less than ε, which results as twice (due to the spin degeneracy) the number of wavevectors smaller than $k_{\max} = \sqrt{2m\varepsilon}/\hbar$. As the spacing between energy levels is $2\pi/L$, which is small, the number of states bounded by $k_{\max}$ can be accurately approximated by an integral expression,

$$G(\varepsilon) = 2 \int_{|\mathbf{k}|<k_{\max}} \left(\frac{L}{2\pi}\right)^3 d^3k = \frac{\mathcal{V}k_{\max}^3}{3\pi^2}. \tag{D.22}$$

Computing $g = \mathcal{V}^{-1} dG/d\varepsilon$ we obtain the density of states for a free electron gas in three dimensions as[6]

$$g_{3\mathrm{D}}(\varepsilon) = \frac{1}{2\pi^2}\left(\frac{2m}{\hbar^2}\right)^{3/2}\sqrt{\varepsilon}. \tag{D.23}$$

In many technological applications, electrons are confined in one or two dimensions, while moving freely in the others, resulting in quasi-2D or quasi-1D systems. The density of states of such confined systems presents singularities at the excitation energies of the transverse modes, as shown in Fig. D.6.

D.3.2   General case in three dimensions

In the general case, the density of states will depend on the band structure. The bands are limited by the maxima or minima of the dispersion relation $\varepsilon(\mathbf{k})$ as shown in Fig. D.5. Close to these extreme values, one can approximate the bands parabolically as $\varepsilon \approx \varepsilon_0 \pm \frac{m^*}{2}(\mathbf{k} - \mathbf{k}_0)^2$, where the upper sign corresponds to a minimum and the lower sign to a maximum. Locally, the dispersion relations are similar to those of free particles, where m^* is called the effective mass.[7] In a bandgap, the density of states vanishes (Fig. D.7), and at the band borders the density of states presents the characteristic van Hove singularities of free particles of the form $g \sim (\varepsilon - \varepsilon_0)^{1/2}$ for a band that starts at ε_0 and an analogous expression when a band terminates.

On many occasions, one needs to sum over the states of the system, for example in Section 8.2 when computing the electrical conductivity. Using the density of states, this sum can be cast into an integral. Also, in the thermodynamic limit, we can integrate in $\mathbf{k}$ in the FBZ as well. Recalling that the distance between successive vectors is $\Delta k = |\mathbf{b}_i|/N_i$, the number of vectors in a volume d^3k is $dN = N_1 N_2 N_3\, d^3k/\hat{v}$, where $\hat{v} = (\mathbf{b}_1 \times \mathbf{b}_2)\cdot\mathbf{b}_3 = (2\pi)^3/v$ is the volume of the FBZ. Combining these two results, we obtain $dN = \mathcal{V} d^3k/(2\pi)^3$, which must be multiplied by the spin degeneracy and summed over the bands. This discussion can be summarised in the following scheme:

$$\sum_{\text{states}} = \mathcal{V}\int g(\varepsilon)\,d\varepsilon = 2\mathcal{V}\sum_{\text{bands}}\int_{\text{FBZ}}\frac{d^3k}{(2\pi)^2} = \mathcal{V}\sum_{s_z}\sum_{\text{bands}}\int_{\text{FBZ}}\frac{d^3k}{(2\pi)^2}. \tag{D.24}$$

[6]See Exercise D.1 for the density of states in one and two dimensions.

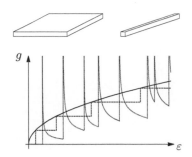

Fig. D.6 Top: quasi-2D and quasi-1D geometries. The electrons move freely in the long directions, while they are confined in the complementary ones, generating an effective band structure. Bottom: density of states $g(\varepsilon)$ for a free electron gas in three dimensions (solid line), quasi-2D (dashed line), and quasi-1D (dotted lines).

[7]The case with the minus sign becomes relevant in semiconductors, where m^* is associated with the mass of holes.

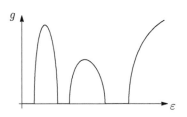

Fig. D.7 Schematic representation of the density of states for a solid in three dimensions presenting bands. The gaps between bands are associated with vanishing density of states.

Exercises

(D.1) **Density of states in one and two dimensions.**
Following the same procedure as in Section D.3, derive the density of states for free electrons moving in one and two dimensions.

(D.2) **Density of states in quasi-1D and quasi-2D systems.** In many technological applications, electrons are confined in one or two dimensions, while moving freely in the others, resulting in quasi-2D or quasi-1D systems, Fig. D.8.

Fig. D.8 Quasi-2D and quasi-1D geometries. The electrons move freely in the long directions, while they are confined in the complementary ones, generating an effective band structure.

In quasi-2D systems the electrons move in a shallow box of dimensions $L_x \times L_y \times a$, where L_x and L_y are large (tending to infinity in the thermodynamic limit), while a is finite. In quasi-1D systems the box dimensions are $L_x \times a \times a$. The box is periodic in the long directions, while fixed boundary conditions should be considered in the confining direction.

Compute the density of states in both cases and show that, if the confinement length tends to infinity, the three-dimensional density of states is recovered.

References

Abergel, D.S.L., Apalkov, V., Berashevich, J., Ziegler, K., and Chakraborty, T. (2010). Properties of graphene: a theoretical perspective. *Adv. Phys.*, **59**, 261.

Adam, S., Hwang, E.H., Galitsky, V.M., and Das Sarma, S. (2007). A self-consistent theory for graphene transport. *Proc. Natl. Acad. Sci. U.S.A.*, **104**, 18392–18397.

Andreotti, B., Forterre, Y., and Pouliquen, O. (2013). *Granular Media: Between Fluid and Solid*. Cambridge University Press, Cambridge.

Aranson, I.S. and Tsimring, L.S. (2005). Pattern formation of microtubules and motors: inelastic interaction of polar rods. *Phys. Rev. E*, **71**, 050901.

Ashcroft, N.W. and Mermin, N.D. (1976). *Solid State Physics*. Holt, Rinehart and Winston, New York.

Astumian, R.D. and Hänggi, P. (2002). Brownian motors. *Phys. Today*, **55**(11), 33–39.

Barnes, J. and Hut, P. (1986). A hierarchical O(N log N) force-calculation algorithm. *Nature*, **324**, 446–449.

Berg, H.C. (1993). *Random Walks in Biology*. Princeton University Press, New Jersey.

Berg, H.C. (2004). E. coli *in Motion*. Springer, New York.

Bernstein, J. (2004). *Kinetic Theory in the Expanding Universe*. Cambridge Monographs on Mathematical Physics. Cambridge University Press, Cambridge.

Binney, J. and Tremaine, S. (2008). *Galactic Dynamics* (2nd edn). Princeton Series in Astrophysics. Princeton University Press, New Jersey.

Bird, G.A. (1994). *Molecular Gas Dynamics and the Direct Simulation of Gas Flows* (2nd edn). Oxford University Press, Oxford.

Blaizot, J.-P., Wu, B., and Yan, L. (2014). Quark production, Bose-Einstein condensates and thermalization of the quark-gluon plasma. *Nuclear Physics A*, **930**, 139–162.

Brilliantov, N.V. and Pöschel, T. (2010). *Kinetic Theory of Granular Gases*. Oxford University Press, Oxford.

Bruus, H. and Flensberg, K. (2004). *Many-Body Quantum Theory in Condensed Matter Physics: An Introduction*. Oxford Graduate Texts. Oxford University Press, Oxford.

Callaway, J. (1991). *Quantum Theory of the Solid State* (2nd edn). Academic Press, London.

Campa, A., Dauxois, T., Fanelli, D., and Ruffo, S. (2014). *Physics of Long-Range Interacting Systems*. Oxford University Press, Oxford.

Campa, A., Dauxois, T., and Ruffo, S. (2009). Statistical mechanics and dynamics of solvable models with long-range interactions. *Phys. Rep.*, **480**, 57–159.

Cercignani, C. (2013). *Mathematical Methods in Kinetic Theory*. Springer, USA.

Chaikin, P.M. and Lubensky, T.C. (2000). *Principles of Condensed Matter Physics*. Cambridge University Press, Cambridge.

Chapman, S. and Cowling, T.G. (1970). *The Mathematical Theory of Non-Uniform Gases*. Cambridge University Press, Cambridge.

Chavanis, P.-H. (2002). Statistical mechanics of two-dimensional vortices and stellar systems. In *Dynamics and Thermodynamics of Systems with Long-Range Interactions*, pp. 208–289. Springer, Berlin Heidelberg.

Chen, F.F. (1984). *Introduction to Plasma Physics and Controlled Fusion: Volume 1: Plasma Physics* (Second edn). Springer, USA.

Cohen, E.G.D. (1993). Fifty years of kinetic theory. *Physica A*, **194**(1-4), 229–257.

Cohen-Tannoudji, C., Diu, B., and Laloe, F. (1992). *Quantum Mechanics, Volume 2*. Wiley, New York.

Das Sarma, S., Adam, S., Hwang, E.H., and Rossi, E. (2011). Electronic transport in two-dimensional graphene. *Rev. Mod. Phys.*, **83**, 407–470.

de Groot, S.R. and Mazur, P. (1984). *Non-equilibrium Thermodynamics*. Dover, New York.

Dodelson, S. (2003). *Modern Cosmology*. Academic Press, London.

Dorfman, J.R. (1999). *An Introduction to Chaos in Nonequilibrium Statistical Mechanics*. Volume 14. Cambridge University Press, Cambridge.

Einstein, A. (1956). *Investigations on the Theory of the Brownian Movement*. Dover, New York.

Ernst, M.H. (1998). Bogoliubov Choh Uhlenbeck theory: Cradle of modern kinetic theory in progress in statistical physics. In *Proceedings of the International Conference on Statistical Physics in Memory of Soon-Takh Choh*. World Scientific Publishing Company.

Evans, D.J. and Morris, G.P. (1990). *Statistical Mechanics of Nonequilibrium Liquids*. Academic Press, London.

Ferziger, J.H. and Kaper, H.G. (1972). *Mathematical Theory of Transport Processes in Gases*. North-Holland, Amsterdam.

Forster, D. (1995). *Hydrodynamic Fluctuations, Broken Symmetry, and Correlation Functions*. Perseus Books, Boulder.

Fox, M. (2006). *Quantum Optics: An Introduction*. Oxford Master Series in Physics. Oxford University Press, Oxford.

Garcia, A.L. (2000). *Numerical Methods for Physics* (2nd edn). Prentice Hall, New Jersey.

Garcia-Ojalvo, J. and Sancho, J. (1999). *Noise in Spatially Extended Systems*. Springer, New York.

Gardiner, C. (2009). *Stochastic Methods* (4th edn). Springer, Berlin.

Goldstein, H., Poole, C.P., and Safko, J.L. (2013). *Classical Mechanics* (3rd edn). Pearson Education Limited, Harlow.

Grad, H. (1949). On the kinetic theory of rarefied gases. *Commun. Pure Appl. Math.*, **2**, 331.

Grad, H. (1958). Principles of the kinetic theory of gases. In *Handbuch der Physik, Vol. XII*.

Greiner, W., Rischke, D., Neise, L., and Stöcker, H. (2012). *Thermodynamics and Statistical Mechanics*. Classical Theoretical Physics. Springer New York.

Hansen, J.-P. and McDonald, I.R. (1990). *Theory of Simple Liquids*. Elsevier.

Hockney, R.W. and Eastwood, J.W. (1988). *Computer Simulation Using Particles*. Institute of Physics, Bristol.

Huang, K. (1987). *Statistical Mechanics* (2nd edn). Wiley, New York.

Ichimaru, S. (1973). *Basic Principles of Plasma Physics: a Statistical Approach*. W.A. Benjamin, Reading, Mass.

Jacoboni, C. (2010). *Theory of Electron Transport in Semiconductors: a Pathway from Elementary Physics to Nonequilibrium Green Functions*. Volume 165. Springer Science & Business Media, Berlin Heidelberg.

Jones, W. and March, N.H. (1973). *Theoretical Solid State Physics, Volume 2: Non-equilibrium and Disorder*. Dover, New York.

Jülicher, F., Ajdari, A., and Prost, J. (1997). Modeling molecular motors. *Rev. Mod. Phys.*, **69**, 1269–1282.

Kadanoff, L.P. and Baym, G. (1962). *Quantum Statistical Mechanics*. Benjamin Cummings, Reading.

Kubo, R., Toda, M., and Hashitsume, N. (1998). *Statistical Mechanics II. Nonequilibrium Statistical Mechanics* (2nd edn). Springer, Berlin.

Landau, L. (1946). On the vibration of the electronic plasma. *J. Phys. USSR*, **10**, 25.

Landau, L.D. and Lifshitz, E.M. (1976). *Mechanics* (3rd edn). Butterworth-Heinemann, Oxford.

Landau, L.D. and Lifshitz, E.M. (1980). *Statistical Physics. Part 1* (3rd edn). Butterworth-Heinemann, Oxford.

Landau, L.D. and Lifshitz, E.M. (1981). *Physical Kinetics*. Pergamon, Oxford.

Landau, L.D. and Lifshitz, E.M. (1986). *Theory of Elasticity* (3rd edn). Butterworth-Heinemann, Oxford.

Landau, L.D. and Lifshitz, E.M. (1987). *Fluid Mechanics* (2nd edn). Butterworth-Heinemann, Oxford.

Lebowitz, J.L. (1993). Boltzmann's entropy and time's arrow. *Physics Today*, **46**, 32.

Lebowitz, J.L. and Montroll, E.W. (1983). *Nonequilibrium Phenomena I: the Boltzmann Equation*. North-Holland, Amsterdam.

Liboff, R.L. (2003). *Kinetic Theory: Classical, Quantum, and Relativistic Descriptions*. Springer, New York.

Lifshitz, E.M. and Pitaevskii, L.P. (1980). *Statistical Physics. Part 2* (3rd edn). Butterworth-Heinemann, Oxford.

Maznev, A.A. and Wright, O.B. (2014). Demystifying umklapp vs normal scattering in lattice thermal conductivity. *Am. J. Phys.*, **82**, 1062.

Messiah, A. (2014). *Quantum Mechanics*. Dover, New York.

Perrin, J. (1991). *Les Atoms*. Gallimard, Paris.

Piasecki, J. (1993). Time scales in the dynamics of the Lorentz electron gas. *Am. J. Phys.*, **61**(8), 718–722.

Pomeau, Y. and Resibois, P. (1975). Time dependent correlation functions and mode-mode coupling theories. *Phys. Rep.*, **19**, 63.

Press, H.W., Teukolsky, S.A., Vetterling, W.T., and Flannery, B.P. (2007). *Numerical Recipes. The Art of Scientific Computing* (3rd edn). Cambridge University Press, New York.

Pulfrey, D.L (2010). *Understanding Modern Transistors and Diodes*. Cambridge University Press, Cambridge.

Purcell, E.M. (1977). Life at low Reynolds number. *Am. J. Phys.*, **45**, 3–11.

Reimann, P. (2002). Brownian motors: noisy transport far from equilibrium. *Phys. Rep.*, **361**(24), 57–265.

Richardson, R. A., Peacor, S. D., Uher, C., and Nori, Franco (1992). $YBa_2Cu_3O_{7-\delta}$ films: Calculation of the thermal conductivity and phonon meanfree path. *J. of Appl. Phys.*, **72**, 4788–4791.

Risken, H. (1996). *The Fokker–Planck Equation: Methods of Solutions and Applications*. Springer, Berlin.

San Miguel, M. and Toral, R. (2000). Stochastic effects in physical systems. In *Instabilities and Nonequilibrium Structures VI* (ed. E. Tirapegui, J. Martínez, and R. Tiemann), Volume 5, Nonlinear phenomena and complex systems, pp. 35–127. Springer, Netherlands.

Semikoz, D.V. and Tkachev, I.I. (1997). Condensation of bosons in the kinetic regime. *Phys. Rev. D*, **55**, 489–502.

Singleton, J. (2001). *Band Theory and Electronic Properties of Solids*. Oxford University Press, Oxford.

Szász, D. (ed.) (2013). *Hard Ball Systems and the Lorentz Gas.* Encyclopaedia of Mathematical Sciences. Springer, Berlin Heidelberg.

Ter Haar, D. (1965). *Collected Papers of LD Landau.* Pergamon, Oxford.

van Kampen, N.G. (2007). *Stochastic Processes in Physics and Chemistry.* North Holland, Amsterdam.

Vasko, F.T. and Raichev, O.E. (2006). *Quantum Kinetic Theory and Applications*: *Electrons, Photons, Phonons.* Springer New York.

Wilson, M. (2006). Electrons in atomically thin carbon sheets behave like massless particles. *Phys. Today*, **59**, 21.

Yagi, K., Hatsuda, T., and Miake, Y. (2005). *Quark-Gluon Plasma*: *From Big Bang to Little Bang.* Cambridge Monographs on Particle Physics, Nuclear Physics and Cosmology. Cambridge University Press, Cambridge.

Zakharov, V.E., L'vov, V.S., and Falkovich, G. (1992). *Kolmogorov Spectra of Turbulence I*: *Wave Turbulence.* Springer Series in Nonlinear Dynamics. Springer, Berlin Heidelberg.

Ziman, J. (2001). *Electrons and Phonons*: *the Theory of Transport Phenomena in Solids.* Oxford University Press, Oxford.

Zwanzig, R. (2001). *Nonequilibrium Statistical Mechanics.* Oxford University Press, New York.

Index